Communications
in Computer and Information Science

Editorial Board Members

Joaquim Filipe ⓘ, *Polytechnic Institute of Setúbal, Setúbal, Portugal*
Ashish Ghosh ⓘ, *Indian Statistical Institute, Kolkata, India*
Lizhu Zhou, *Tsinghua University, Beijing, China*

Rationale

The CCIS series is devoted to the publication of proceedings of computer science conferences. Its aim is to efficiently disseminate original research results in informatics in printed and electronic form. While the focus is on publication of peer-reviewed full papers presenting mature work, inclusion of reviewed short papers reporting on work in progress is welcome, too. Besides globally relevant meetings with internationally representative program committees guaranteeing a strict peer-reviewing and paper selection process, conferences run by societies or of high regional or national relevance are also considered for publication.

Topics

The topical scope of CCIS spans the entire spectrum of informatics ranging from foundational topics in the theory of computing to information and communications science and technology and a broad variety of interdisciplinary application fields.

Information for Volume Editors and Authors

Publication in CCIS is free of charge. No royalties are paid, however, we offer registered conference participants temporary free access to the online version of the conference proceedings on SpringerLink (http://link.springer.com) by means of an http referrer from the conference website and/or a number of complimentary printed copies, as specified in the official acceptance email of the event.

CCIS proceedings can be published in time for distribution at conferences or as postproceedings, and delivered in the form of printed books and/or electronically as USBs and/or e-content licenses for accessing proceedings at SpringerLink. Furthermore, CCIS proceedings are included in the CCIS electronic book series hosted in the SpringerLink digital library at http://link.springer.com/bookseries/7899. Conferences publishing in CCIS are allowed to use Online Conference Service (OCS) for managing the whole proceedings lifecycle (from submission and reviewing to preparing for publication) free of charge.

Publication process

The language of publication is exclusively English. Authors publishing in CCIS have to sign the Springer CCIS copyright transfer form, however, they are free to use their material published in CCIS for substantially changed, more elaborate subsequent publications elsewhere. For the preparation of the camera-ready papers/files, authors have to strictly adhere to the Springer CCIS Authors' Instructions and are strongly encouraged to use the CCIS LaTeX style files or templates.

Abstracting/Indexing

CCIS is abstracted/indexed in DBLP, Google Scholar, EI-Compendex, Mathematical Reviews, SCImago, Scopus. CCIS volumes are also submitted for the inclusion in ISI Proceedings.

How to start

To start the evaluation of your proposal for inclusion in the CCIS series, please send an e-mail to ccis@springer.com.

Gulchohra Mammadova · Telman Aliev ·
Kamil Aida-zade
Editors

Information Technologies and Their Applications

Second International Conference, ITTA 2024
Baku, Azerbaijan, April 23–25, 2024
Proceedings, Part II

Editors
Gulchohra Mammadova
Azerbaijan University of Architecture
and Construction
Baku, Azerbaijan

Telman Aliev
Azerbaijan University of Architecture
and Construction
Baku, Azerbaijan

Kamil Aida-zade
Institute of Control Systems
Baku, Azerbaijan

ISSN 1865-0929　　　　　　ISSN 1865-0937 (electronic)
Communications in Computer and Information Science
ISBN 978-3-031-73419-9　　　ISBN 978-3-031-73420-5 (eBook)
https://doi.org/10.1007/978-3-031-73420-5

© The Editor(s) (if applicable) and The Author(s), under exclusive license
to Springer Nature Switzerland AG 2025

This work is subject to copyright. All rights are solely and exclusively licensed by the Publisher, whether the whole or part of the material is concerned, specifically the rights of translation, reprinting, reuse of illustrations, recitation, broadcasting, reproduction on microfilms or in any other physical way, and transmission or information storage and retrieval, electronic adaptation, computer software, or by similar or dissimilar methodology now known or hereafter developed.
The use of general descriptive names, registered names, trademarks, service marks, etc. in this publication does not imply, even in the absence of a specific statement, that such names are exempt from the relevant protective laws and regulations and therefore free for general use.
The publisher, the authors and the editors are safe to assume that the advice and information in this book are believed to be true and accurate at the date of publication. Neither the publisher nor the authors or the editors give a warranty, expressed or implied, with respect to the material contained herein or for any errors or omissions that may have been made. The publisher remains neutral with regard to jurisdictional claims in published maps and institutional affiliations.

This Springer imprint is published by the registered company Springer Nature Switzerland AG
The registered company address is: Gewerbestrasse 11, 6330 Cham, Switzerland

If disposing of this product, please recycle the paper.

Preface

These conference proceedings are a collection of the best papers accepted by ITTA 2024—the 2nd International Conference on Information Technologies and Their Applications, held on April 23–25, 2024 in Baku, Azerbaijan. The conference was organized by leading educational and research organizations from five countries: Ministry of Science and Education of the Republic of Azerbaijan, Azerbaijan University of Architecture and Construction, Institute of Control Systems, Azerbaijan, Institute of Information Technology, Azerbaijan, Bydgoszcz University of Science and Technology, Poland, Ege University, Türkiye, V. M. Glushkov Institute of Cybernetics, Ukraine, and Institute of Information and Computational Technologies, Kazakhstan. The conference was chaired by Gulchohra Mammadova, Rector of the Azerbaijan University of Architecture and Construction.

The conference topics were as follows:

1. Information technologies in construction and architecture (including smart building, smart village, smart city)
2. Artificial intelligence and its applications
3. Problems and methods of big data processing
4. Parallel computing
5. Problems and methods of signal processing
6. Information technologies in the management of complex objects and processes
7. Intelligent control systems
8. Intelligent information systems
9. Computer networks and cyber security issues
10. Application of information technologies in education

The conference work was organized in the following sections:

- Information Technology in Intelligent Systems
- Information Technology in Construction, Industry, and Engineering
- Information Technology in Modeling
- Information Technology in Decision Making

324 participants (including presenters) registered to attend the conference. Totally 200 papers were submitted. The papers were submitted and reviewed through the EasyChair system. All the papers were single-blind peer reviewed by leading information technology experts, with submissions receiving three reviews on average. The reviewers evaluated the compliance of research papers with the conference topics, their scientific novelty and practical value. Out of 200 papers submitted to the conference, 108 were included in the conference program. Presentations were made in a hybrid format: 68 presentations were delivered in-person, and 40 presentations were delivered online. 60 papers, evaluated by the reviewers as the best of the presented papers and approved by the Program Committee, were selected to be included in the Conference Proceedings.

Five leading information technology researchers were invited as plenary speakers: Telman Aliev (Azerbaijan University of Architecture and Construction), Boris Mordukhovich (Wayne State University, USA), Sedat Akleylek (University of Tartu, Estonia), Efendi Nasiboglu (Dokuz Eylul University, Türkiye), and Adil Bagirov (Federation University Australia, Australia).

April 2024

Gulchohra Mammadova
Telman Aliev
Kamil Aida-zade

Organization

General Chair

Gulchohra Mammadova — Azerbaijan University of Architecture and Construction, Azerbaijan

Program Committee Chairs

Ali Abbasov — Institute of Control Systems, Azerbaijan
Rasim Alguliyev — Institute of Information Technology, Azerbaijan
Telman Aliev — Azerbaijan National Academy of Sciences, Azerbaijan
Kamil Aida-zade — Institute of Control Systems, Azerbaijan

Program Committee Members

Abdullayev, Vagif — Azerbaijan State Oil and Industry University, Azerbaijan
Ahmedzade, Perviz — Ege University, Türkiye
Akleylek, Sedat — Ondokuz Mayis University, Türkiye
Aliguliyev, Ramiz — Institute of Information Technology, Azerbaijan
Aliyev, Fikret — Baku State University, Azerbaijan
Amirgaliyev, Yedilkhan — Institute of Information and Computing Technologies, Kazakhstan
Ashrafova, Yegana — Baku State University, Azerbaijan
Aslanova, Nigar — Azerbaijan University of Architecture and Construction, Azerbaijan
Aytaç, Aysun — Ege University, Türkiye
Bagirov, Adil — Federation University Australia, Australia
Bashirov, Agamirza — Eastern Mediterranean University, Northern Cyprus
Bashirov, Rza — Eastern Mediterranean University, Northern Cyprus
Boruzanlı Ekinci, Gülnaz — Ege University, Türkiye
Chikriy, Arkadiy — V. M. Glushkov Institute of Cybernetics, Ukraine
Gasimov, Vagif — Azerbaijan Technical University, Azerbaijan

Gasimov, Yusif	Azerbaijan University, Azerbaijan
Górecki, Jarosław	Bydgoszcz University of Science and Technology, Poland
Guliyev, Samir	Azerbaijan State Oil and Industry University, Azerbaijan
Gürsoy, Arif	Ege University, Türkiye
Hajiyev, Chingiz	Istanbul Technical University, Türkiye
Ibrahimov, Bayram	Azerbaijan Technical University, Azerbaijan
Jin, Jionghua	University of Michigan, USA
Kasımzade, Azer	Ondokuz Mayıs University, Türkiye
Kasimbeyli, Refail	Eskishehir Technical University, Türkiye
Kırlangıc, Alpay	Ege University, Türkiye
Kiedrowski, Piotr	Bydgoszcz University of Science and Technology, Poland
Mahmudov, Elimhan	Istanbul Technical University, Türkiye
Mammadova, Masuma	Institute of Information Technology, Azerbaijan
Mamyrbayev, Orken	Institute of Information and Computing Technologies, Kazakhstan
Mardani, Ali	Bursa Uludag University, Türkiye
Melikov, Agassi	National Aviation Academy, Azerbaijan
Molina-Moreno, Valentín	University of Granada, Spain
Mordukhovich, Boris	Wayne State University, USA
Musaeva, Naila	Azerbaijan University of Architecture and Construction, Azerbaijan
Nabiyev, Vasif	Karadeniz Technical University, Türkiye
Nasiboglu, Efendi	Dokuz Eylül University, Türkiye
Normatov, Ibrokhimali	National University of Uzbekistan, Uzbekistan
Nunez-Cacho, Pedro	University of Jaén, Spain
Nuriyev, Urfat	Ege University, Türkiye
Oproescu, Mihai	University of Pitesti, Romania
Ordin, Burak	Ege University, Türkiye
Polyak, Roman	George Mason University, USA
Rahimov, Anar	Institute of Control Systems, Azerbaijan
Rahimov, Rustam	Tashkent State Transport University, Uzbekistan
Ramyar, Kambiz	Ege University, Türkiye
Rustamov, Samir	ADA University, Azerbaijan
Sadigov, Aminaga	Institute of Control Systems, Azerbaijan
Sattarova, Ulkar	Ministry of Science and Education, Azerbaijan
Sidorenko, Sergii	National Technical University of Ukraine, Ukraine
Sladkowski, Aleksander	Silesian University of Technology, Poland
Stetsyuk, Petro	V. M. Glushkov Institute of Cybernetics, Ukraine

Organizing Committee Chair

Reshad Ismibayli   Azerbaijan University of Architecture and
                                    Construction, Azerbaijan

Organizing Committee Members

Abbaszade, Narmin	Institute of Control Systems, Azerbaijan
Abdullayev, Nijat	Azerbaijan University of Architecture and Construction, Azerbaijan
Abdullayeva, Gulchin	Institute of Control Systems, Azerbaijan
Ahmadli, Nigar	Institute of Control Systems, Azerbaijan
Akbarli, Reyhan	Azerbaijan University of Architecture and Construction, Azerbaijan
Aliyev, Alakbar	Baku State University, Azerbaijan
Aliyeva, Aygun	Institute of Control Systems, Azerbaijan
Alizada, Tahir	Institute of Control Systems, Azerbaijan
Azimov, Rustam	Institute of Control Systems, Azerbaijan
Chaudhary, Prince Shiva	Intel, USA
Gasimov, Akif	Azerbaijan University of Architecture and Construction, Azerbaijan
Gelashvili, Otar	Georgian Technical University, Georgia
Huseynova, Rana	Azerbaijan University of Architecture and Construction, Azerbaijan
Imanov, Kamran	Intellectual Property Agency, Azerbaijan
Isayev, Mazahir	Azerbaijan University of Architecture and Construction, Azerbaijan
Kasımzade, Azer	Azerbaijan University of Architecture and Construction, Azerbaijan
Kazim-zade, Aydin	Azerbaijan University of Architecture and Construction, Azerbaijan
Khurshudov, Dursun	Azerbaijan University of Architecture and Construction, Azerbaijan
Kılıç, Elgin	Ege University, Türkiye
Kuspangaliyev, Bolat	Institute of Architecture and Civil Engineering, Kazakhstan
Mamedov, Israfil	Azerbaijan University of Architecture and Construction, Azerbaijan
Mamedov, Sabir	Azerbaijan University of Architecture and Construction, Azerbaijan
Mammadli, Maryam	Azerbaijan University of Architecture and Construction, Azerbaijan

Mammadov, Vusal	Institute of Control Systems, Azerbaijan
Nagdiyev, Orxan	Azerbaijan University of Architecture and Construction, Azerbaijan
Orujov, Orkhan	Azerbaijan University of Architecture and Construction, Azerbaijan
Pashayev, Adalat	Institute of Control Systems, Azerbaijan
Rzayev, Ramin	Institute of Control Systems, Azerbaijan
Rzayeva, Narmin	Azerbaijan Technical University, Azerbaijan
Sabziev, Elkhan	Institute of Control Systems, Azerbaijan
Talibzada, Sarkhan	Azerbaijan University of Architecture and Construction, Azerbaijan

Contents – Part II

Information Technology Applied in Construction, Industry, and Engineering

Optimizing the Performance Characteristics of SBS - Geopolymer Modified Bitumen: A Data-Driven Approach with ANN and Design Expert Software 3
    Perviz Ahmedzade, Burak Yiğit Katanalp, Murat Tastan, Çiğdem Canbay Türkyılmaz, and Emrah Türkyılmaz

Innovative Construction Technologies for Smart Villages: Case Study Karabakh 18
    Gulchohra Mammadova and Samira Akbarova

Turkish Legal Single-Document Summarizing 32
    Maha Ahmed Abdullah Albayati and Oğuz Fındık

Technologies and Systems for Monitoring the Onset of Accidents at Strategic Construction 42
    Telman Aliev, Naila Musaeva, Narmin Rzayeva, and Ana Mammadova

Optimizing Feature Distributions for Unsupervised Deep Learning-Based Fabric Defect Detection and Localization 52
    Eissa Alzabidi and Oğuz Fındık

Conservation of Wildlife in Hunting Tourism Using Artificial Intelligence and Image Processing in Smart Firearms 67
    Ufuk Asil and Efendi Nasibov

Optimizing Microarray Gene Selection in Colon Cancer: An Enhanced Metaheuristic Algorithm for Feature Selection 76
    Salsabila Benghazouani, Said Nouh, and Abdelali Zakrani

Securing Vehicular Ad-Hoc Networks: A Blockchain and Software Defined Network Approach for Enhanced Efficiency, Safety, and Security 87
    F. Ajesh, Felix M. Philip, T. Triwiyanto, Danyalov Shafi, Mammadov Sabir, Vusala Abuzarova, Samira Aliyeva, Dursun Khurshudov, Vugar Hacimahmud Abdullayev, Latafat Mikailzade, and Taleh Asgarov

Influence of Process Noise Biases to Satellite Attitude Filters' Estimates 100
  Chingiz Hajiyev and Demet Cilden-Guler

Information-Retrieval System of Rock Pictures of Different Countries
and Continents 111
  Aydin Kazim-zada and Hayat Huseynova

Convergence of Sharpness-Aware Minimization with Momentum 123
  *Pham Duy Khanh, Hoang-Chau Luong, Boris S. Mordukhovich,
  Dat Ba Tran, and Truc Vo*

Vulnerability Assessment and Penetration Testing of University Network 133
  *Dursun Khurshudov, Akif Imanov, Jamalladdin Nuraliyev,
  Malahat Nagiyeva, and Samira Aliyeva*

Preliminary Orbit Estimation for Turksat 5B via Extended and Unscented
Kalman Filtering 144
  Hasan Kinatas and Chingiz Hajiyev

Three-Axis Attitude Determination of a Nanosatellite Using Optimized
TRIAD Algorithms 159
  Orhan Kirci and Chingiz Hajiyev

Computer System of Evaluation of the Mass Exam Results Based
on Recognition of Handprinted Azerbaijani Characters 171
  Elshan Mustafayev and Rustam Azimov

Empirical Analysis of Conformer Impact on the CTC-CRF Model
in Kazakh Speech Recognition 184
  *Dina Oralbekova, Orken Mamyrbayev, Keylan Alimhan,
  and Nurdaulet Zhumazhan*

Ontological Approach to Describing the Interaction of Information
Processes in Intelligent Systems 198
  Valerian Ivashenko and Daniil Shunkevich

LANet: Lightweight Attention Network for Medical Image Segmentation 213
  *Yi Tang, Dmitry Pertsau, Di Zhao, Dziana Kupryianava,
  and Mikhail Tatur*

Lightweight Hybrid CNN Model for Face Presentation Attack Detection 228
  Uğur Turhal, Asuman Günay Yilmaz, and Vasif Nabiyev

Disaggregation Analyses of the External/Internal Parameters for Chloride Diffusion in Concrete Structures: Artificial Intelligence by k-means Algorithm 241
    Enrico Zacchei, Emilio Bastidas-Arteaga, and Ameur Hamani

K-means for Small Earthquakes. Alternative Disaggregation Analyses by Considering Wave Components and Soil Types 252
    Enrico Zacchei and Reyolando Brasil

Algorithm for Distributing Incoming Messages Among Handlers 262
    Ivan Zverev and Anna Zykina

Comparison of Machine Learning Based Anomaly Detection Methods for ADS-B System 275
    Nurşah Çevik and Sedat Akleylek

Image Processing Based Wood Defect Detection 287
    Merve Özkan and Caner Özcan

Anomaly Detection with Machine Learning Models Using API Calls 298
    Varol Sahin, Hami Satilmis, Bilge Kagan Yazar, and Sedat Akleylek

Information Technology in Decision Making

Finding a Minimum Distance Between Two Smooth Curves in 3D Space 313
    Majid Abbasov and Lyudmila Polyakova

On Numerical Solution of Block-Structured Discrete Systems 325
    Jamila Asadova

Detection of Leaks in the Water Distribution Network 336
    Kamil Aida-zade and Yegana Ashrafova

Numerical Solution to a Problem of Optimizing Placement and Flow Rates of Wells 348
    Arzu Bagirov, Tatiana Gunkina, and Alexander Handzel

Time Dilation Principle to Solve Game Problems of Control 360
    Arkadii Chikrii, Greta Chikrii, and Viktor Kuzmenko

Synthesis of Power Control of Moving Sources with Optimization of Measurement Points Location on Heating of the Rod 372
    Vugar Hashimov

"Schedule" System for Universities Under the Bologna Education Process 386
  Gulchohra Mammadova, Reshad Ismibayli, and Sona Rzayeva

Author Index 401

Contents – Part I

Information Technology in Intelligent Systems

2-bit Multiplication Quantum Circuit with Addition 3
    Saadi Andishmand and Sedat Akleylek

Techniques for Determining the Operational and Non-operation Condition of a Technical Facility Based on Estimates of the Relationship Between the Useful Signal and the Noise 17
    Telman Aliev and Naila Musaeva

Unsupervised Attack Detection on MIL-STD-1553 Bus for Avionic Platforms 28
    Mustafa Evcil, Zaliha Yüce Tok, and Mehmet Tahir Sandıkkaya

The Principle of Development of an Intelligent System for Automatic Calibration of Control-Measuring Devices of Aircraft 44
    Mazahir Isayev, Majid Gurbanov, Leyla Mahmudbeyli, Natavan Khasayeva, and Sevil Isgandarova

Research on the Effectiveness of Wireless Networks Cellular Communications Based on Digital Technologies 55
    Bayram G. Ibrahimov, Almaz Aliyev, and Ulvi R. Rafizade

A WABL-Based Two-Dimensional Representative of Fuzzy Numbers 66
    Resmiye Nasiboglu and Efendi Nasibov

Performance Analysis of Queuing-Inventory System with Catastrophes Under (s, Q) Policy 78
    Agassi Melikov, Serife Ozkar, and Laman Poladova

A Comprehensive Comparison of Lattice-Based Password Authenticated Key Exchange Protocols Defined on Modules 91
    Kübra Seyhan and Sedat Akleylek

Computer Technology for Parametric Research of the Hierarchical Structure of a Multicommodity Communication Network with Discrete Flows 106
    Volodymyr Vasyanin and Liudmyla Ushakova

Information Technology in Modeling

Application of Active Experiment Planning Methods for Determining the Coefficients of the Regression Equation to Determine the Optimal Composition of the Mixture in the Production of Ceramic Products 125
   Nigar Abasova and Maryam Mammadli

Determination of the Nonlinear Resistance Coefficient of a Linear Section of a Water Supply Network 142
   Kamil Aida-zade and Samir Quliyev

Selection of Facts Placements for Optimal Control of Power System Modes 157
   Ashraf Balametov, Elman Halilov, Tarana Isayeva, and Tofiq Yaqublu

Optimal Tuning of the Nanosatellite Attitude Controller Using TRIAD-Aided Kalman Filter and Particle Swarm Optimization 172
   Mehmet Fatih Ertürk and Chingiz Hajiyev

Towards Comparison of Various Variants of CKKS 187
   Nargiz Khankishiyeva Hati

Investigation of Observation Conditions and Detection Distance of Aircraft in the Thermal Range 199
   Huseynova Rana and Aliyeva Gunel

Pythagorean Fuzzy Pattern Recognition Model in the Assessment of Social Inclusion Index for Azerbaijan 210
   Gorkhmaz Imanov and Asif Aliyev

An Interactive Intelligent System of Creating a Class Schedule 221
   Reshad Ismibayli and Sona Rzayeva

System Identification for Non-destructive Assessment of Structures: Lessons Learned 235
   Azer A. Kasimzade, Emin Nematli, Eldaniz Mahmudov, and Nicat Huseynli

Calculation of Direction Angles of Video Camera to UAV with Stable and Uniform Motion 251
   E. N. Sabziev and L. N. Nabadova

Algorithm for Circuit Covering the Euclidean Complete Graph 260
   Fidan Nuriyeva

Bridge Model for Individualized Digital NasoAlveolar Molding Using
Uniform Cross-Section Elliptic Segment 275
    Hathaichanok Parakarn, Buddhathida Wangsrimongkol,
    Nawapak Eua-Anant, and Tatpong Katanyukul

Automated and Automatic Systems of Management of a Programs
Package for Making Optimal Decisions 287
    Samir Quliyev

Numerical Solution to a Problem of Determining Parameters of Oil Field
and Boundaries of Their Constancy Domains 300
    Anar Rahimov

Application of the Simulated Annealing Algorithm for Finding the Optimal
Trajectory in the Sense of Construction Cost 313
    Andrey Rychkov and Majid Abbasov

Modeling Quadcopter Stabilization 324
    Adalat Pashayev and Elkhan Sabziev

Determining the Minimum Operational Period of the Refinery Production
Operation Plan 338
    Mikhail Saveliev and Anna Zykina

Using of Ellipsoid Method for Finding Linear Regression Parameters
with L_1-Regularization 350
    Petro Stetsyuk, Viktor Stovba, and Mykola Korablov

On the Calculation of the Boiler Thermal Diagram 363
    Mykola Zhurbenko and Tamara Bardadym

Author Index 375

Information Technology Applied in Construction, Industry, and Engineering

Optimizing the Performance Characteristics of SBS - Geopolymer Modified Bitumen: A Data-Driven Approach with ANN and Design Expert Software

Perviz Ahmedzade[1(✉)] 🆔, Burak Yiğit Katanalp[2] 🆔, Murat Tastan[3] 🆔, Çiğdem Canbay Türkyılmaz[1] 🆔, and Emrah Türkyılmaz[1] 🆔

[1] Samarkand State Architectural and Civil Engineering University, 140147 Samarkand, Uzbekistan
p.ahmedzade@samdaqi.edu.uz
[2] Cukurova University, 01100 Adana, Turkey
[3] Ege University, 35100 İzmir, Turkey

Abstract. In this study, the optimization of rutting and fatigue performance characteristics in SBS and geopolymer composite-modified bitumen was achieved by using an experimental design approach and an Artificial Neural Network (ANN) model. The Design Expert 10.0.3 software was employed to construct an optimal mixture design, taking into account the ratios of SBS and Geopolymer within the bitumen matrix. A reference binder of B 50/70 penetration grade bitumen was used, and modified binders were prepared considering various ratios of SBS and Geopolymer. The factors are defined as SBS and Geopolymer concentrations, and the responses include SHRP rutting ($G^*/\sin \delta$) and fatigue ($G^*\sin \delta$) factors, non-recoverable creep compliance (Jnr), and number of cycles to failure (Nf), which were measured through Dynamic Shear Rheometer (DSR), Multiple Stress Creep Recovery (MSCR), and Linear Amplitude Sweep (LAS) tests. In experimental design, four central composite design (CCD) models were constructed, and variance analysis was performed to investigate the statistical significance of the model results. In ANN model, the multilayer perceptron was employed. Findings indicated that a higher percentage of Geopolymer in SBS modified bitumen is considered a promising option for reducing the SBS content, thus offering a potential pathway for the enhancement of the performance characteristics of the modified bitumen.

Keywords: SBS · Geopolymer · Central Composite Design · ANN · MSCR · LAS

1 Introduction

Ecofriendly technologies, such as geopolymerization, facilitate the recycling of byproducts, enabling the reutilization of these waste materials in diverse applications. Geopolymerization is considered a clean technology based on the procedure of activating aluminosilicate-containing materials in an alkaline environment, first introduced by Davidovits [1]. The aluminosilicate sources of geopolymer materials are industrial and agricultural wastes such as fly ash [2], furnace slag or dust [3], metakaolin [4], red mud [5], rice husk [6], etc. Alkaline activator typically considered as NaOH, Na_2SiO_3, KOH and K_2SiO_3 or combinations thereof in different proportions [7]. Previously, the microstructural properties of geopolymer were found to be highly mesoporous and amorphous or semi-crystalline, further attributing the rapid dissolution of glassy contents in alkaline environments [8]. The unique 3D-structure consisting of linked oxygen atoms with tetrahedral AlO_4 units provides many advantages including fire resistance, superior durability, and high comprehensive strength [9]. Reduced CO_2 release during geopolymer production, fast hardening of geopolymer gel and strong interface bonding ability are the most common benefits of the geopolymer when compared with traditional cement [10]. On the other hand, the lack of standardized geopolymer production and evaluation techniques and synthesis difficulties arising from reaction speed prevent the using high amount of geopolymer concrete in building industry.

Over the last several years, scholars have focused on analyzing the potential usage of geopolymer materials in asphalt applications. Prior studies have shown that adding geopolymer to the base binder reduces mixing and compaction temperature requirements, improves thermal stability, and provides high strength under loading conditions especially moderate test temperatures [11]. Tang, Deng [12] demonstrated that the inclusion of 10% geopolymer in the bituminous binder increases its mechanical strength by 2.7%. Previous works showed that the introduction of geopolymers into bitumen significantly influences its viscoelastic characteristics. The improvement in rutting resistance of the asphalt binder becomes notably pronounced when the modification ratio exceeds 5% [13]. In a more recent study by Gao, Hao [14], the impact of geopolymers derived from coal liquefaction waste on emulsified bitumen was explored. Findings revealed significant improvements in the high-temperature rheological characteristics of the geopolymer modified binder, along with an extended fatigue life. Katanalp and Ahmedzade [13] illustrated that the geopolymer-modified binder exhibits significantly higher values for complex modulus (G^*) and rutting factor ($G^*/\sin\delta$) within the temperature range of 58 °C to 70 °C when compared to the reference bitumen. Hamid, Baaj [15] noted that a reduction in dissipated energy, indicative of improved asphalt rutting resistance, was attainable within different frequencies through the incorporation of geopolymers.

Geopolymer modification not only improves the engineering characteristics of asphalt but also offers the potential for sustainable solutions to mitigate environmental challenges. Tang, Yang [16] investigated the particulate matter (PM) and volatile organic compound emissions (VOCs) of the geopolymer incorporated bitumen. In addition, studies have indicated that geopolymer modification can lead to lower viscosity values, enhancing workability and reducing energy requirements [13]. Milad, Ali [11] asserted that when compared to conventional binder, geopolymer-modified asphalt has the potential to offer economic benefits of up to 50%. Katanalp and Ahmedzade [13] stated that incorporating 15% geopolymer by weight into a 3.5% styrene butadiene styrene (SBS) modified binder results in a minimal increase of only 0.05% in the overall production costs. Hamid, Baaj [15] revealed that binders modified with 2% SBS and 8%12% geopolymer modified binders displayed comparable performance. Katanalp and

Ahmedzade [13] conducted research on the combined utilization of SBS and geopolymer, with their findings indicating that the most noteworthy improvements were achieved through hybrid modification. Within the frequency range of 10^{-3} Hz to 10^2 Hz and at a test temperature of 20 °C, Hamid, Baaj [17] observed that the utilization of 8% and 12% geopolymer modification rates resulted in higher G* values and lower δ values when compared to both unmodified bitumen and bitumen modified with 2% SBS.

Overall, the application of geopolymers in the asphalt industry as an alternative or complementary to widely used polymer modifiers like SBS offers numerous benefits. However, as the number of components in the designed structure increases, there is a greater demand for specimens, thereby requiring a larger number of test repetitions [18]. The precise determination of raw material percentages for modified bitumen or asphalt mixtures often necessitates a significant number of repetitive experiments. In this regard, computer-aided methodologies such as Design Expert or Machine Learning (ML) applications emerge as valuable tools for predicting experimental outcomes with a reduced reliance on a large number of data points [19].

1.1 Research Objective

The biggest challenges in the asphalt industry are the rutting and fatigue deteriorations of the pavement due to the repetitive traffic loads, environmental conditions, and inherent oxidation tendency of the bitumen. The previous studies by Hamid, Baaj [15], Hamid, Baaj [17] and Katanalp and Ahmedzade [13] demonstrated the effectiveness of blending geopolymer and SBS on bitumen modification to prevent permanent deformation. However, the optimal combination of SBS and geopolymer in bitumen modification still needs to be determined. Accordingly, a central composite design (CCD) method and an artificial neural network (ANN) model were used to analyze the impact of factors like SBS, geopolymer, and bitumen ratios in the resulting blends, along with responses including Superpave rutting (G*/sinδ) and fatigue (G*sinδ) factors, non-recoverable creep compliance (Jnr), as well as number of cycles to failure (Nf), from Dynamic Shear Rheometer (DSR), Multiple Stress Creep Recovery (MSCR) and Linear Amplitude Sweep (LAS) analyses. To achieve this, CCD models for each response variable were constructed based on abovementioned features. Subsequently, the data extracted from the CCD results were provided as a training set to the ANN. The trained ANN model was then tested with the initial laboratory data to assess the efficiency and compliance of both CCD and ANN models.

2 Material and Method

2.1 Raw Materials, Geopolymerization and Bitumen Modification

The neat bitumen utilized was PG 64–22, obtained from TUPRAS in İzmir, Turkey, with specific gravity, penetration, and softening point of 1.035, 62 mm^{-1}, and 47°C, respectively. Geopolymerization involved the use of waste Electric Arc Furnace Fume (EAFF) and Fly Ash (FA) as aluminosilicate sources, along with an alkaline activator prepared from NaOH and Na_2SiO_3. The production of SBS-geopolymer composites incorporated the SBS copolymer Kraton D1101. Geopolymers were formulated with a 1.5 SiO_2/Na_2O ratio (MS). This was achieved by blending 10.2 g of Na_2O with every 100 g of Na_2SiO_3. The ratio of Electric Arc Furnace Fume (EAFF) to Fly Ash (FA) was maintained at 2:1 [29]. Following a 10-min mechanical mixing process, the geopolymer mortars were molded, cured at 60 °C for 3 days, and subsequently milled using a RETSCH PM100 ball mill to obtain geopolymer dust. The modified binders were categorized into three groups: geopolymer-only contained binders (GMB), SBS-geopolymer contained binders (S-GMB), and only SBS-contained binders (SMB). GMBs were produced at a temperature of 150°C, while SMB and S-GMB groups were prepared at 180°C and a shear rate of 2500 rpm.

2.2 Sample Characterization

Rutting and Fatigue Performance with DSR Analyses

DSR analyses were conducted following AASHTO T315 [20]. The tests were performed at constant frequency of 1.59 Hz. The test temperature of 64°C and 25-mm diameter parallel plate test geometry was used on unaged binders to obtain G*/sinδ SHRP rutting factor. Additionally, tests were conducted at 25 °C using an 8-mm diameter test geometry on long-term aged specimens (in accordance with AASHTO T 391–20 [21]) to obtain G*sinδ SHRP fatigue factor.

Rutting Performance with MSCR Analyses

MSCR tests are employed in accordance with AASHTO T 350–14 [22] at 64 °C test temperature and under 0.1 kPa and 3.2 kPa stress levels. Short-term aged bitumen samples with respect to AASHTO T240 [23] were used for analyses. Analyses were conducted at three phases. During analyses, specimens underwent 20 creep and recovery cycles at 0.1 kPa (with the first 10 cycles used to reach a steady-state condition), followed by an additional 10 cycles at 3.2 kPa stress levels. Following that, the Jnr parameter was determined as the ratio of the deformation caused by each loading cycle to the corresponding stress level as $(\varepsilon c - \varepsilon 0)/\tau$.

Fatigue Performance with LAS Analyses

LAS tests were utilized according to the AASHTO T 391–20 [21] specification. Long-term aged bitumen samples were used in the analyses and the test temperature was set at 25°C. Analyses were consisted of two phases. In the first phase, samples were subjected 0.1% strain level at the frequency range of 0.2 Hz–30 Hz to investigate undamaged properties using G* and δ. In the second phase, the strain level elevated from 0 to 30% at a constant frequency of 10 Hz to analyze the samples under accumulated damage effect. Using the material structural integrity (A) as a function of accumulated damage and material response of the strain changes (B), the number of cycles to failure was obtained as $Nf = A \times (\gamma)^{-B}$, at different strain levels (γ).

2.3 Central Composite Design

Design Expert 10.0.3 software was employed to implement the CCD. Based on CCD, the design process unfolded in three main steps. The initial step involved defining the factors and response variables to construct the model. Subsequently, the statistical analysis of variance (ANOVA) was applied to assess model significance. Afterward, regression coefficients for each variable were derived, and the interpretation of the factor-response relationship, identified using Eq. 1, was undertaken.

$$Y = b_o + \sum_{i=1}^{k} b_i X_i + \sum_{i=1}^{k} b_{ii} X_i^2 \sum_{i=1}^{k-1} \sum_{j=i+1}^{k} b_{ij} X_i X_j + e \quad (1)$$

where Y is the dependent variable, b is regression coefficient, X is the independent variable, k is number of parameters and e denotes random error. The independent variables are SBS and Geopolymer ratio in the composite blend. The responses include G*/sinδ, G*sinδ, Jnr (at τ = 3.2 kPa), and Nf (at γ = 5%). Details of the factors (X_i) and responses (Y_i) are presented in Table 1.

Table 1. Details of the design parameters.

Parameter	Unit	Min	Max	Mean	Std. Dev
SBS (X_1)	%	2.5	3.5	3	0.35
Geopolymer (X_2)	%	5	15	10	3.53
G*/sinδ (Y_1)	kPa	3.3	7.02	4.84	1.08
G*sinδ (Y_2)	kPa	2181	2943	2495.23	221.71
Jnr @ τ = 3.2 kPa (Y_3)	kPa−1	0.78	1.65	0.96	0.23
Nf @ γ = %5 (Y_4)		685.07	905.29	812.67	60.74

2.4 Artificial Neural Network

To ensure comparable results with CCD, a separate model was developed for each output variable. This study employed a multilayer perceptron (MLP) with 1 input layer, 2 hidden layers with 5 neurons, and 1 output layer to analyze the aforementioned responses of the test results. For the MLP, the activation function was chosen as the rectified linear unit function (relu), the learning rate was set at 0.5, momentum coefficient was 0.9, and the stochastic gradient-based optimizer 'Adam' was used to update weights after each iteration.

3 Results

3.1 Results of the CDD Analyses

Experimental dataset structured using DSR, MSCR and LAS test results regarding the different proportions of the modified binders. As outlined in Table 2, a total of 13 laboratory experiments were carried out in a randomized order, including five replications of the central point to enhance the precision of estimating experimental errors. Prima facie, it can be inferred from the dataset that the SBS (X_1) at a high factorial level is associated with an increase in the $G^*/\sin\delta$ (Y_1) and Nf (Y_4) responses, while the $G^*\sin\delta$ (Y_2) and Jnr (Y_3) values show a decrease. This indicates that the interaction between SBS and bitumen enhances the viscoelastic response capacity of the binder, contributing to higher rutting and fatigue resistance [24]. Examining the geopolymer content (X_2), it becomes evident that the low factorial level (-1) is associated with a lower rutting factor and higher Jnr values, as observed in experimental runs 6, 7, and 2.

Table 2. CDD matrix and values of the response variables.

Run	Space	X_1	X_2	Y_1	Y_2	Y_3	Y_4
1	Factorial	+1	−1	6.02	2181	0.90	865.36
2	Axial	0	+1	4.88	2516	0.78	741.63
3	Axial	+1	0	6.48	2224	0.85	884.86
4	Center	0	0	4.69	2457	0.89	824.17
5	Factorial	−1	−1	3.30	2734	1.65	768.92
6	Center	0	0	4.72	2487	0.89	831.17
7	Axial	0	−1	4.44	2403	0.92	806.47
8	Axial	−1	0	3.62	2822	1.17	758.56
9	Center	0	0	4.65	2398	0.90	829.23
10	Factorial	−1	+1	3.73	2943	1.01	685.07
11	Center	0	0	4.71	2481	0.89	833.6
12	Factorial	+1	+1	7.02	2314	0.77	905.29
13	Center	0	0	4.72	2478	0.88	830.44

Table 3. ANOVA results of quadratic polynomial model for $G^*/\sin\delta$.

Source	SS	df	MS	F-value	p-value	Observation
Model ($G^*/\sin\delta$)	14.10	5	2.82	539.23	< 0.0001	significant
X_1	13.05	1	13.05	2494.5	< 0.0001	-
X_2	0.595	1	0.59	113.77	< 0.0001	-
$X_1 X_2$	0.087	1	0.087	16.63	0.0047	-
X_1^2	0.342	1	0.342	65.48	< 0.0001	-
X_2^2	0.003	1	0.003	0.752	0.4144	-
Residual	0.036	7	0.005	-	-	-
Lack of Fit	0.001	4	0.0004	9.201	0.0269	significant

To gain a better understanding of the relationship between factors and responses, ANOVA tests were conducted. A total of 4 quadratic polynomial models were fitted each one representing one of the response variables. Models were chosen based on the highest determination coefficient (R^2) and lowest sequential p-value. ANOVA findings of the models are presented in Tables 3, 4, 5 and 6. The confidence interval (CI) was utilized as %95. As can clearly be seen in Table 3, general model for $G^*/\sin\delta$ and X_1, X_2, $X_1 X_2$, and X_1^2 terms were found statistically significant with lower p-values. The model F-value was found as 539.23, which indicates that there is %0.01 chance than an F-value this large occur due to noise. In Table 4, regarding the $G^*\sin\delta$ model, the terms of X_2^2 and $X_1 X_2$ are excluded from the final model due to the greater p-value than 0.05.

Table 4. ANOVA results of quadratic polynomial model for $G^*\sin\delta$.

Source	SS	df	MS	F-value	p-value	Observation
Model ($G^*\sin\delta$)	583078	5	116615	120.40	< 0.0001	significant
X_1	528066	1	528066	545.20	< 0.0001	-
X_2	34504	1	34504	35.62	0.0006	-
$X_1 X_2$	1444	1	1444	1.490	0.2616	-
$X_1 2$	14634	1	14634	15.10	0.0060	-
X_2^2	238.52	1	238.52	0.2467	0.6349	-
Residual	6779.9	7	968.56	-	-	-
Lack of Fit	1433	3	477.72	0.357	0.7878	not significant

Table 5. ANOVA results of quadratic polynomial model for Jnr.

Source	SS	df	MS	F-value	p-value	Observation
Model (Jnr)	0.59	5	0.12	23.23	0.0003	significant
X_1	0.28	1	0.28	55.36	0.0001	-
X_2	0.14	1	0.14	27.02	0.0013	-
$X_1 X_2$	0.067	1	0.067	13.00	0.0087	-
$X_1 2$	0.084	1	0.084	16.39	0.0049	-
X_2^2	0.0007	1	0.0007	0.15	0.7084	-
Residual	0.036	7	0.005	-	-	-
Lack of Fit	0.035	3	0.011	211.21	< 0.0001	significant

The "lack of fit" p-value for the $G^*\sin\delta$ model was 0.7878 with an F-value of 0.357, suggesting that there is approximately a 78% chance of such a large F-value occurring. In this case, all models had a significant "lack of fit" p-value except for $G^*\sin\delta$. Nevertheless, it was observed that all models exhibited reasonable agreement between

Table 6. ANOVA results of quadratic polynomial model for Nf.

Source	SS	df	MS	F-value	p-value	Observation
Model (Nf)	42163	5	8432.74	27.93	0.0002	significant
A	32702	1	32702.26	108.30	< 0.0001	-
B	1971.4	1	1971.46	6.53	0.0378	-
AB	3830.3	1	3830.37	12.68	0.0092	-
A2	356.25	1	356.25	1.18	0.3134	-
B^2	3639.8	1	3639.89	12.05	0.0104	-
Residual	2113.7	7	301.97	-	-	-
Lack of Fit	2065.0	3	688.35	56.52	0.0010	significant

the predicted and adjusted R^2 values. This indicates that the proposed models for all responses can be effectively utilized to explore the design space, as discussed in [25]. The model equations for responses are given in Eq. 2, 3, 4 and 5. The synergistic and antagonistic effects of the model terms on response variable can be identified with positive or negative signs of the terms.

$$G^*/sin\delta = 4.70 + 1.47X_1 + 0.31X_2 + 0.15X_1X_2 + 0.35X_1^2 - 0.038X_2^2 \quad (2)$$

$$G^*sin\delta = 2457.3 - 296.6X_1 + 75.83X_2 - 19X_1X_2 + 72.7X_1^2 + 9.29X_2^2 \quad (3)$$

$$Jnr = 0.88 - 0.22X_1 - 0.15X_2 + 0.13X_1X_2 + 0.17X_1^2 - 0.018X_2^2 \quad (4)$$

$$Nf = 824.1 + 73.83X_1 - 18.13X_2 + 30.94X_1X_2 + 11.36X_1^2 - 36.3X_2^2 \quad (5)$$

A higher regression coefficient is indicative of a more substantial impact of the related factor on response prediction. The degree of fit for the models was assessed using R^2 values, calculated as 0.97, 0.98, 0.94, and 0.95 for $G^*/sin\delta$, $G^*sin\delta$, Jnr, and Nf models, respectively. A R^2 value closer to 1 signifies a well-fitted model, indicating reasonable agreement between dependent and independent variables. To further assess the predictive capabilities of the models, graphs comparing actual and predicted data were plotted.

In Figs. 1, 2, 3, and 4, corresponding values, along with lines of equality and surface graphs illustrating multifactor relationships with responses, are presented. It is evident that all points of the response data align well with the equality lines, signifying enhanced precision between model predictions and experimental results. In addition, the fitting lines suggest that small differences between actual and predicted values were observed.

The 3D-surface graphs within the 95% CI, facilitated the assessment of the impact of composite component ratios on the response variable beyond the tested points. Note that surface spaces in the graphs can be applicable to navigate the data within related ranges, with -adequate precision (AP) values greater than 4. In this case AP values were

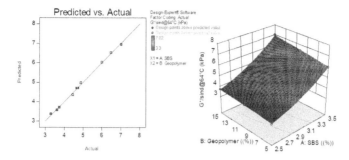

Fig. 1. Actual vs. Predicted data for G*/sinδ model.

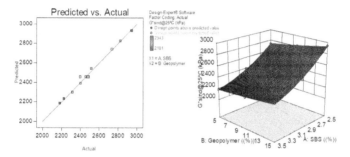

Fig. 2. Actual vs. Predicted data for G*sinδ model.

obtained as 72.84, 35.23, 16.95 and 19.73 for G*/sinδ, G*sinδ, Jnr, and Nf models, respectively. Prior studies have indicated that higher geopolymer modification leads to a stiffer binder, as also inferred from the G*/sinδ graphs in Fig. 1. For G*sinδ, it is evident that SBS content has a significantly decreasing effect on the fatigue factor, while the presence of geopolymer slightly increases it. In addition, when the geopolymer content exceeded 10%, the elastic capability imparted by SBS became less conspicuous. During cyclic loading, when the strain rebound in response to applied stress unfolds over a more extended duration and at a lower magnitude than anticipated, it signifies that viscous component of the total strain energy surpasses the elastic part.

In Fig. 3, it is seen that both geopolymer and SBS content in composite blend have decreasing effect on Jnr values, indicating that the ratio of the deformation after creep portion to applied stress is lower in SBS-geopolymer modified binders. The lower Jnr represents higher deformation resistance at high temperatures related with rutting. When compared the surface graphs between G*/sinδ and Jnr models, it is evident that the increasing geopolymer effect is more apparent at Jnr values.

Regarding the Nf (cycles to failure) in Fig. 4, an increase in SBS content in the resulting blend is observed to enhance the binder's fatigue life under accumulated damage, while geopolymer content reduces the Nf values. An interesting finding is that, despite the negative effect of geopolymer on Nf values, this trend is not observable at elevated SBS ratios. The highest Nf values are observed in modified bitumen containing 8% to

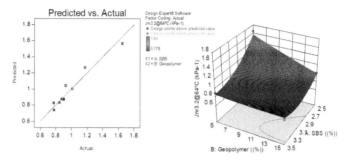

Fig. 3. Actual vs. Predicted data for Jnr model.

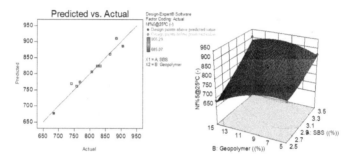

Fig. 4. Actual vs. Predicted data for Nf model.

11% geopolymer and 3.5% SBS. Consequently, it can be suggested to use geopolymer with a high amount of SBS to achieve a longer fatigue life.

The multi-objective numerical optimization was carried out to determine the best combination of SBS and geopolymer in modified binder considering enhanced permanent deformation resistance and fatigue life. To do so, the desired goal of for SBS was chosen "minimize" and for geopolymer was chosen as "in range", while maximizing the G*/sinδ, Nf, and minimizing the Jnr and G*sinδ responses. Ordering of the alternatives and determining the best option were done by utilizing desirability function. Accordingly, geometric mean of free scale values of responses (d_i), was calculated by the Eq. 6, where the N is the total number of response variables.

$$D = \frac{1}{\left[\prod_{n=1}^{N} d_n\right]^N} \tag{6}$$

Desirability is considered as a critical criterion that evaluates expected outcomes within a range of 0–1, depending on the proximity of predicted responses to the model's lower and upper limits. Different solutions obtained from the CCD with desirability values are presented in Table 7.

As can be seen in Table 7, the best solution with desirability value of 0.773 was obtained at 3.39% SBS and 11.50% geopolymer blend. The other options have different geopolymer concentrations, while SBS content were relatively similar in composite

Table 7. Optimized CDD solutions with desirability values.

Solution	SBS	Geopolymer	G*/sinδ	G*sinδ	Jnr	Nf	D
1	3.39	11.50	6.19	2289.1	0.798	887.3	0.773
3	3.31	7.13	5.51	2665.3	0.807	862.7	0.717
2	3.30	11.38	5.85	2320.3	0.788	871.6	0.697
4	3.26	14.18	5.88	2381.1	0.751	840.1	0.682
5	3.13	11.05	5.18	2398.5	0.805	840.5	0.604

modifiers. The optimization results were found acceptable with higher fatigue life and permanent deformation resistance.

3.2 Results of the ANN Analyses.

The ANN model was implemented using the Python programming language. The training data, comprising 971 samples, was derived from optimized CDD surfaces. To achieve this, all constraints were eliminated from the optimization process, and responses and factors were assessed at the "in-range" level. As a result, the desirability values for each collected data point were set to 1.00. Descriptive statistics of the training dataset are given in Table 8. During the data preprocessing phase, all values were normalized to fall within the range of 0 to 1. Following normalization, the training data was partitioned into two sets, with 80% allocated for fitting the network and 20% reserved for network validation.

Table 8. Details of the design parameters.

Parameter	Min	Max	Mean	Std. Dev
SBS	2.500	3.498	2.991	0.295
Geopolymer	5.010	14.99	10.03	2.921
G*/sinδ	3.422	6.873	4.781	0.888
G*sinδ	2188.0	2927.7	2491.8	183.7
Jnr	0.729	1.495	0.942	0.170
Nf	679.0	909.8	813.9	47.98

The models were structured with one of the variables (G*/sinδ, G*sinδ, Jnr, or Nf) designated as the output, while the remaining were outlined as input variables. Thus, the composition of each MLP model consists of 5 input variables and 1 output variable. In Fig. 5, the graph depicts the variation in error concerning iteration. Notably, the maximum iteration number was fixed at 1000, and the momentum coefficient, representing the rate at which the weight factors are adjusted, was set to 0.9.

As illustrated in Fig. 5, each model achieved the minimum MSE value at distinct iterations. In Figs. 5c and 5d, the chosen momentum coefficient effectively eliminated local minima for the error term. Moreover, the models exhibited MSE values very close to zero, signifying well-fitted models during the training process. In Fig. 6, it is evident that the actual vs. predicted data points in validation process closely align with line of equality. Take note that data points falling below the line suggest predicted values that are lower than the actual values, whereas data points above the line signify the converse. Despite achieving high R2 values in validation, the models exhibited slightly diminished performance when tested on laboratory data. Notably, the R2 values for G*/sinδ, Jnr, Nf, and G*sinδ predictions during testing were recorded as 0.87, 0.83, 0.84, and 0.90, respectively.

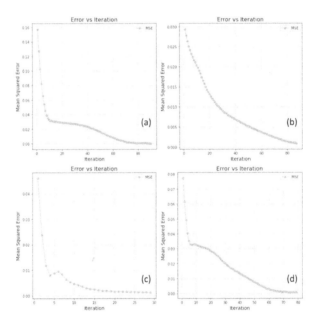

Fig. 5. Error vs. Iteration at training process for (a) G*sinδ (b) Jnr (c) Nf (d) G*/sinδ.

The CCD and ANN prediction performances were presented in Table 9. The predictive capability of CCD surpassed that of ANN, as evidenced by higher R^2 values and diminished error terms across all responses. Notably, the disparities in R^2 values between CCD and ANN models for G*/sinδ, G*sinδ, Jnr, and Nf parameters were determined to be 9.83%, 12.44%, 11.82%, and 12.01%, respectively. Considering that the ANN training data was structured from CCD results and the testing data was the CCD design matrix, it can be stated that the most compatible results were obtained for optimizing the G/sinδ parameter.

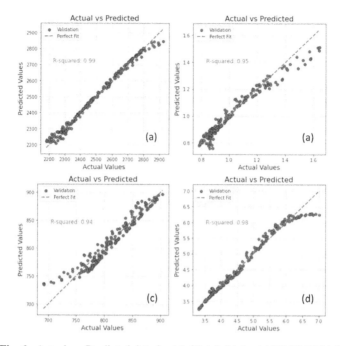

Fig. 6. Actual vs. Predicted data for (a) G*sinδ (b) Jnr (c) Nf (d) G*/sinδ.

Table 9. Comparison of the CDD and ANN performances.

Parameter	R2		MAE		RMSE	
	CCD	ANN	CCD	ANN	CCD	ANN
G*/sinδ	0.997	0.899	0.044	0.221	0.053	0.331
G*sinδ	0.988	0.865	17.32	62.45	22.83	78.12
Jnr	0.939	0.828	0.039	0.075	0.052	0.091
Nf	0.949	0.835	9.443	18.467	12.75	23.67

4 Conclusion

Building upon the achievements discussed in the previous section, the key findings of the study are outlined below.

- ANOVA results demonstrated the statistical significance of models structured using the CCD method, with p-values < 0.001.
- The model coefficients indicated that the combination of geopolymer and SBS had a positive effect on rutting and fatigue deteriorations. Furthermore, the impact of SBS on the changes in G*/sinδ and G*sinδ SHRP parameters was notably more pronounced when compared to geopolymer.

- According to the optimization results, the most effective combination for enhancing bitumen fatigue life and rutting resistance was achieved by blending 3.40% SBS with 11.50% geopolymer, yielding a desirability value of 0.77.
- The best model performance was achieved in G*/sinδ prediction for both the, CCD and ANN models, attaining R^2 values of 0.997 and 0.899, respectively.
- Based on the evaluation using MAE, R^2, and RMSE goodness-of-fit metrics, the CCD demonstrated better prediction capability in comparison to the ANN model for all responses.

Overall, the prediction results and model comparisons indicated that both the ANN and CCD computer-aided methods are capable of predicting the asphalt binder's performance, considering parameters such as G*/sinδ, G*sinδ, Jnr, and Nf. This study specifically focused on fatigue and rutting performance indicators to identify the optimal combination of SBS and geopolymer blends in bitumen modification. Given the complex nature of bitumen, characterized by unique temperature, time, and load-dependent behavior, there is a need for comprehensive exploration. Multiple indicators for asphalt binder performance can be considered to optimize both modified asphalt binder performance and component proportions. Consequently, for future studies, it is recommended to explore various test parameters, encompassing low-temperature performance, adhesion capability, and asphalt mixture performance tests.

References

1. Davidovits, J.: Geopolymers and geopolymeric materials. J. Therm. Anal. **35**, 429–441 (1989)
2. Ahmad, A., et al.: Compressive strength prediction of fly ash-based geopolymer concrete via advanced machine learning techniques. Case Studies in Construction Mat. **16**, e00840 (2022)
3. Khater, H.: Hybrid slag geopolymer composites with durable characteristics activated by cement kiln dust. Constr. Build. Mater. **228**, 116708 (2019)
4. Fu, C., et al.: Alkali cation effects on chloride binding of alkali-activated fly ash and metakaolin geopolymers. Cement Concr. Compos. **114**, 103721 (2020)
5. Bai, B., et al.: A high-strength red mud–fly ash geopolymer and the implications of curing temperature. Powder Technol. **416**, 118242 (2023)
6. Pham, T.M.: Enhanced properties of high-silica rice husk ash-based geopolymer paste by incorporating basalt fibers. Constr. Build. Mater. **245**, 118422 (2020)
7. Shilar, F.A., et al.: Optimization of alkaline activator on the strength properties of geopolymer concrete. Polymers **14**(12), 2434 (2022)
8. Cong, P., Cheng, Y.: Advances in geopolymer materials: a comprehensive review. J. Traffic and Transportation Eng. (English Edition) **8**(3), 283–314 (2021)
9. Ahmed, H.U., et al.: Compressive strength of geopolymer concrete composites: a systematic comprehensive review, analysis and modeling. Eur. J. Environ. Civ. Eng. **27**(3), 1383–1428 (2023)
10. Zhao, J., et al.: Eco-friendly geopolymer materials: a review of performance improvement, potential application and sustainability assessment. J. Clean. Prod. **307**, 127085 (2021)
11. Milad, A., et al.: Utilisation of waste-based geopolymer in asphalt pavement modification and construction—a review. Sustainability **13**(6), 3330 (2021)
12. Tang, N., et al.: Geopolymer as an additive of warm mix asphalt: preparation and properties. J. Clean. Prod. **192**, 906–915 (2018)

13. Katanalp, B.Y., Ahmedzade, P.: Rheological evaluation and life cycle cost analysis of the geopolymer produced from waste ferrochrome electric arc furnace fume as a composite component in bitumen modification. J. Mater. Civ. Eng. **35**(11), 04023401 (2023)
14. Gao, Y., et al.: Interaction, rheological and physicochemical properties of emulsified asphalt binders with direct coal liquefaction residue based geopolymers. Constr. Build. Mater. **384**, 131444 (2023)
15. Hamid, A., Baaj, H., El-Hakim, M.: Rutting behaviour of geopolymer and styrene butadiene styrene-modified asphalt binder. Polymers **14**(14), 2780 (2022)
16. Tang, N., et al.: Reduce VOCs and PM emissions of warm-mix asphalt using geopolymer additives. Constr. Build. Mater. **244**, 118338 (2020)
17. Hamid, A., Baaj, H., El-Hakim, M.: Temperature and aging effects on the rheological properties and performance of geopolymer-modified asphalt binder and mixtures. Materials **16**(3), 1012 (2023)
18. Katanalp, B.Y., et al.: Atik kömür katkili asfalt betonu performans karakteristiklerinin yapay sinir ağlari ve merkezi kompozit tasarim yöntemleri kullanilarak karşilaştirilmasi. Mühendislik Bilimleri ve Tasarım Dergisi **7**(3), 680–688 (2019)
19. Ndepete, C.P., et al.: Exploring the effect of basalt fibers on maximum deviator stress and failure deformation of silty soils using ANN, SVM and FL supported by experimental data. Adv. Eng. Softw. **172**, 103211 (2022)
20. AASHTO T315, Standard method of test for determining the rheological properties of asphalt binder using a dynamic shear rheometer (DSR). American Association of State Highway and Transportation Officials (2012)
21. TP101, A., Standard method of test for estimating fatigue resistance of asphalt binders using the linear amplitude sweep. AASHTO: Washington, DC, USA (2012)
22. Aashto, T.: Standard method of test for multiple stress creep recovery (MSCR) test of asphalt binder using a dynamic shear rheometer (DSR). American Association of State Highway and Transportation Officials, Washington, DC (2018)
23. Aashto, T.: Effect of heat and air on a moving film of asphalt binder (rolling thin-film oven test). AASHTO Standard Specifications for Transportation Materials and Methods of Sampling and Testing. American Association of State Highway and Transportation Officials: Washington DC (2011)
24. Kim, T.W., et al.: Fatigue performance evaluation of SBS modified mastic asphalt mixtures. Constr. Build. Mater. **48**, 908–916 (2013)
25. Nassar, A.I., Thom, N., Parry, T.: Optimizing the mix design of cold bitumen emulsion mixtures using response surface methodology. Constr. Build. Mater. **104**, 216–229 (2016)

Innovative Construction Technologies for Smart Villages: Case Study Karabakh

Gulchohra Mammadova and Samira Akbarova[✉]

Azerbaijan University of Architecture and Construction, 11 A. Sultanova,
Baku Az1073, Azerbaijan
samira.akbarova@azmiu.edu.az

Abstract. Currently, Azerbaijan is rebuilding the once-destroyed settlements of Karabakh, implementing the concept of a "smart" village. Despite a significant number of such projects around the world, the level of their theoretical and methodological validity is not sufficient. This paper examines the construction technologies for sustainable smart villages, focusing on innovative energy-efficient engineering systems and measures. The subject of the study is a set of theoretical and practical aspects of the realization of the smart village concept to justify the effectiveness of the applied innovative construction technologies in the realized pilot project of smart village Agali in Karabakh. The methodological basis of the work was the comparative, statistical, and logical analysis of foreign academic studies on the problems of rural development. The main conceptual components of the Azerbaijani model of a smart village are described, the effectiveness of applied innovative construction technologies is substantiated, and scientific-practical recommendations for further realization of the smart village concept in the socio-economic realities of Azerbaijan are developed. It is proposed to create a catalog of energy-saving and energy-efficient innovative technologies adaptive for the region; a map of the village competencies with the indication of economic specialization; and a model concept of integrated development of the village with the help of advanced IT solutions and green energy sources, taking into account the availability of resources, specific features, and the historical development path of a village.

Keywords: Green Energy · Efficient Systems · Scientific-practical Recommendations · Catalog of Energy-Saving Technologies · Map of Competencies · Economic Specialization

1 Introduction

In Azerbaijan, the large-scale works are currently underway to restore the once-destroyed villages of the Karabakh region through the implementation of the popular trend of the current century - the concept of a "smart" village [1], which means the most efficient use of the available resources of a particular region through the digitalization of all spheres of human activity and settlement infrastructure with a fully automated an ecosystem that includes automation of electricity, heat, and water supply; the use of innovative construction technologies that reduce resource consumption due to their advanced technical features, as well as a lot of digital trends without a negative impact on the environment [2, 3].

The term smart village is interpreted in different ways, but the key role is given to information-telecommunication technologies, innovations, including construction ones, and green energy supply, which help solve the problems of villages within the framework of a multilateral partnership between citizens, business and government (Fig. 1). This understanding was formed in 1993 in Silicon Valley, USA, where the concept of a smart community appeared [4]. Today, the concept smart village is complemented by the mandatory requirement to ensure sustainable development of rural area, i.e. the concept of a smart sustainable village is more relevant, where integrated intelligent and innovative solutions are used in all spheres of village life in real time through the formation of a unified environmentally friendly system of technologies and services [5]. Currently, there is no a unified list of technologies and innovations that unambiguously characterizes the village as smart and sustainable [6]. But it is possible to highlight the digital transformations and technological improvements in infrastructure management that are most characteristic of a smart village: the use of intelligent metering systems and remote network management for energy delivery, thermal supply, water distribution, and electricity supply systems; the use of innovative control systems allows real-time monitoring of buildings, objects, structures, and infrastructure, promptly detecting and preventing accidents and emergencies, increasing resource use efficiency, and optimizing consumption (Table 1). Also, the concept of a smart village implies a focus on stimulating the rural economy through high-speed internet access, the introduction of green technologies, and the transfer of public services to digital format [7, 8].

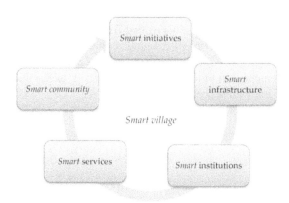

Fig. 1. Conceptual major parts of smart village [9].

In this study, the main conceptual components of the Azerbaijani smart village model are described; the list of the implemented innovative technologies in homes and buildings in the first smart village Agali is given; the energy efficiency of the applied innovative construction technologies and engineering systems is justified; scientific and practical recommendations developed on the issues of adapting strategies of the smart village concept in the context of the socio-economic realities of Azerbaijan.

Table 1. Main priorities for smart village development.

Smart environment (natural resources): • energy efficiency • renewable energy sources • environment protection • saving resources	Smart lifestyle (quality of life): • smart consumption • convenient layout • social interaction • healthy lifestyle
Smart people (social and human capital): • qualified users of information-communication technologies (ICT) • affordable education • engagement in public life and entrepreneurship	Smart economy (competitiveness): • productivity • new products, services, business models • the international cooperation • flexibility
Smart mobility (transport and ICT): • integrated transport systems • environmentally friendly transportation	Smart management (participation): • involving citizens in making the decisions • convenient services • open data

2 Methods and Materials

Despite a significant quantity of publications and implemented projects of smart villages around the world (Fig. 2), the theoretical-methodological validity of this concept is not sufficiently researched and scientifically justified. The object of this article is the concept of a smart village in the conditions of a digital economy, using the example of Agali village in the Karabakh region. The subject of the study is a set of theoretical and practical aspects of the implementation of the smart village concept in Karabakh to substantiate the effectiveness of innovative construction technologies.

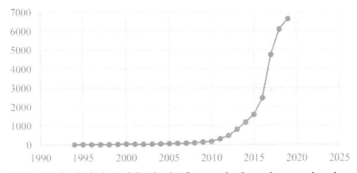

Fig. 2. The number of scholarly articles in the Scopus database that mention the term "smart village" between 1994 and 2020 [10].

The methodological basis of the article was a comparative, statistical, and logical analysis of foreign academic research on the problems of rural development and the reports of international organizations on their sustainable development.

As the smart village concept is not fully developed in theoretical and methodological aspects and is reflected in a relatively small quantity of scientific research, this concept is often criticized due to the absence of scientific reasons, although some authors have tried to place it in some theoretical framework [11]. Komorowski L. places the smart village concept in the theory of regional development as "center-periphery" [12]. Crump J. is considering the connection between smart technologies, policy strategies, and rural viability in the context of the Internet of Things (IoT) [13]. Another approach when the concept of a smart village as an alternative strategy for sustainable development of rural areas is proposed in [14]. A similar approach implies improving the quality of rural residents' lives through economic benefits and ensuring the protection of nature and landscape, i.e., maintaining ecological order [15].

3 Results

3.1 Conceptual Components of the Azerbaijani Smart Village Model

The first pilot project smart village, implemented in Azerbaijan, is Agali village (Fig. 3), rightly considered the Azerbaijani analog of the same world trend and the first in the post-Soviet space. The village is located in the south of Karabakh, a few kilometers south of the village runs the Araks River, which marks the natural border with the Republic of Iran.

Fig. 3. Smart Agali village, 2023, (https://caliber.az/print/179595).

It took eight months to construct the first row of village amenities, which include 200 energy-efficient homes spread across 110 hectares, a school with 360 seats, a kindergarten with 60 spots, a clinic, administrative and public buildings, transportation routes, and trails for bicyclists and pedestrians. There are plans to build an additional 150 homes in the subsequent phase.

The key strategy point for the creation of smart villages is their self-sufficiency by using innovative technologies based on local resources in four development spheres [16, 17]:

- green energetics;
- smart infrastructure and services;
- energy efficiency of construction projects;
- full employment of local residents.

All construction structures and buildings—social facilities, homes, administrative buildings, catering establishments, and buildings for processing and production of agricultural products—are provided with alternative energy sources, equipped with an environmentally friendly heating system, smart external and internal lighting, and have a smart waste management system.

Innovative technologies cover five components of rural life: energy supply, housing sector, manufacturing, social services, and agriculture. Engineering communications and systems are created on the basis of smart technologies. Plans for further development of the village include the formation of a tourism infrastructure. The second village in line to implement the smart village concept is Dovletyarly in Fizuli district, and the third is Bash Garvand in Agdam district.

3.2 Innovative Technologies of the Smart Village Concept, Adapted in Agali

1. The main source of electricity for the village is a small hydroelectric power plant built on an artificial water reservoir, where electricity is produced by a hydrodynamic turbine, which has a certificate of CO_2 emissions' minimization. Another method of obtaining electricity is using of solar photovoltaic panels.
2. The source of the domestic hot water supply is solar vacuum collectors (Fig. 4) and geothermal wells.
3. Wind turbines with a 212 kW capacity are used for public and administrative buildings.
4. Applied environmental monitoring sensors indicate a reduction of CO_2 emissions by up to 70–80% compared to baseline.
5. Electrode water boilers and anodized radiators are used for homes' heating;
6. All buildings and houses are equipped with meters for water, thermal heat, and electricity consumption, operating using wireless technologies.
7. All houses and buildings are designed in compliance with environmental construction standards using innovative external thermal insulation such as basalt wool.
8. Energy-saving windows with a metal-plastic frame have double glazing with an internal silver coating.
9. For non-residential buildings, mechanical ventilation with heat recovery and sewerage water heat recovery are used.
10. Into architectural and construction design digital Building Information Modeling (BIM) technologies are introduced.

Fig. 4. Residential home with a solar vacuum collector on the roof.

11. A smart waste collection system is used: the waste containers are equipped with ultrasonic sensors called "SmartBin" that give a signal when the container is full (Fig. 5).

Fig. 5. SmartBin waste collection system.

12. A smart system for recycling waste and emissions is applied.
13. Recycled waste and emissions are used as valuable fuels for energy generation.
14. All energy-efficient measures provide for 80% efficiency compared to the baseline.
15. There is a smart street lighting automated system with local and centralized sensors.
16. Smart video surveillance cameras for transport, bicycle, and pedestrian paths are used. Automatic gates based on video analytics for recognizing license vehicle numbers are installed on transport roads.
17. Smart school education system combines distance and interactive learning.
18. Smart medicine uses remote diagnostics and online consultations.
19. Smart agriculture uses drones, innovative irrigation systems, treated wastewater, environmental monitoring, and stations for controlling agricultural pests.

20. Disposal of household and industrial waste is carried out: waste from a buffalo farm is used as raw material for the production of biogas and fertilizers.
21. To facilitate the use of government online services, "ASAN", "DOST", Azerpost, the Agency for the Development of Small and Medium Businesses, and the State Center for Agricultural Development operate.

3.3 Energy Efficiency of Implemented Construction Technologies

Hydraulic Auger Turbine. For the continuous supply of electricity to the village, a hydroelectric power plant was built, operating on the principle of the hydraulic auger turbine (Fig. 6). The hydroelectric power plant with three generators is able to fully meet the electricity needs of the population and future industries in Agali. The advantage of the auger turbine compared to other systems is that it can generate energy with minimal water pressure, is easy to install, and is cost-effective. It is also important that the fish pass through the auger turbine without any risk [18, 19].

Fig. 6. Photo of hydraulic auger turbine.

The main element of the power plant is a screw turbine, which consists of a screw rotor inserted into a trough and suspended on upper and lower bearings. The rotor is connected through a gearbox to an asynchronous generator, typical for use in small hydroelectric power plants [20]. The rotor converts the potential energy of the water flow into its own rotational motion and thereby drives the asynchronous generator into rotational motion. Water flows freely from the lower end of the turbine. An efficient gearbox and control system ensure optimal turbine efficiency in the range of 10–100% of water flow. One turbine can operate in the following range of basic technical parameters:

– water consumption: 0.1 - 10 m^3/sec;

- difference in water levels: 1 - 8 m;
- inclination: 22 - 36^0;
- screw diameter - from 0.7–2 m;
- screw rotation speed - 20–80 rev/minute;
- average efficiency - up to 85%;
- generation capacity: 1–500 kW.

The auger turbine has no equal in its simplicity and reliability of operation. Due to its design, i.e. the large rotor diameter and very low rotation speed, the turbine is safe and convenient for the free movement of aquatic fauna through it. Also, leaves and branches flow freely through the rotor, as it does not interfere with power generation, there is no need to install dense grids and cleaners at the top of the turbine. The hydraulic auger turbine is used in places where there are strict requirements for power plant equipment regarding environmental protection. The hydraulic auger turbine is the most practical source of electricity in modern low-power hydroelectric power plants due to its following technical advantages:

1. it has a single, compact, and simple design, long service life, significant efficiency, and significantly low construction, operation, maintenance costs compared to traditional hydraulic turbines;
2. it can be used on small inclinations and slight differences in the levels upper and lower levels of water;
3. it efficiently operates even at very low water flow rates, for example, at a water flow rate of 20%, the efficiency is 74%;
4. no cavitation;
5. it has self-cleaning effect;
6. it does not harm the biological resources of the reservoir, it is environmental friendliness and safety for fish and the environment;
7. it saturates water with oxygen and helps improve the quality of water in the reservoir;
8. it works effective even in floods.

Trend Solutions for Smart Home. According to [21], a smart home is implemented due to innovative technologies with stable access to high-speed Internet, it includes the concepts of environmental friendliness and sustainability, based on saving and efficient use of resources with constant monitoring and data collection. A smart home (apartment, building) is a system of devices and sensors interconnected using platform solutions, where the end-user interacts with the system through a specially designed interface (Fig. 7).

Key global trends of the smart home concept are:

- development of complex software, hardware and platform products and services of home, with the help of which the user can assemble the necessary functional comfortable life scenario at home according to his needs;
- increasing the importance of security and protection systems against attacks and penetration into the smart home system;
- reducing the cost of smart home solutions by 1.5–5 times and increasing their availability for the end users.

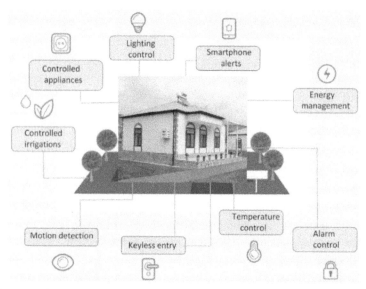

Fig. 7. Smart sensing devices used for various purposes in the Internet of Things for smart home.

Fundamental and rapidly growing smart home technologies are the efficient production, storage and use of energy; applying of computer 3D modeling and 3D printing of home- BIM technologies, sensors and robotics components, innovative technologies and energy-efficient building materials that reduce consumption resources due to its improved technical characteristics and thermal properties.

The implemented innovative technologies in homes and buildings in Agali village are:

- monitoring and control system for energy, heat and water consumption- automation work systems for housing and communal services, also accounting and control resources;
- lighting control system with smart sockets- setting up scene management, control of electricity consumption;
- various options for smart lamps: LED lamp with motion and light sensors (works without additional smart devices); smart Wi-Fi lamps, which are controlled from a mobile phone, with the ability to change lighting shades;
- smart system ventilation, cooling and heating- smart air conditioners, thermostats;
- climate-control sensors- temperature, relative humidity, carbon dioxide level, air pressure;
- climate-control equipment- Magic Air station and Tion Breer for single-room air purification, ventilation and heating;
- comprehensive monitoring and forecasting of the condition of housing and communal services facilities, prevention of emergency situations;
- smart security sensors- motion, smoke, broken glass, gas leaks, leaks from open taps and so forth;

- smart intercoms - automatic opening doors through application on mobile phone, signals about breakdown to management services;
- remote payment services and systems;
- closed-circuit television systems (CCTV), smart gates, cameras and barriers.

The result of the applied innovative technologies and measures for homes and buildings is:

- reduction in: energy consumption by 30–40%;

    water consumption by 20–30%;
    operating costs by 10–30%.

- an increase comfort and opportunity of individual "settings" of homes and buildings in compliance with lifestyle, working routine and needs;
- improving the quality of services for residents and users;
- reduction in operating and maintaining costs;
- prevention emergency situations.

Innovative Heating for Smart Home. For home heating, electrode water boilers and anodized radiators are used (Fig. 8), equipped with universal remote control thermostats by a given algorithm [22]. Electrode boilers are highly efficient and have an efficiency rate of 96–100%. They utilize a progressive heating technology based on the movement of ions in the heat medium when a current is passed through it. One of the key features of these boilers is that they heat up water immediately, without requiring a heat exchanger [23]. As a result, electrode boilers are more than twice as efficient as other types of water boilers and can heat a much larger area. They also consume much less electricity, as they do not require additional equipment to be installed [24]. Other benefits of electrode boilers include automatic control of temperature parameters, simple design to avoid frequent breakdowns and costly repairs, easy installation with minimal costs, soft start and stop power control functions, and electronic control that allows to choose the most optimal operation mode and reduce electricity consumption by 30–40% [25]. With a heated room area of 60–350 m^2 and a height of no more than 2.5–3 m, the boiler's power is 4.5–25 kW.

Anodized aluminum radiators, due to anodic oxidation of the inner surface, have improved strength properties, a number of thermal advantages and their technical characteristics are superior to all available analogs:

- the power of one radiator section is 110–210 W;
- the ability to heat 2 m^2 of room area with one radiator section;
- operating pressure of heated water- up to 2.5 MPa;
- heated water temperature - up to 130 ^{0} C.

The heat transfer rate of anodized radiators is 2 times higher than traditional aluminum radiators due to their improved design and completely smooth internal surface.

Fig. 8. Electrode boiler and anodized radiator, photo.

4 Discussion

The specific strategy for sustainable development of rural areas and carrying out restoration works in the villages of Karabakh is to adapt them to the global trend of smart villages. Based on the results of this study, the author offers the following set of scientific-practical recommendations for tackling the problem with tools selected from world practice [26, 27]. It is proposed to develop:

1- catalog of energy-saving and energy-efficient innovative technologies adaptive for a certain region, based on the formation of more expanded agglomerations around the rural center, i.e. "business center—residential districts." According to [28] villages with more dense populations have higher rates and possibilities of innovation, productivity, and entrepreneurship. The concentration of human, creative, business, and material resources will allow for an increased synergistic effect and create additional pulse development villages through the formation of a common economic space [29].
2- map of the village's competencies, indicating its economic specialization. "Smart" specialization of rural areas is a promising approach to their sustainable development. It is necessary to proceed based on the available resources, the potential of rural areas, and the needs of the economy [30]. The initial attempts to identify prospective economic specializations were made in Agali:
3- concrete enterprise with a fully automated production process;
4- buffalo farm whose waste is used to produce fertilizer and biogas [31, 32]. It is also important to create a database on the village specializations, which will optimize the mechanism of support from the state. Regular monitoring and evaluation of the effectiveness of applied innovations will increase an investment attractiveness in the region [33].
5- model concept of integrated village development with cutting-edge IT solutions based on resource availability and characteristics, taking into account the historical path of a particular village with the active involvement of local residents as the primary

development resource. For any region, especially rural ones, the development of engineering systems and information and communication infrastructure is essential [34]. In order to effectively integrate cutting-edge digital technologies into the management of socio-economic processes in rural areas, highway broadband Internet generation of at least 4G should be utilized. This is particularly valid in this age of rapid technology advancement and the shift to the digital economy [35, 36].

5 Conclusion

The study attempted to concretize the principles of implementing the smart village concept for the sustainable development of rural areas through the introduction of innovative technologies, including construction ones, based on the experience of Agali village in the Karabakh region of Azerbaijan. Using ICT, high-speed internet, green energy, and other resources are examples of strategic growth directions [37, 38]. It is impossible to create and implement rural development programs using a template or universal method [38]. An examination of Agali's experience reveals that attempts to tackle village socio economic development from a single perspective, ignoring its unique characteristics, the availability of natural resources, and the course of past development, ultimately result in the rejection of innovations. The locals must take an active role in the smart restoration and development of the village in order to create a comfortable living environment that satisfies the demands of the younger generation while preserving the distinctive originality of the surrounding area.

References

1. Akbarova, S., Mammadov, N., Rustamov, V.: Evaluation of thermal energy production by solar panels for Karabakh "green" energy zone. Reliability: Theory and Appl. **4**(70), 200–206 (2022)
2. Adamowicz, M., Zwolinska-Ligaj, M.: The "smart village" as a way to achieve sustainable development in rural areas of Poland. Sustainability **12**(16), 65–83 (2020)
3. Bell, D., Jayne, M.: The creative countryside: policy and practice in the UK rural cultural economy. J. Rural. Stud. **26**(3), 209–218 (2010)
4. Bijker, R., Haartsen, T., Strijker, D.: Migration to less-popular rural areas in the Netherlands: exploring the motivations. J. Rural. Stud. **24**(8), 490–498 (2012)
5. Kneafsey, M.: Rural cultural economy: tourism and social relations. Ann. Tour. Res. **28**(3), 762–783 (2001)
6. Deller, S., Tsai, T., Marcouiller, D.: The role of amenities and quality of life in rural economic growth. Am. J. Agr. Econ. **83**(2), 352–365 (2001)
7. Guzal-Dec, D.: Intelligent development of the countryside- the concept of smart villages: assumptions, possibilities and implementation limitations. Economic and Regional Stud. **3**, 32–49 (2018)
8. Haider, M., Siddique, A., Alam, S.: An approach to implement frees space optical technology for smart village energy autonomous systems. Far East J. Electronics and Communication **18**, 439–456 (2018)
9. Komorowski, L., Stanny, M.: Smart villages: where can they happen? Land **9**(5), 151–169 (2020)

10. Meyn, M.: Digitalization and its impact on life in rural areas: exploring the two sides of the atlantic: USA and Germany. Smart Village Technol. **17**, 99–116 (2020)
11. Mammadova, G., Akbarova, S.: Experimental study of air cavity thermal performance of opaque ventilated facades under extreme wind conditions: case study Baku. Construction Reports **73**(561), 63–76 (2021)
12. Komorowski, L., Stanny, M.: Smart village laboratory: a visit to finnish smart villages. Land **9**, 151–165 (2023)
13. Crump, J.: Finding a place in the country: exurban and suburban development in Sonoma County California. Environ. Behav. **35**, 187–202 (2003)
14. Merino, F., Prats, M: Why do some areas depopulate? the role of economic factors and local governments. Cities **97**, 102506 (2020)
15. Zwolinska-Ligaj, M., Guzal-Dec, D., Adamowicz, M.: The concept of smart development of local territorial units in peripheral rural areas: the case of Lublin Voivodeship. Wie´s I Rol **179**, 247–280 (2018)
16. Mammadova, G., Akbarova, S.: Certification methods as a mechanism for estimation of building sustainability. E3S Web of Conferences **458**(7) (2023)
17. Musaeva, N., Mammadova, G., Aliyev, E., Sattarova, U.: Methods, technologies and means of control of seismic stability of complexes of construction structures in operation. In: 4th International Conference "Problems of Cybernetics and Informatics", p. 648635 (2012)
18. Nelson, L., Nelson, P.: The global rural: gentrification and linked migration in the rural USA. Prog. Hum. Geogr. **35**(4), 441–459 (2011)
19. Paniagua, A.: Counter urbanisation and new social class in rural Spain: the environmental and rural dimension revisited. Scottish Geographical J. **118**, 1–18 (2002)
20. Paniagua, A.: The environmental dimension in the constitution of new social groups in an extremely depopulated rural area of Spain. Land Use Policy **25**(1), 17–29 (2008)
21. Stolojescu-Crisan, C., Calin, C., Bogdan-Petru, B.: An IoT-based smart home automation system. Sensors **21**(11), 3784 (2021)
22. Aliev, T., Musaeva, N., Suleymanova, M., Gazizade, B.: Density function of noise distribution as an indicator for identifying the degree of fault growth in sucker rod pumping unit. J. Autom. Inf. Sci. **49**(4), 1–11 (2017)
23. Paniagua, A.: Urban–rural migration, tourism c and rural restructuring in Spain. Tour. Geogr. **4**, 349–371 (2002)
24. Philip, L., Williams, F.: Healthy ageing in smart villages? Observations from the field. European Countryside **11**, 616–633 (2019)
25. Stockdale, A.: The diverse geographies of rural gentrification in Scotland. J. Rural Studie **26**(1), 31–40 (2010)
26. Wolski, O.: Smart villages in EU Policy: how to match innovativeness and pragmatism? Wies Roln **181**, 163–179 (2018)
27. Musaeva, N.: Robust correlation coefficients as initial data for solving a problem of confluent analysis. Autom. Control. Comput. Sci. **2**, 76–87 (2007)
28. Nathan, M., Overman, H.: Agglomeration, clusters and industrial policy. Oxf. Rev. Econ. Policy **29**(2), 383–404 (2013)
29. Aliev, T., Musaeva, N.: Technologies for early monitoring of technical objects using the estimates of noise distribution density. J. Autom. Inf. Sci. **5**, 12–23 (2019)
30. Seppanen, O.: Modern heating systems for residential buildings: expert opinions. J. Sustainable Building Technologies **5**, 4–10 (2019)
31. Tabunschikov, Y.: Road map of green construction: problems and growth perspectives. J. Sustainable Building Technologies **3**, 5786 (2014)
32. Rayak, M.: Development of Foreign and Domestic Heating and Ventilation Systems for Civil and Industrial Buildings. News of heat supply, Moscow, p. 183 (2007)

33. Aliev, T., Musaeva, N., Sattarova, U.: The technology of forming the normalized correlation matrices of the matrix equations of multidimensional stochastic objects. J. Autom. Inf. Sci. **45**(1), 1–15 (2013)
34. Zavratnik, V., Kos, A., Duh, E.: Smart villages: comprehensive review of initiatives and practices. Sustainability **10**, 2559 (2018)
35. Visvizi, A., Lytras, M.: It's not a fad: smart cities and smart villages research in European and global contexts. Sustainability **10**, 2727 (2018)
36. Aliev, T., Musaeva, N., Sattarova, U.: Robust technologies for calculating normalized correlation functions. Cybern. Syst. Anal. **46**(1), 153–166 (2010)
37. Philip, L., Williams, F.: Healthy ageing in smart villages? Observations from the field. European Countries **11**, 616–633 (2019)
38. Aliev, T., Musaeva, N., Gazizade, B.: Algorithms for calculating high-order moments of the noise of noisy signals. J. Autom. Inf. Sci. **50**(6), 1–13 (2018)
39. Visvizi, A., Lytras, M., Mudri, G.: Smart villages: comprehensive review of initiatives and practices. Sustainability **10**(7), 2559 (2019)

Turkish Legal Single-Document Summarizing

Maha Ahmed Abdullah Albayati[✉] and Oğuz Fındık

Karabuk University, Karabuk 78000, Turkey
mahaa6022@gmail.com

Abstract. Text summarization is the compression of source text into a condensed version while preserving the information content and overall meaning. Due to the large amount of information research, articles, documents, books, etc., and the development of Internet technologies, text summarization has become an important tool by extracting the most important parts and clarifying the basic purpose of the text. It is very necessary for lawyers and ordinary citizens to conduct thorough research related to their case before answering questions in court. For some time, they had to read very long rulings and try to pick out useful information from them or hire legal editors to create summaries. Due to the lack of research related to summarizing Turkish legal texts, we propose an automated text summarization system by using two novel hybrid pretrained approach models. Our new dataset is collected from official government sites that included Supreme Court, high court and district court cases. We achieved satisfactory results according to the legal professionals (lawyers) evaluation and rouge evaluation metrics, our summary text is Readable and understandable even by non-specialized people. In the future, we aim to obtain better summaries and build an application available to everyone, and open the way for further research using the dataset used in our study.

Keywords: Turkish legal text summarize · pretrained model · hyper approach · text summarization

1 Introduction

In the era of digital information, the legal domain faces the formidable challenge of efficiently managing and processing vast quantities of complex documents. The Turkish legal system, with its unique linguistic and structural characteristics, presents an additional layer of intricacy in legal document analysis. To address this challenge, our study introduces a pioneering approach to legal document summarization, combining the strengths of both extractive and abstractive methodologies within a hybrid framework powered by Hyper AI's advanced pre-trained models.

The need for accurate and concise summaries of legal documents is paramount for legal professionals, enabling them to quickly grasp the essence of cases and make informed decisions. Traditional methods of manual summarization are time-consuming and prone to human error, underscoring the necessity for an automated solution that can navigate the intricacies of legal jargon and the specificity of Turkish language constructs.

Our research breaks new ground by curating a novel dataset comprising diverse Turkish legal documents, which serves as the foundation for training and evaluating our AI models. This dataset reflects the broad spectrum of document types found within the Turkish legal system, including but not limited to court decisions, contracts, and legislation. By tailoring our dataset to the domain-specific requirements of Turkish legal documents, we ensure that our models are well-equipped to handle the nuances of legal language and the specificities of the Turkish legal context.

Furthermore, we employ two of Hyper AI's pre-trained models, which have been fine-tuned to the Turkish legal domain. The first model, an extractive summarizer, is adept at identifying and extracting key information and salient points from the original texts. The second model, an abstractive summarizer, excels at generating coherent and contextually relevant summaries that encapsulate the core meaning of the documents while rephrasing content to achieve brevity and clarity.

To achieve unparalleled efficiency in summarizing Turkish legal documents, our research deploys two sophisticated Hyper AI pre-trained models, each consisting of both extractive and abstractive components tailored to process legal texts effectively. The first model integrates a Bag of Words (BOW)-TF-IDF extractive algorithm with an abstractive 'pipeline' model, which together form a powerful summarization tool. Here, the BOW-TF-IDF component is responsible for sifting through the text to pinpoint and pull out the most substantive elements. These elements are then seamlessly passed into the 'pipeline' model, which is designed to reconstruct the extracted information into an abstractive summary that is not only concise but also maintains the essence and clarity of the original content.

The second Hyper AI model employs a similar dual-approach strategy, where Text-Rank serves as the extractive mechanism, deftly identifying the key points within the text, while the Transformer-based T5 model operates as the abstractive agent. In this model, the Text-Rank algorithm excels at zeroing in on the salient features of the legal documents, which are then artfully re-envisioned by the T5 model, producing summaries that are both contextually rich and linguistically sharp.

Our innovative "Hyper Approach" synthesizes these two models into a cohesive hybrid framework, strategically utilizing the extractive components to prepare the foundational layer of the summary. The abstractive components of both models then take over, transforming the distilled content into a final summary that is not only succinct and easy to understand but also faithful to the original legal text's purpose and detail. This hybrid extractive-abstractive dynamic ensures that the summaries generated are of the highest caliber, providing legal professionals with reliable and quick access to the critical information embedded within dense legal documentation.

1.1 Related Work

[11] Luhn developed the first automatic text summary in 1958 using phrase frequency. [9] Document summaries were created in the 1990s using statistical approaches and machine learning in natural language processing. Recently, trained language models have become a crucial tool for attaining remarkable improvements in a broad range of natural language activities. By learning contextual representations from large-scale corpora with a language modeling purpose, these models expand on the concept of

word embedding. Legal judgment documents are now available digitally, providing a wealth of options for information extraction and application. The unique structure and great complexity of these legal writings make automatic summarization of them both an important and difficult endeavor. Prior efforts in this area have concentrated their attention on a small sub-domain for maximum efficacy, used enormous labeled datasets, hand-engineered features, and leveraged domain expertise.

[12] Suggested utilizing neural networks to do extractive legal document summarizing tasks for Indian legal judgment documents. These techniques were easy and generic. For this objective, they investigate two neural network architectures (LSTM and feed forward neural networks), using word and phrase embedding to capture the semantics. They address the issue of labelled data not being available for the task by classifying or scoring sentences in the training set according to how well they match human-produced reference summaries. These experimental analyses demonstrate how successful these suggested strategies are.

[13] Introduced previous case retrieval models for Turkish courts. They achieved a good score with a micro-F1 score of 57.28% using two retrieval methods: RNN auto encoders and the combination of RNN auto encoders with BM25.

The automatic summary of Greek legal texts was discussed by [14]. In order to facilitate automatic legal document summation, they created a new, metadata-rich dataset that included specific rulings from the Greek Supreme Civil and Criminal Courts together with their category tags and reference summaries. They implemented a number of cutting-edge techniques for extractive and abstractive summarization and carried out a thorough assessment of the techniques utilizing both automatic and human metrics. LexRank, Biased LexRank, and BERT were the algorithms utilized for extractive and abstractive summaries, respectively. The findings indicated the need for metrics that better capture the coherence, relevance, and consistency of a legal document summary. They also showed that BERT models can perform noticeably better when fine-tuned for a particular upstream task. Extractive methods perform averagely, while abstractive methods generate moderately fluent and coherent text, but they typically receive low scores in relevance and consistency metrics.

[15] Proposed an NLP-based method for summarizing legal texts. For the extractive approach, they used Naive Bayes, Decision Trees, and Random Forest; for the abstractive approach, they used LSTM and an improved PEGASUSLARGE model. They used Justia, a legal information platform, to produce Labeled Corpus. They evaluated their models with a relatively excellent performance in recall and F1 score for the particular task of automatically summarizing the legal views.

[16] Study looked into whether summaries of court rulings may be produced more effectively by shrewdly mixing the outputs of several summarizing algorithms than by using any one of the underlying techniques alone. Utilizing two datasets of Indian Supreme Court case judgment documents: one dataset had abstractive gold standard summaries, while the other had extractive summaries. Simple voting-based, ranking-based, and graph-based ensembling techniques are the ensembling techniques used. They demonstrated that, in terms of ROUGE and METEOR scores, many of their assembling techniques create summaries that outperform those generated by any one of the individual base algorithms.

[17] Introduced LegalSumm, a technique that generates a candidate summary by dividing the legal document into preset chunks that the model can handle. This improves abstractive summarization of lengthy judicial opinions. By choosing the best summaries from these candidate summaries, it can also eliminate irrelevant themes. LegalSumm has a benefit over other strategies for preventing "hallucination" in that it does not require NER or POS taggers. They used ROUGE F-scores in a variety of evaluation experiments. The initial trial showcases LegalSumm's ability to select well from the candidate summaries produced by conventional Transformer models. In the second experiment, BertSumExt, BertSumAbs (Liu and Lapata 2019), and BART (Lewis et al. 2020)—all of which required modifications to function with Portuguese texts—are compared with LegalSumm as summarizing baselines. Additionally, we present a comparison between LegalSumm and Feijo and Moreira's (2019) findings using the same dataset. They asked legal professionals to judge the coverage, coherence, fidelity, and replace ability of the machine generated summaries.

From the original dataset, [18] constructed a modified abstractive dataset. This guarantees that each document-summary pair's length is reasonable and compatible with the most advanced summarizing techniques (like BART). In order to overcome the data scarcity issue, they extracted multiple extractive summaries from each sample in the original dataset. These extractive summaries were then assigned ground-truth summary sentences, resulting in new training samples that could be used to improve summarization models. The methodology was assessed on two distinct legal datasets: BillSum by (3 - 8) %, Forum of Information Retrieval Evaluation (FIRE) by (1 - 2) % for BERTScore metrics, and the ROUGE metrics were able to outperform the pre-trained BART model that was fine-tuned on the original dataset by 1–3% for BillSum test sets and by 3–8% for FIRE test sets.

For the purpose of extractive summarizing, [10] supplied a new corpus known as EUR-LexSum, which is made up of roughly 4.5K documents regarding European rules and the summaries that go along with it. They used this corpus to assess the effectiveness of sophisticated transformer-based models in extractive summarization tasks. When their studies demonstrate that even with a moderate quantity of data for fine-tuning, optimized transformer models outperform Text Rank, a conventional baseline model. Given the encouraging Rouge1-F1 scores for the basic hybrid strategy, they recommended that future study focus on the application of advanced hybrid systems. They planned to score the projected summaries by human judges and determine the correlation between the evaluations—automated and human—in order to conduct additional quality assessment.

2 Problem Statement

The need for automatic summarization of legal documents and other word processing has been felt for a long time, but has only come to the focus of computer scientists very recently. This prompted us to research more deeply into this field and we discovered that there were no legal summaries in the Turkish language. We obtained a set of legal documents from the Turkish civil courts. We will list the set of problems we attempted to solve in this paper:

1. The Turkish language problem: The morphological structure of the Turkish language is very complex, as one word contains a large number of affixes that make it difficult for the models used to distinguish them and know the root of the word. In addition, the Turkish language, like other languages, may have more than one meaning for a single word, and here the problem of disambiguation appears.
2. the characteristics of the Legal text:
   a. Size: Since many domains still rely on collections of abstracts rather than the complete text of the documents, legal documents typically have a larger file size than documents in other domains.
   b. Structure: The internal organization of legal documents varies. They adhere to a hierarchical framework and have status and administrative norms.
   c. Vocabulary: Legal documents have a distinct vocabulary. In addition to the common language, legal language employs a variety of domain-specific terms.
   d. Ambiguity: Text that is legal may be unclear since the same term, phrase, or statement can have more than one interpretation. If the identical text appeared in a district court opinion as opposed to a high court opinion, it might be construed differently.
   e. Citations: Generally speaking, citations highlight the major points of the case and are more common in the legal realm than in other disciplines.
3. Hyper approach method problem: which combines the extractive and abstractive approaches, were most researchers employed the extractive method, with a small number touching on the abstractive approach, while a bit research was done on the usage of the hyper approach method.
4. The suitable model: identifying a model capable of handling and resolving each of the aforementioned issues.

3 The Main Part

3.1 Dataset

Our data consists of legal judgments issued by the Turkish judiciary system. Supreme Court, high court and district court data was collected from official sites having extensions of.gov. To retrieve this data, a web crawler was created in python using the Requests and Json libraries. This method of gathering data is referred to as web scraping. More than 1 hundred documents were collected, consisting of documents related to "consumer rights", this group of documents belong to civil cases type. Every document includes the court's name, the case number and name, the laws belong to the case, the court decision and all details related to the case. This group formed our new novel dataset.

3.2 Models

Preprocessing. The first step in pre-processing is cleaning the input document which is in text format from unwanted punctuations except the single full stop because it is refer to the end of the sentence and saving the cleaned text in text format too. The cleaning process is successfully done by means of using regular expressions. The next step is word, sentence tokenization and stemming by using Zemberek library which is made specially to deal with Turkish language.

Processing. In our research, we used two hybrid pre-training models which they are (BOW)-TF-IDF with pipeline and TextRank with T5. (BOW)-TF-IDF and TextRank represented the step of extractive summarize then the output of those models will pass to the input of the abstractive models which they are pipeline and T5. Together, extractive and abstract models constitute hybrid models:

the bag-of-words (BOW) technique. It is a natural language processing (NLP) feature representation technique that can be used to represent each remaining word as a document feature [1]. It entails counting the frequency of words in a document or corpus to create a vector representation of the text. For tasks like text classification and information retrieval, machine learning algorithms frequently employ the bag-of-words model. It offers a quick and easy method of encoding text input, collecting crucial details needed for linear classifiers to generate precise predictions. However, the bag-of-words representation might become sparse in environments with a large vocabulary and brief texts, which can impact classifier accuracy in addition to This method treats every document as a "bag" of words, ignoring the words' structure and order. Techniques like phrase count, term frequency, and term frequency–inverse document frequency can be used to change the values of the features in BOW. We can now extract features from individual documents into a feature vector by using a master dictionary of features. The simplest method bases the value on the presence of a term; that is, a "0" or a "1" is assigned when a phrase (or characteristic) is present or absent, respectively. We combined term frequency–inverse document frequency (TF-IDF) with this method to make it better.

term frequency–inverse document frequency (TF-IDF) technique. It is the process of determining a word's relevance to a text within a corpus or series [2]. The meaning of a word grows in direct proportion to how frequently it occurs in the text; however, this is offset by the word frequency in the corpus (data-set).

$$TF\_IDF = TF \times IDF$$

term frequency (TF) technique. The term frequency in a document indicates how many times a certain word appears [3]. As a result, it makes sense that when a word appears in the text, it becomes more relevant. We can utilize a vector to represent the text in the bag of term models since the term ordering is not important. There is an entry with the term frequency as the value for each unique term in the document.

$$TF = \frac{number\ of\ times\ the\ term\ appears\ in\ the\ document}{total\ number\ of\ terms\ in\ the\ document}$$

inverse document frequency (IDF) technique. It primarily assesses the word's relevance. Finding the relevant records that meet the demand is the main goal of the search. It is also not simply able to utilize term frequencies to measure a term's weight in the article, since tf regards all phrases as equally relevant. To get a term's document frequency, first count the number of documents that contain the term. Put another way, the word's IDF is the total number of documents in the corpus divided by the text's frequency.

$$IDF = \log\left(\frac{number\ of\ the\ documents\ in\ the\ corpus}{number\ of\ documents\ in\ the\ corpus\ contain\ the\ term + 1}\right)$$

TextRank algorithm. It is a graphical technique that divides pre-processed data into sentences and words, which serve as graph vertices [4, 5]. The weight of the connections connecting the phrases and words is determined by how similar they are to one another. A similarity matrix is created in order to calculate this similarity. TextRank's main benefit is that it is an unsupervised graph-based algorithm, meaning it can determine the key elements of a textual content without the need for a human summary or training dataset. Compared to comparable supervised algorithms, unsupervised algorithms require less manual data preparation, which saves time overall [6]. TextRank is an ATS system that operates on word occurrence and is language-independent, in addition to being an unsupervised algorithm. Only the most enlightening sentences are chosen for the output summary once each sentence's relative importance is determined.

Text to Text Transfer Transformer (T5). It is a transformer model based on encoder-decoders that Colin Raffel suggested [7]. This universal transformer-based architecture produces state-of-the-art results on numerous natural language processing tasks, including text summarization, question answering, machine translation, and sentiment classification. It takes "text" as input and outputs "text." Text summarization is a strong suit for this pre-trained language model, which is well-known for a number of NLP tasks. Hugging Face API allows seamless text summarization using T5. However, numerous more features can be unlocked by optimizing T5 for text summarization.

3.3 Evaluations and Results

Text summarization is a critical task in NLP, with the need for robust evaluation metrics to evaluate the quality of generated summaries. In our study, we employ the Recall-Oriented Understudy for Gisting Evaluation (ROUGE) [8] is the industry-standard evaluation metric for text summarization (Lin 2004). This metric's fundamental idea is to quantify the number of units that overlap between the machine-generated summary and one or more reference summaries. A good summary should, ideally, employ the same terms as the reference summaries, and preferably in the same sequence.

ROUGE-N, ROUGE-L, ROUGE-W, ROUGE-S, and ROUGE-SU are the five different forms of the ROUGE metrics. We will assess the summarizing job in our studies using the two most popular forms, ROUGE-N and ROUGE-L to evaluate the performance of our models. ROUGE-N counts the number of n-gram units that separate a candidate from a group of reference summaries. ROUGE-L(RL) is referred to the longest common subsequence (LCS) between the output of our model and reference, in another words the longest in order sequence of words that is common to both. We computed precision(P), recall(R) and F1-score(F) for ROUGE-N and ROUGE-L as well. Below are the mathematical equations we used to calculate ROUGE values: (Table 1)

$$ROUGE - N_{precision} = \frac{overlapping\ number\ of\ n - grams}{number\ of\ n - grams\ in\ the\ candidate}$$

$$ROUGE - N_{recall} = \frac{overlapping\ number\ of\ n - grams}{number\ of\ n - grams\ in\ the\ reference}$$

$$ROUGE - L_{precision} = \frac{number\ of\ words\ in\ LCS}{number\ of\ the\ words\ in\ candidate}$$

$$ROUGE - L_{recall} = \frac{number\ of\ words\ in\ LCS}{number\ of\ the\ words\ in\ reference}$$

$$ROUGE - (NorL)_{F1-score} = 2 \times \frac{precision \times recall}{precision + recall}$$

Table 1. Present the result of the ROUGE evaluation metrics for the two models tested in our study.

Model name	R1-P	R1-R	R1-F	R2-P	R2-R	R2-F	RL-P	RL-R	RL-F
(BoW)-TF-IDF + pipeline	0.92	0.27	0.41	0.82	0.23	0.36	0.57	0.16	0.25
TextRank + T5	0.8	0.14	0.24	0.62	0.10	0.18	0.8	0.14	0.24

We successfully applied our summary model and produced outcomes:

Experimental Setup. Two pre-trained super-models for summarization were built using Python. Those forms used to summarize individual Turkish legal documents. The documents used are related to one issue, "Consumer Rights". Using the first model, which is ((BoW)-TF-IDF with pipeline), the summary was approximately 37% of the original text, it is containing the court's decision with the legal aspect of the decision. The second model, which is (TextRank with T5), summarized the original document by an approximate percentage of 13% of the original text, mainly emphasizing the final court decision without going into complicated legal details.

Evaluation of the System. The results were evaluated based on ROUGE metrics (Rouge-1(R1), Rouge-2(R2), Rouge-L(RL)), the results showed in table (1). From the results it is obvious the second model achieve better results from the first one in the terms of Precision and F1-Score where they showed medium to high scores, while the retrieval showed low scores, which means a lack of deep understanding of the texts. We attribute this divergence partly to our intentional constraint on the number of sentences in the summaries, limiting them to a single sentence. We believe that expanding the summary length could potentially improve overall ROUGE scores for both models.

We would like to point out here that we did not improve or train the models on our own data. The other aspect of the evaluation was relying on human evaluation by specialists in the legal field (lawyers). It is worth noting here that human evaluation differs from one person to another. While ROUGE metrics provide valuable quantitative assessments, human evaluation adds a qualitative dimension, offering nuanced feedback that may not be captured by automated metrics alone.

However, because terms in machine-generated summaries do not match words in human-generated summaries, the system's effectiveness cannot be adequately assessed using ROGUE metrics data. In addition, there is a significant variation in length between summaries produced by systems and humans, even when it comes to trying to fit too many ideas into a single line. In light of these details, our ROGUE result is more than satisfactory. Yet, upon closer examination by legal experts, the conceptual alignment between the machine-generated and human-generated summaries was found to be nearly identical, indicating the success of our models in capturing key information.

Subsequently, our evaluation using both automated metrics and human judgments provides a comprehensive understanding of the performance of our summarization models. While automated metrics offer valuable quantitative insights, human evaluation adds qualitative depth, contributing to a more holistic assessment of the system's effectiveness (Fig. 1).

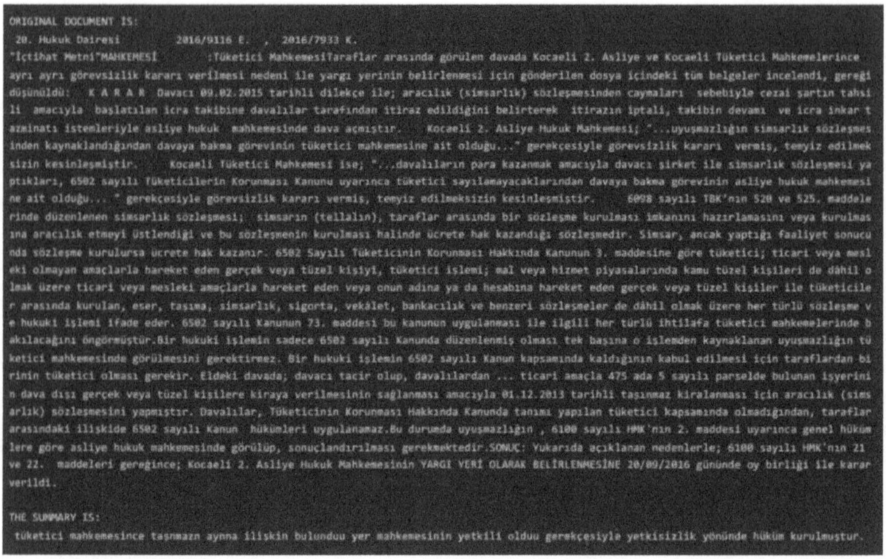
ORIGINAL DOCUMENT IS:
20. Hukuk Dairesi 2016/9116 E. , 2016/7933 K.
"İçtihat Metni"MAHKEMESİ :Tüketici MahkemesiTaraflar arasında görülen davada Kocaeli 2. Asliye ve Kocaeli Tüketici Mahkemelerince ayrı ayrı görevsizlik kararı verilmesi nedeni ile yargı yerinin belirlenmesi için gönderilen dosya içindeki tüm belgeler incelendi, gereği düşünüldü: K A R A R Davacı 09.02.2015 tarihli dilekçe ile; aracılık (simsarlık) sözleşmesinden caymaları sebebiyle cezai şartın tahsi li amacıyla başlatılan icra takibine davalılar tarafından itiraz edildiğini belirterek itirazın iptali, takibin devamı ve icra inkar t azminatı istemleriyle asliye hukuk mahkemesinde dava açmıştır. Kocaeli 2. Asliye Hukuk Mahkemesi; "...uyuşmazlığın simsarlık sözleşmes inden kaynaklandığından davaya bakma görevinin tüketici mahkemesine ait olduğu..." gerekçesiyle görevsizlik kararı vermiş, temyiz edilmek sizin kesinleşmiştir. Kocaeli Tüketici Mahkemesi ise; "...davalıların para kazanmak amacıyla davacı şirket ile simsarlık sözleşmesi ya ptıkları, 6502 sayılı Tüketicilerin Korunması Kanunu uyarınca tüketici sayılamayacaklarından davaya bakma görevinin asliye hukuk mahkemesi ne ait olduğu... " gerekçesiyle görevsizlik kararı vermiş, temyiz edilmeksizin kesinleşmiştir. 6098 sayılı TBK'nın 520 ve 525. maddele rinde düzenlenen simsarlık sözleşmesi; simsarın (tellalın), taraflar arasında bir sözleşme kurulması imkanını hazırlamasını veya kurulmas ına aracılık etmeyi üstlendiği ve bu sözleşmenin kurulması halinde ücrete hak kazandığı sözleşmedir. Simsar, ancak yaptığı faaliyet sonucu nda sözleşme kurulursa ücrete hak kazanır. 6502 Sayılı Tüketicinin Korunması Hakkında Kanunun 3. maddesine göre tüketici; ticari veya mesl eki olmayan amaçlarla hareket eden gerçek veya tüzel kişiyi, tüketici işlemi; mal veya hizmet piyasalarında kamu tüzel kişileri de dâhil o lmak üzere ticari veya mesleki amaçlarla hareket eden veya onun adına ya da hesabına hareket eden gerçek veya tüzel kişiler ile tüketicile r arasında kurulan, eser, taşıma, simsarlık, sigorta, vekâlet, bankacılık ve benzeri sözleşmeler de dâhil olmak üzere her türlü sözleşme v e hukuki işlemi ifade eder. 6502 sayılı Kanunun 73. maddesi bu kanunun uygulanması ile ilgili her türlü ihtilafa tüketici mahkemelerinde b akılacağını öngörmüştür.Bir hukuki işlemin sadece 6502 sayılı Kanunda düzenlenmiş olması tek başına o işlemden kaynaklanan uyuşmazlığın tü ketici mahkemesinde görülmesini gerektirmez. Bir hukuki işlemin 6502 sayılı Kanun kapsamında kaldığının kabul edilmesi için taraflardan bi rinin tüketici olması gerekir. Eldeki davada; davacı tacir olup, davalılardan ... ticari amaçla 475 ada 5 sayılı parselde bulunan işyerini n dava dışı gerçek veya tüzel kişilere kiraya verilmesinin sağlanması amacıyla 01.12.2013 tarihli taşınmaz kiralanması için aracılık (sims arlık) sözleşmesini yapmıştır. Davalılar, Tüketicinin Korunması Hakkında Kanunda tanımı yapılan tüketici kapsamında olmadığından, taraflar arasındaki ilişkide 6502 sayılı Kanun hükümleri uygulanamaz.Bu durumda uyuşmazlığın , 6100 sayılı HMK'nın 2. maddesi uyarınca genel hüküm lere göre asliye hukuk mahkemesinde görülüp, sonuçlandırılması gerekmektedir.SONUÇ: Yukarıda açıklanan nedenlerle; 6100 sayılı HMK'nın 21 ve 22. maddeleri gereğince; Kocaeli 2. Asliye Hukuk Mahkemesinin YARGI YERİ OLARAK BELİRLENMESİNE 20/09/2016 gününde oy birliği ile karar verildi.

THE SUMMARY IS:
tüketici mahkemesince taşnmazn aynna ilişkin bulunduu yer mahkemesinin yetkili olduu gerekçesiyle yetkisizlik yönünde hüküm kurulmuştur.

Fig. 1. An example showing the result of the second summary model (TextRank + T5)

4 Conclusion

We succeeded in implementing a Turkish legal text summary using our proposed model that uses two hybrid models and relied on new dataset from Turkish courts that, to our knowledge, have not been used in previous research. Thus, we broke the traditional methods of summarizing texts that rely on pre-prepared data with their summaries and training algorithms on this data. A short summary has been created keeping important ideas from the original document. We were able to achieve an average ROGUE-1 score of 0.92 in addition to the human evaluation by specialists. However, this evaluation method is not completely effective, and in the future, we aim to build a more comprehensive model for the purpose of deploying our system on the mobile phone platform or electronic browser to ensure the maximum benefit for all of the community layers.

References

1. Juluru, K.: Bag-of-words technique in natural language processing: a primer for radiologists. RadioGraphics (2021)
2. Wang, W.: Improvement and Application of TF-IDF Algorithm in Text Orientation Analysis. Atlantis Press (2016)
3. Capitalone. https://www.capitalone.com/tech/machine-learning/understanding-tf-idf/, 1 Oct 2024
4. Gulati, V.: Extractive Article Summarization Using Integrated TextRank and BM25+ Algorithm. MDPI (2023)
5. Zha, P.: An Efficient Improved Strategy for the PageRank Algorithm . IEEE Xplore (2011)
6. Moratanch, N.: A Survey on Extractive Text Summarization. IEEE Xplore (2017)
7. Raffel, C.: Exploring the limits of transfer learning with a unified text-to-text transformer. The Journal of Machine Learning Research (2020)
8. Lin, C.Y.: ROUGE: a package for automatic evaluation of summaries. Scientific Research (2004)
9. Gholamrezazadeh, S., Salehi, M.A.: A comprehensive survey on text summarization systems. IEEE Xplore (2009)
10. Klaus, S.: Summarizing legal regulatory documents using transformers. ACM Digital Library (2022)
11. Turtle, H.: Text Retrieval in the Legal World. Artif Intell Law (1995)
12. Anand, D.: Effective Deep Learning Approaches for Summarization of Legal Texts. Elsevier (2022)
13. Öztürk, C.E.: Prior Case Retrieval for the Court of Cassation of Turkey. IEEE Xplore (2022)
14. Koniaris, M.: Evaluation of Automatic Legal Text Summarization Techniques for Greek Case Law. MDPI (2023)
15. Ghimire, A.: Too Legal; Didn't Read (TLDR): Summarization of Court Opinions. IEEE Xplore (2023)
16. Deroy, A.: Ensemble Methods for İmproving Extractive Summarization of Legal Case Judgements. SpringerLink (2023)
17. Feijo, D.V.: Improving Abstractive Summarization of Legal Rulings Through Textual Entailment. Springer (2023)
18. Jain, D.: Summarization of Lengthy Legal Documents via Abstractive Dataset Building: An Extract-then-Assign Approach. ELSEVIER (2024)

Technologies and Systems for Monitoring the Onset of Accidents at Strategic Construction

Telman Aliev[1,2], Naila Musaeva[1,2]($\boxtimes$), Narmin Rzayeva[1,3], and Ana Mammadova[1,2]

[1] Azerbaijan University of Architecture and Construction, Ayna Sultanova 11, Baku, Azerbaijan
director@cyber.az, musanaila@gmail.com
[2] Institute of Control Systems of Ministry of Science and Education, 68 B.Vahabzade, Baku AZ1141, Azerbaijan
[3] Azerbaijan Technical University, G. Javid Avenue, 25, AZ1148 Baku, Azerbaijan

Abstract. Currently, the beginning of transition to emergency states is not reflected in the readings of measurement equipment of control systems of many essential technical facilities. The signaling of the beginning of the latent period of an emergency state is not implemented either. There are no tools for monitoring malfunctions of the railroad bed, or for controlling the beginning of the emergence of various typical malfunctions in the equipment of urban water supply systems, which are operated in continuous rotary motion under high pressure. Control systems of said facilities detect malfunctions only when they already have a clearly pronounced form. Nevertheless, specialists and operating personnel use this information to take action and prevent various accidents. However, in some cases, it is too late and an accident becomes unavoidable. At the same time, the emergence of a malfunction at these facilities is accompanied with the emergence of noise in the vibration signals at the outputs of the vibration sensors in their control systems. This noise contains diagnostic information. In this study, we propose technologies for generating informative attributes that can be used to control the beginning of the latent period of accidents. We also show the possible applications of the proposed technologies in developing intelligent systems for controlling the onset of accidents at facilities under investigation and other strategic construction.

Keywords: control · signaling · noise · informative attribute · construction facilities · pumping stations · malfunction · seismic stability · railroad track · vibration signal

1 Introduction

Currently, automatic control systems are used for control and monitoring of pumping facilities in all technological water supply and wastewater disposal systems, as well as in municipal cold and hot water supply systems, sewage pumping stations of wastewater pumping and treatment facilities. This enables the facility to stay operational for an extended period without maintenance [1–3]. They process discrete and analog data in using predetermined algorithms, forming the signals required for technological equipment control, displaying information about the parameters and state of the technological process, preparing the transfer of information about the current state of the technological process, detecting emergencies or malfunctions in the equipment being used, and automatically connecting additional pumps in the event that the plant's output is insufficient. As a result, the population's critical issues with water supply, wastewater pumping, agricultural crop irrigation, etc. are appropriately and consistently addressed [4–10]. A high degree of accident-free operation is assured since monitoring and control systems are utilized as well to control the earthquake resistance of strategic buildings, infrastructure, urban construction, and other essential assets. [1, 2].

Railroads remain the most economical and profitable means of transportation, with pipeline and water transport coming in second. There is no seasonality or weather reliance in rail transportation. Rail transport works well because of its high speed, capacity to carry heavy loads, adaptability, and other factors. The highly affordable energy cost of moving steel wheels along steel rails is one of the primary benefits of railroads.

As high-speed train traffic grows, so do the requirements for railroad infrastructure components and equipment. These requirements pertain to the quality of determining track occupancy as well as the state of the rail line, specifically the track superstructure (ballast), which is essential for maintaining train performance, safety, and uninterrupted operation. Since it is thought that there are no appreciable changes between the checks when no control is performed, the technical condition of the railroad bed at each track haul is currently practically controlled "in turns" as per the schedule with the aid of railroad test cars, geometry cars, and flaw detector cars. Meanwhile, in real life, some changes occur even a day after control as a result of a variety of causes. Consequently, besides the current ones, it would be beneficial to develop straightforward and reasonably priced intelligent technical monitoring tools that can be mounted on one of each rolling stock cars for continuous management of the onset of changes in the track's technical state. In this instance, the "Safety center" can decide whether or not to exercise "out of turn" control based on data gathered from trains in the relevant hauls [1, 11–14].

2 Problem Statement

Controlling the start of the latent period of accidents affecting economic facilities is crucial since each accident is preceded by the appearance of specific faults, which mark the beginning of the accident's latent period. It takes some time to develop, and only then does it show up in the measurements taken by the control and management systems' measuring equipment. The dynamics of the defect development determine how long the latent period lasts. As a result, the onset of an emergency is identified by the measurement equipment of the control systems of the aforementioned facilities at the precise instant when it manifests itself. Because of this, there are instances in real life where it is too late to prevent the accident. Of course, the development of new technologies and intelligent systems that enable the control and signaling of the onset of the latent period of the critical status of these facilities is required in order to address that shortcoming [1, 12–16].

Research findings have revealed that continuous operation under high load causes vibration processes to develop in many of the most significant economic infrastructures.

Consequently, as the onset of these facilities' malfunction is mostly reflected in the vibration signals, it is advised to utilize vibration sensors to control the start of the latent period of malfunction $g(i\Delta t)$. This process typically manifests itself as the noises $\varepsilon(i\Delta t)$, which are correlated with the useful signals $X(i\Delta t)$ [1, 2]. As a result, during this time, the total noise $\varepsilon(i\Delta t)$ is generated, as the sum of the noise $\varepsilon_1(i\Delta t)$ caused by outside factors and the noise $\varepsilon_2(i\Delta t)$ produced by numerous faults.

Experiments have shown [4, 5] that the latent period of accidents in the process of operation of the equipment of the facilities under investigation is caused by the fact that the operating conditions in the equipment are accompanied by fractures, fatigue cracks, residual stresses, fatigue damage, fatigue, wear, friction and abrasion. Hence, the estimates of the cross-correlation function $R_{X\varepsilon}(\mu)$ between the useful signal and the noise and the noise variance $R_{\varepsilon\varepsilon}(\mu)$ of the vibration signals are appreciable values, i.e. the following inequality takes place:

$$\begin{cases} R_{X\varepsilon}(\mu) \gg 0 \\ R_{\varepsilon\varepsilon}(\mu) \gg 0 \end{cases}$$

This makes it difficult to guarantee the adequacy of vibration control at the onset of the latent period of accidents when using traditional technologies. Therefore, at the aforementioned key economic facilities, it is necessary to develop alternative, efficient technologies as well as smart technical tools for controlling and signaling the start of the latent period of accidents.

3 The Algorithm for Finding the Estimates of the Informative Attributes to Control the Beginning of Accidents at Economic Facilities

The results of an analysis of causes of accidents have revealed [1] that the latent period of malfunction is caused by fatigue damage, wear, friction, abrasion, and fractures from the equipment's continuous rotary motion under high pressure during the operation of the facilities under consideration. Due to their prolonged, continuous operation, estimates of different features of noisy vibration signals clearly show the start of the latent period of these defects of these facilities.

According to analysis, it is most practical to employ the following technologies in order to estimate the most significant informative features that appear at the onset of the specified defects in the facilities under consideration:

The technology for determining the variance $D_{\varepsilon\varepsilon}$ of the noise $\varepsilon(i\Delta t)$ of the noisy vibration signal $g(i\Delta t)$:

$$D_{\varepsilon\varepsilon} \approx R_{\varepsilon\varepsilon}(0) \approx \frac{1}{N}\sum_i =1^N[g^2(i\Delta t) + g(i\Delta t)g((i+2)\Delta t) - 2g(i\Delta t)g((i+1)\Delta t)]$$

The technology for determining the cross-correlation function $R_{X\varepsilon}$ between the useful vibration signal $X(i\Delta t)$ and the noise $\varepsilon(i\Delta t)$:

$$\frac{1}{N}\sum_{i=1}^{N} g'^2(i\Delta t)g'(i\Delta t) + \sum_{i=1}^{N} g'^2(i\Delta t)g'((i+2)\Delta t)$$

$$= \sum_{i=1}^{N} 2g'^2(i\Delta t)g'((i+1)\Delta t),$$

The technology for determining the relay cross-correlation function $R^*_{X\varepsilon}$ between the useful vibration signal $X(i\Delta t)$ and the noise $\varepsilon(i\Delta t)$:

$$R^*_{X\varepsilon} \approx \frac{1}{N} \sum_{i=1}^{N} [\text{sgng}(i\Delta t)g((i+m-1)\Delta t) - 2\text{sgng}(i\Delta t)g((i+m)\Delta t) + +\text{sgng}(i\Delta t)g((i+m+1)\Delta t)]$$

The technology for determining the variance of the useful signal $.X(i\Delta t)$
$D_X = D_g - D_{\varepsilon\varepsilon}$, where
$D_g = \frac{1}{N} \sum_{i=1}^{N} g^2(i\Delta t)$.

Our analysis demonstrates that it is advisable to also use estimates of coefficients $K_1, K_2, K_3, K_4, K_5, K_6$ as informative attributes indicating the onset of malfunction of these facilities. These coefficients are determined from the formulas

$$K_1 = \frac{D_{\varepsilon\varepsilon}}{D_X} \quad K_4 = \frac{R_{X\varepsilon}}{D_{\varepsilon\varepsilon}}$$

$$K_2 = \frac{D_{\varepsilon\varepsilon}}{D_g} \quad K_5 = \frac{R^*_{X\varepsilon}}{D_{\varepsilon\varepsilon}}$$

$$K_3 = \frac{D_X}{D_g} \quad K_6 = \frac{R^*_{X\varepsilon}}{R_{X\varepsilon}}$$

As the previous paragraph makes clear, the noise $\varepsilon(i\Delta t)$ is regarded as a diagnostic information carrier in the vibration control of the onset of the malfunction of facilities with the used algorithms of analysis of noisy vibration signals. Here, to ensure the adequacy of control results, the proposed control systems in analog to digital conversion of the vibration signals $g(i\Delta t)$ must use algorithms and technologies for adaptive determination of the sampling interval of the noise Δt_ε in real time. This is because, for instance, depending on the speed of the rolling stock, the spectrum of vibration signals varies in time within a wide range, affecting the adequacy of control results. Therefore, give the variation of both the spectrum of the useful signals $X(i\Delta t)$ and the noise $\varepsilon(i\Delta t)$ in time caused by the speed of the train, to obtain sought estimates with required accuracy, the sampling interval must be determined adaptively in real time. Only then the estimates of the sought informative attributes can be determined with sufficient accuracy.

As we know [1], the vibration signal $X(t)$ in the process of analog to digital conversion can take any value within a range of $X_{min} \ldots X_{max}$. At the same time, in practice, during signal measurement, there is a minimum value of the increment that can be determined by the instrument being used and depends on its resolution. Denote this minimum value of the increment as ΔX. Therefore, in the measurement of the signal, the number of ita discrete values is finite

$$n = X/\Delta X + 1$$

and during the signal measurement its amplitude sampling takes place, i.e. the range of its possible variation (X_{max}, X_{min}) is spli into n sampling intervals, and at

$$S\Delta X - \frac{\Delta X}{2} \leq X(t) \leq S\Delta X + \frac{\Delta X}{2}$$

the value of the signal that ends up in the S-th interval belongs to the center of the interval $S\Delta X$.

It is shown 3 [6, 7] that adaptive determination of the sampling interval can be achieved by using the frequency properties of the low order bit $q_0(i\Delta t)$, of the samples $g(i\Delta t)$ of vibration signals during its analog to digital conversion with excessive frequency f_v that significantly exceeds the traditional sampling frequency f_c.

$f_{q0} \approx \frac{N_{q0}}{N} f_v$, where N_{q0} is the number of transitions of the low order bit $q_0(i\Delta t)$ of the sample $g_V(i\Delta t)$ from one to zero, N is the total number of samples of the analyzed signal $g(i\Delta t)$, f_{q0} is the frequency of the low order bit $q_0(i\Delta t)$, which is the recommended sampling frequency of the vibration signal $g(i\Delta t)$.

Due to this, in the process of analog to digital conversion of the vibration signal $g(t)$, conversion by means of module 2 with excessive frequency f_v in the proposed signaling system the inequalities $f_v \gg f_c$ will take place.

4 The Possibility of Control of the Beginning of the Latent Period of Seismic Stability of Critical Facilities in Seismically Active Regions

In countries located in seismically active zones, it is necessary to regularly monitor the seismic stability of important facilities and residential buildings in order to guarantee public safety. When there is a potential for a landslide in addition to seismic danger, the significance of this issue multiplies. Research has indicated that severe, catastrophic earthquakes are uncommon and occur more than ten years apart in many seismically active regions. Nonetheless, weak earthquakes typically happen in these areas for a short while (one to two months), which might be used to build a system that controls changes in the seismic stability or technical condition of buildings [1, 9]. This may result in a notable decrease in the amount of damage caused by catastrophic earthquakes. In light of this, it makes sense to develop technology and a control system that can be used to regularly monitor the onset of the latent period of change in the seismic stability of critical buildings during each mild earthquake.

Research has demonstrated that, in seismically active areas, stable conditions for socially important construction facilities are indicated by the invariability of the estimates of informative attributes of seismic signals received from seismic sensors installed on these facilities during weak earthquakes. Meanwhile, facilities that exhibit changes in the estimates of informative attributes can be classified as belonging to the group where the latent period of changes in seismic stability is beginning. Considering this unique characteristic of seismic zones, a schematic representation of a possible approach to building an intelligent city-wide system for detecting the onset of changes in seismic stability during mild earthquakes is given in Fig. 1.

In this version of the system, each building $O_1, O_2, ..., O_N$ is equipped with a controller-based local unit and with corresponding seismic sensors $D_1, D, ..., D_m$, installed in the most vulnerable parts of the building. During weak earthquakes the signals $g_1(i\Delta t), g_2(i\Delta t), ..., g_n(i\Delta t)$ are transmitted from seismic sensors of each building to the system through communication means. Therefore, during weak earthquakes, the

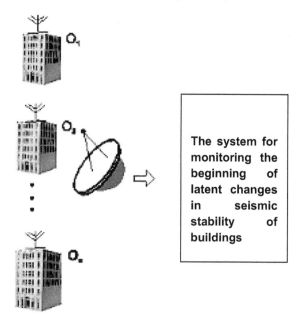

Fig. 1. Diagram of the system for monitoring the beginning of latent changes in seismic stability of buildings during weak earthquakes.

system receives seismic signals from monitoring objects, which are practically vibration signals that can be used to determine the technical condition of the buildings.

It is shown in literature [1, 2] that the seismic stability of an object can be considered stable if the following conditions are satisfied for the seismic signals $g(i\Delta t)$:

$$\frac{1}{N}\sum_{i=1}^{N} g'(i\Delta t)g'(i\Delta t) + \sum_{i=1}^{N} g'(i\Delta t)g'((i+2)\Delta t) = \sum_{i=1}^{N} 2g'(i\Delta t)g'((i+1)\Delta t), \quad (1)$$

$$\frac{1}{N}\sum_{i=1}^{N} g'(i\Delta t)g(i\Delta t) + \sum_{i=1}^{N} g'(i\Delta t)g'((i+2)\Delta t) = \sum_{i=1}^{N} 2g'(i\Delta t)g'((i+1)\Delta t), \quad (2)$$

$$\frac{1}{N}\sum_{i=1}^{N} g^2(i\Delta t)g(i\Delta t) + \sum_{i=1}^{N} g^2(i\Delta t)g((i+2)\Delta t) = \sum_{i=1}^{N} 2g^2(i\Delta t)g((i+1)\Delta t), \quad (3)$$

$$[\frac{1}{N}\sum_{i=1}^{N} g'^2(i\Delta t)g'(i\Delta t) + \sum_{i=1}^{N} g'^2(i\Delta t)g'((i+2)\Delta t)$$
$$= \sum_{i=1}^{N} 2g'^2(i\Delta t)g'((i+1)\Delta t), \quad (4)$$

$$\frac{1}{N}\sum_{i=1}^{N} \operatorname{sgn} g(i\Delta t)g(i\Delta t) + \sum_{i=1}^{N} \operatorname{sgn} g(i\Delta t)g((i+2)\Delta t)$$
$$= \sum_{i=1}^{N} 2\operatorname{sgn} g(i\Delta t)g((i+1)\Delta t), \quad (5)$$

$$\frac{1}{N}\sum_{i=1}^{N} \operatorname{sgn} g'(i\Delta t)g'(i\Delta t) + \sum_{i=1}^{N} \operatorname{sgn} g'(i\Delta t)g'((i+2)\Delta t)$$
$$= \sum_{i=1}^{N} 2\operatorname{sgn} g'(i\Delta t)g'((i+1)\Delta t), \quad (6)$$

where $g(i\Delta t)$ is a centered seismic vibration signal, $g\prime(i\Delta t)$ is a non-centered seismic vibration signal, which show that the seismic properties of the monitored facilities are stable and there is no cause for concern.

However, when the seismic stability of objects is compromised, during the analysis of seismic signals, these equalities are broken and the sought estimates are non-zero, i.e.

$$R_{X\varepsilon}^{*1}(0) = \frac{1}{N}\sum_{i=1}^{N} g'(i\Delta t)g'(i\Delta t) + \sum_{i=1}^{N} g'(i\Delta t)g'((i+2)\Delta t)$$
$$- \sum_{i=1}^{N} 2g'(i\Delta t)g'((i+1)\Delta t) \neq 0,$$

$$R_{X\varepsilon}^{*2}(0) = \frac{1}{N}\sum_{i=1}^{N} g'(i\Delta t)g(i\Delta t) + \sum_{i=1}^{N} g'(i\Delta t)g'((i+2)\Delta t)$$
$$- \sum_{i=1}^{N} 2g'(i\Delta t)g'((i+1)\Delta t) \neq 0,$$

$$R_{X\varepsilon}^{*3}(0) = \frac{1}{N}\sum_{i=1}^{N} g^2(i\Delta t)g(i\Delta t) + \sum_{i=1}^{N} g^2(i\Delta t)g((i+2)\Delta t)$$
$$- \sum_{i=1}^{N} 2g^2(i\Delta t)g((i+1)\Delta t) \neq 0,$$

$$R_{X\varepsilon}^{*4}(0) = \frac{1}{N}\sum_{i=1}^{N} g'^2(i\Delta t)g'(i\Delta t) + \sum_{i=1}^{N} g'^2(i\Delta t)g'((i+2)\Delta t)$$
$$- \sum_{i=1}^{N} 2g'^2(i\Delta t)g'((i+1)\Delta t) \neq 0,$$

$$R_{X\varepsilon}^{*5}(0) = \frac{1}{N}\sum_{i=1}^{N} \operatorname{sgn} g(i\Delta t)g(i\Delta t) + \sum_{i=1}^{N} \operatorname{sgn} g(i\Delta t)g((i+2)\Delta t)$$
$$- \sum_{i=1}^{N} 2\operatorname{sgn} g(i\Delta t)g((i+1)\Delta t) \neq 0,$$

$$R_{X\varepsilon}^{*6}(0) = \frac{1}{N}\sum_{i=1}^{N} \operatorname{sgn} g'(i\Delta t)g'(i\Delta t) + \sum_{i=1}^{N} \operatorname{sgn} g'(i\Delta t)g'((i+2)\Delta t)$$
$$- \sum_{i=1}^{N} 2\operatorname{sgn} g'(i\Delta t)g'((i+1)\Delta t) \neq 0.$$

Studies have shown that using this technology it is possible to build an intelligent monitoring system, where by analyzing the seismic signals $g_1(i\Delta t)$, $g_2(i\Delta t)$, $g_3(i\Delta t),\ldots, g_m(i\Delta t)$ received from the sensors of respective objects, the estimates $R_{X\varepsilon}^{*1}(0)$, $R_{X\varepsilon}^{*2}(0)$, $R_{X\varepsilon}^{*3}(0)$, $R_{X\varepsilon}^{*4}(0)$, $R_{X\varepsilon}^{*5}(0)$, $R_{X\varepsilon}^{*6}(0)$ are calculated during periods of time when low magnitude earthquakes occur, which are used as informative attributes.

Current estimates are also calculated during the periods of the next low magnitude earthquake. Equality (1)-(6) is checked. If the difference in Eqs. (1)-(6) does not exceed the accepted minimum ranges, then it is considered that the seismic stability and technical condition of respective objects $O_1, O_2,\ldots, O_N$ have not changed. Otherwise, information about the beginning of changes in the technical condition of the corresponding object is generated. In this case, the difference between the current and reference estimates determines the range of deviation and the severity of the situation.

Received from different objects, is taken as a sign of the beginning of violation of seismic stability.

In this case, those of the estimates $R_{X\varepsilon}^{*1}(0)$, $R_{X\varepsilon}^{*2}(0)$, $R_{X\varepsilon}^{*3}(0)$, $R_{X\varepsilon}^{*4}(0)$, $R_{X\varepsilon}^{*5}(0)$, $R_{X\varepsilon}^{*6}(0)$ that are non-zero show that seismic properties of the objects under control are unstable, and it is necessary to control their technical condition using standard equipment and technologies.

5 The Possibilities of Developing Intelligent Systems for Control of Railroad Track Malfunctions

Modern geometry cars, flaw detector cars and other railroad test cars reliably control the technical condition of railroad track hauls at "certain intervals". Their limited quantity makes "continuous control" of all track hauls practically impossible. In actual life, however, problems with the railroad tracks can still arise even a day after the control due to a variety of causes, such as the impact of seismic processes, hurricane winds, or intense rainfall. According to the results of our analysis, one approach for "continuous" monitoring of the onset of changes in the track's technical condition is to analyze the noise in the signal that is received from ground vibration caused by the impact of the rolling stock. This allows for the formation of informative attributes that can be used to identify the track's technical condition. Using traditional correlation and spectral analysis technologies and other methods here does not give adequate results because of the influence of noise on useful vibration signals. It was therefore deemed appropriate to use the technology of selection and analysis of useful vibration signal, noise vibration signal, and the relationship between them, using noise as the main source of diagnostic information.

Our analysis of the possibilities of controlling the technical condition of the railroad bed shows the possibility of creating simple and low-cost intelligent tools to be mounted on one of the cars of each rolling stock to identify the hauls to be controlled "out of turn" [1–5].

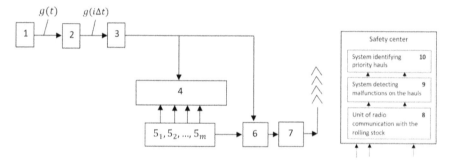

Fig. 2. The intelligent system for identifying railroad track hauls in need of out-of-turn control.

Figure 2 is the block diagram of a possible version of an intelligent system of malfunction control based on the analysis of vibration signals from the railroad bed that emerge at the beginning of typical accidents during train movement.

The system is comprised of the following modules:

1- Vibration sensor.
2- Module of adaptive analog-to-digital conversion of vibration signals $g(t) = g(i\Delta t) = X(i\Delta t) + \varepsilon(i\Delta t)$.
3- Module for determining the estimates of current informative attributes
4- Module for generating reference informative attributes.
5- 5_1-5_m - Module for memorizing reference informative attributes.
6- Module for identifying the onset of the latent period of malfunctions.
7- Module for generating and transmitting information via radio channel.

During the operation of the control system, the vibration signal $g(t)$ from the vibration sensor 1 arrives at the input of the module 2, i.e. the adaptive analog-to-digital converter to be converted into digital code $g(i\Delta t)$ with adaptive determination of the sampling interval. In practical application of the system, the process states with a training stage that goes on for a certain period of time. During the movement of trains in all hauls, when there is a malfunction on the track, the sets $\{D_X, D_{\varepsilon\varepsilon}, K_1, K_2, K_3, K_4, K_5, K_6, R_{X\varepsilon}, R^*_{X\varepsilon}\}$ consisting of relevant estimates of informative attributes are determined and memorized. For a certain period of time, all possible reference informative attributes are determined and memorized by analyzing vibration signals on different trains. After the training stage, railroad malfunction monitoring and identification of "problematic hauls" begins. During the movement of all rolling stocks, the current estimates $\{D_X, D_{\varepsilon\varepsilon}, K_1, K_2, K_3, K_4, K_5, K_6, R_{X\varepsilon}, R^*_{X\varepsilon}\}$ are determined by analyzing vibration signals $g(i\Delta t)$ at the current time Instant to be compared with the reference ones. If they are greater than the threshold values set on the modules $5_1, 5_2, \ldots 5_m$, module 6 generates a signal about the beginning of a malfunction. After a second comparison, if the current estimates exceed the respective reference estimates, then module 7 generates a signal about the beginning of a malfunction. The signal is transmitted to the "Safety Center" via radio. As a result of the operation of the "system detecting control hauls", the information about the beginning of a malfunction is transmitted on the faulty hauls in each rolling stock.

Thus, during the movement of trains on the faulty hauls, the information through the "Unit of radio communication with the rolling stock" is successively received by the "System establishing the presence of malfunctions on the hauls" of the "Safety Center". The set of informative attributes indicating the presence of malfunctions from these "problematic" hauls in all rolling stocks for the day will have the following form.

They are periodically analyzed as they are received, and the final judgment on the occurrence of a malfunction at the relevant railroad haul, where it is necessary to send "test cars" out of turn, is produced based on the received combination of a set of informative attribute estimates. At the same time, the reliability of the control results is achieved by parallelizing the control process with the use of many algorithms. The degree of reliability is determined by the number of coinciding results regarding the occurrence of a malfunction from the rolling stock in each haul in the "system identifying priority hauls." Thus, in the simplest instance, the "problematic" hauls are finally established from those sections of the tracks from which the rolling stock communicates alerts in the "system identifying priority hauls" and it is proposed to alter the schedule of their control. If several such problematic hauls exist at the same time, schedules for their out-of-turn control can be supplied.

6 Conclusion

The existing monitoring and control systems ensure continuous operation of critical infrastructure facilities. However, they are unable to monitor the onset of accidents. At the same time their accident-free operation requires that this process be detected in the initial latent period. Vibration diagnostics is of great importance in solving this problem, as vibration signals have a great diagnostic information potential. Nevertheless, the adequacy of vibration control results is not ensured at this stage, as these systems do not use noise of vibration signals as a carrier of diagnostic information. At the same time in vibration signals, noise emerges precisely at the inception of a malfunction and is correlated with the useful signal. Therefore, only by using the noise estimates as an informative attribute it is possible to ensure the adequacy of vibration diagnostics results. Experimental studies have shown that when the ranges of deviations of estimates of informative attributes from vibration sensors are greater than the threshold values, the initial stages of malfunction are detected with sufficient reliability, which allows taking action and avoiding catastrophic accidents.

References

1. Aliev, T.: Noise control of the Beginning and Development Dynamics of Accidents. Springer, p. 201 (2019)
2. Aliyev, T.A., Mamedov, S.I.: Telemetric information system to prognose accident when drilling wells by robust method. Oil Industry J. **2** (2002)
3. Aliyev, T.A., Alizada, T.A., Rzayeva, N.E.: Noise technologies and systems for monitoring the beginning of the latent period of accidents on fixed platforms. Mech. Syst. Signal Process. **87**, 111–123 (2017)
4. Aliev, T.A.: Intelligent seismic-acoustic system for identifying the area of the focus of an expected earthquake. Earthquakes tectonics. Hazard and risk mitigation, edited by Taher Zouaghi, Published by in Tech, Janeza Trdine 9, 51000 Rijeka, Croatia, The Editor(s) and the Author(s), pp. 293–315 (2017)
5. Aliyev, T.A., Rzayev, A.H., Guluyev, G.A., Alizada, T.A., Rzayeva, N.E.: Robust technology and system for management of sucker rod pumping units in oil wells. Mech. Syst. Signal Process. **99**(15), 47–56 (2018)
6. Aliev, T.A., Musaeva, N.A., Babayev, T.A., Mammadova, A.I., Alibayli. E.E.: Technologies and intelligent systems for adaptive vibration control in rail transport. Transport Problems: An International Scientific J. **17**(3), 31–38 (2022)
7. Metin, M., Guclu, R.: Rail Vehicle Vibrations Control Using Parameters Adaptive PID Controller, pp. 1–10. Mathematical Problems in Engineering, Hindawi (2014)
8. Bendat, J.S., Piersol, A.G.: Random Data Analysis & Measurement Procedures. Wiley, New York (2000)
9. Collacott, R.A.: Mechanical Fault Diagnosis and Condition Monitoring, p. 506 (1977)
10. Yakovlev, S.V., Karelin, Y.A., Laskov, Y.M., Kalitsun, V.I.: Water Drainage and Wastewater Treatment. Stroyizdat, Moscow, p. 591 (1996)
11. Lin, C.C., Wang, J.F., Chen, B.L.: Train-induced vibration control of high-speed railway bridges equipped with multiple tuned mass dampers. J. Bridg. Eng. **10**(4), 398–414 (2005)
12. Dudkin, E.P., Andreeva, L.A., Sultanov, N.N.: Methods of noise and vibration protection on urban rail transport. Procedia Eng. **189**, 829–835 (2017)
13. Gvozdkova, S.I., Shvartsburg, L.E.: Analysis of sources and methods for reducing noise by minimizing vibrations of engineering technological processes. Procedia Eng. **206**, 958–964 (2017)
14. Agic, A., Eynian, M., Ståhl, J.-E., Beno, T.: Experimental analysis of cutting-edge effects on vibrations in end milling. CIRP J. Manuf. Sci. Technol. **24**, 66–74 (2019)
15. Shirman, A.R., Soloviev, A.B.: Practical vibration diagnostics and monitoring of mechanical equipment. Mechanical Engineering, Moscow, p. 276 (1996)
16. Anderson, D., Gautier, P., Iida, M., et al.: Noise and vibration mitigation for rail transportation systems. In: Proceedings of the 12th International Workshop on Railway Noise, 12–16 September 2016, Terrigal, Australia (2016)

Optimizing Feature Distributions for Unsupervised Deep Learning-Based Fabric Defect Detection and Localization

Eissa Alzabidi(✉) and Oğuz Fındık

Department of Computer Engineering, Karabuk University, Karabuk, Turkey
1938166016@ogrenci.karabuk.edu.tr, oguzfindik@karabuk.edu.tr

Abstract. Fabric defect detection techniques are significantly important to improve the quality of the fabric industry. However, most existing models face challenges related to overgeneralization due to relying only on the in-distribution, which represents normal samples for training. In this paper, we introduce an improved approach to address overgeneralization issues and augment the accuracy of fabric defect detection and localization. Our approach leverages a U-Net architecture with end-to-end modeling in the context of unsupervised deep learning-based defect detection. We propose the Gamma-Weighted Discrepancy loss function (L_{GWD}) to tackle feature matching and defect detection more effectively. Moreover, we optimized the quality of the synthetic anomalies generated during training to reduce the gap between simulated anomalies and real-world defects. The innovative strategy focuses on enhancing the distinctiveness and dissimilarity of synthetic anomalies compared to normal patterns. Experimental results conducted on the AITEX fabric and Carpet-MVTecAD datasets demonstrate the effectiveness of these enhancements with high average AUC scores of 98.4% at the image level and 98.1% at the pixel level. The proposed model demonstrates superior performance compared to existing state-of-the-art deep learning methods in the detection and localization of fabric defects, significantly contributing to progress in this domain.

Keywords: Defect Detection · Unsupervised Deep Learning · Feature Mapping · Transfer Learning · Synthetic Anomalies

1 Introduction

The anomaly detection techniques are significantly important in improving the quality of industrial products [1]. Despite the variety and complexity of anomalies that can occur in industrial images and the lack of labeled datasets, anomaly detection is a rapidly growing field with the potential to significantly improve the quality of industrial products [2]. Fabric defect detection can be utilized to pinpoint surface defects like holes, cuts, color contamination, etc. that can impact both the performance and aesthetics of a product. In fabrics anomalies can be represented by more than 70 types of fabric defects. Detecting defects early in production contributes to product quality, waste reduction, cost savings, and increases profits for manufacturers. Moreover, it is possible to prevent defects from reaching the customer and causing problems [3]. Fabric defect detection involves scenarios of image-level classification. On the other hand, fabric defect localization refers to the identification of abnormal regions that differ from the normal, which is typically achieved through pixel or patch-level anomaly detection techniques [4]. We can refer to them as defect detection. Fabric defect detection entails the identification of prominent regions within an image that deviates from the surrounding areas. This form of detection is categorized as low-level since it involves recognizing distinct visual characteristics. Conversely, some anomalies may not be immediately apparent throughout the entire image, particularly those associated with functional defect detection entails the identification of irregularities in the overall function or behavior of an image. This type of detection can be employed to detect abnormalities in the product's overall function or behavior, such as malfunctioning parts, improper assembly, or misaligned components. It is regarded as high-level detection as it necessitates an understanding of the image's context and semantics [5].

Deep learning-based defect detection (DLDD) methods have gained immense popularity due to their ability to handle large amounts of data and extract complex features. Most of the articles focused on unsupervised or semi-supervised DLDD, where these approaches do not necessitate labeled data for training the models. Unlike supervised DLDD not as popular and not as widely used for several limitations, such as limited availability of labeled data, high cost of labeling, difficulty in identifying all types of anomalies, and lack of interpretability by providing results without providing detailed information about the specific location of the defect [6]. Therefore, in our study, we focused on the literature that utilizes only normal samples for training, under the assumption of ignoring the abnormal samples because it is very rare, and few compared to the normal ones. Additionally, we utilized a minimal number of normal and abnormal samples for testing [7]. DLDD methods have advanced from just comparing images to comparing features of images. Additionally, the methods used for detecting anomalies have moved from simple defect synthesis to more complex self-supervised methods that utilize anomaly simulation strategies or contrastive learning. Therefore, supervised DLDD algorithms learn from examples without prior knowledge of the anomalies expected appearance by analyzing common features between similar and different features between non-similar images [8, 9]. Moreover, the majority of DLDD methods operate under the assumption that the distribution of testing and training datasets is identical. Nevertheless, this assumption is often not met in practice, which can lead to inaccurate results in the defect detection process [10].

Detecting fabric defects is essential for enhancing product quality in the textile industry. However, most current unsupervised DLDD methods often face difficulties in overgeneralization, primarily because they rely exclusively on in-distribution data during training, where the model learns the training data too well and fails to generalize to new, unseen data. This can result in the model identifying patterns that are not truly indicative of defects, leading to overgeneralization and reduced detection accuracy. Overgeneralization problems can arise when training a model on a set of data that is not sufficiently diverse or representative of the real-world data it will encounter. This can lead to the model performing poorly when presented with new or unexpected inputs. It may have the potential to affect the discrepancy distribution DLDD of the corresponding features between the expert domain and apprentice domain, consequently influencing anomaly detection performance. When overgeneralization occurs, the margin between the DLDDs of normal and abnormal features may be reduced or eliminated, making it difficult for the model to differentiate between normal and abnormal inputs. Similarly, the overlap between the DLDDs may increase, making it more likely for the model to misclassify the inputs. To tackle this issue, this paper introduces the following key contributions:

- We present an innovative approach in the context of unsupervised DLDD by minimizing differences normal between features of normal samples while simultaneously maximizing differences abnormal between features of synthetic abnormal samples. We proposed the Gamma-Weighted Discrepancy loss function (L_{GWD}) is tailored to enhance the distinctiveness and dissimilarity of synthetic anomalies compared to normal patterns. Thereby augmenting the separation between normal and abnormal patterns. For end-to-end Modeling, our approach leverages a U-Net architecture to achieve more efficient feature matching and defect detection.
- We develop an optimization algorithm to generate synthetic anomalies. This includes choosing the best noise generation masks, which are the Perlin noise mask [11], the free-form mask [12], and CutPaste [13]. Next, the mask is combined with one of the various noises to generate different types of synthetic anomalies, including texture noise, structure noise, image blending noise, filling noise, and stain noise. They are randomly selected and applied to the normal image to generate an artificial anomaly image. These masks allow for greater diversification in terms of size, shape, location, and appearance of anomalous patterns and can better generalize to real anomalous image samples.

This paper is organized as follows: Sect. 2 presents a review of the related work methods for defect detection; Sect. 3 details the implementation of our method; Sect. 4 provides extensive experimental demonstrating the effectiveness of our method; Sect. 5 provides a summary of the paper and recommendations for future research.

2 Related Work

This section presents the existing literature on Unsupervised DLDD methods related to our work. We categorize these techniques into three primary categories: reconstruction-based, feature embedding-based, and knowledge distillation-based methods. Each of these categories includes several different techniques, and each has a set of advantages and limitations.

2.1 Reconstruction-Based Methods

In industrial production, this method is useful in identifying defective products or parts along the production line. The neural network undergoes training using a normal image and then it is used to reconstruct new images in real-time. When the deviation between the original image and the reconstructed image exceeds a predefined threshold, the system identifies it as an anomaly, signaling the presence of a potential defect [14]. In this context, some recent studies have presented new algorithms for reconstruction-based image anomaly detection in industrial production. For example, RIAD [15] and DFR [16]. RIAD [15] is intuitive and interpretable. However, it lacks the incorporation of prior knowledge and solely depends on the latent layer's expressive capacity to identify normal features. DFR [16] is computationally efficient. Nevertheless, it may be incomplete localization of anomalous regions and poor visualization of those regions. Additionally, this method may not perform well in certain datasets. Generally, reconstruction-based methods have shown remarkable improvement in the detection and location of anomalies over the last few years. Nevertheless, it may not do well on multiscale complex texture datasets due to the challenges in managing the ability to generalize the bottleneck layer and coarse localization of anomalous regions. Additionally, a lack of strong generalization ability can lead to the inadequate reconstruction of edge regions which can cause over-detection problems [4].

2.2 Feature Embedding-Based Methods

Feature embedding-based methods are widely used in the field of industrial production. These methods typically consist of two stages: The first stage is representation learning by a pre-trained network model using extensive datasets like MNIST [17], CIFAR10 [18], and ImageNet [19] or using self-supervised learning techniques. The second stage is anomaly estimation in new images. This involves calculating a distance or similarity score between the reference normal samples in the training set and the test sample. If the resulting score falls below a specific threshold, the image is classified as anomalous. In addition, anomaly localization is used to identify the specific location of the anomalous regions within the image. This approach eliminates the need for separate teacher and student networks and simplifies the model architecture, which uses a feature space to build a representation of the input image. Many of these current methods still have better performance than reconstruction-based models [5]. This is shown by the results obtained by some algorithms such as SPADE [20], PatchCore [21], PaDiM [22], MemSeg [11], and HFFMM [23]. The SPADE [20] model works by finding correspondences between anomalous images and a set of normal images and creates a mask that indicates which pixels in the anomalous image are anomalous. SPADE methods show excellent performance by directly comparing normal features with input features. However, their time complexity increases with the amount of data due to the KNN algorithm. In HFFMM [23], to detect defects the authors developed a multi-scale high-frequency information extraction module based on the difference between the original and high-frequency features calculated by the transformer encoder. PaDiM [22] model is a distribution of patches in normal images. Although it can detect anomalies at multiple scales, it may lead to rough and non-fine-grained training results. This is attributed to the susceptibility of patch region settings to the size of the cropped patch, and the correlation between neighborhood patches is not always present in complex images.

2.3 Knowledge Distillation-Based Methods

Knowledge distillation (KD) is a technique employed to transfer knowledge from a pre-trained large and more complex teacher model to a small and simple student model. The teacher model is pre-trained to generate a set of features for the normal product images. These features are then used to train the student model. The student model can then be used for anomaly detection on new images. This approach is useful for DLDD in industrial products when the original model is too computationally expensive to run in real-time applications. Several studies have been proposed in the literature using KD-based DLDD methods for industrial products, such as ST-AD [24], STPM [25], and CDO [26]. ST-AD [24], utilizes solely the last layer's output as a feature for KD. Moreover, to improve anomaly localization, it adopts a multi-patch approach, which can lead to a considerable increase in computing time. STPM [25] model calculates the differences between multi-feature layers of a pre-trained teacher network and a student network. Many existing methods typically focus on minimizing discrepancies between normal feature distributions. However, due to the high generalization ability, the apprentice may produce abnormal feature distributions like those in the expert domain. This similarity can result in low discrepancies between abnormal feature distributions, making it more difficult to detect anomalies. To address the impact of overgeneralization and enhance the apprentice model's ability to accurately detect anomalies, the CDO [26] is proposed. Nevertheless, they have several limitations, including being highly sensitive to the hyperparameters, which may not effectively locate anomalies. They als Moreover, they are not able to adapt to multiscale anomalies and datasets that are not aligned. In addition, choosing the wrong layer in KD can significantly impact their effectiveness, leading to inaccurate results. The multi-patch approach can result in longer computation times. Furthermore, they may suffer from incomplete transfer of knowledge, and difficulty in handling scaling.

3 Proposed Method

In this paper, we design an approach that achieves enhanced performance in discerning subtle visual anomalies, reducing false positives, and improving overall detection accuracy inspired by the MemSeg [11] method. In MemSeg, the anomaly simulation strategy is employed to generate synthetic anomalies to enhance the training dataset. The memory module effectively stores general patterns of normal images without significantly increasing computational cost. MSFF and spatial attention modules are used to prevent overlapping features. The decoder reconstructs input images based on extracted features and completes the end-to-end pixel-level defect localization, which allows the identification and localization of surface defects effectively. However, the model still suffers from a gap between anomalies simulated and real anomalies, as well as the potential overlap between normal features and simulated anomalies.

In our proposed method, together with the proposed anomaly simulation strategy, we propose an innovative approach in the context of unsupervised DLDD by minimizing differences normal (D_n) between features of normal samples while simultaneously, maximizing differences abnormal (D_a) between features of synthetic abnormal samples. The proposed method is designed to render synthetic anomalies more distinct and dissimilar to normal patterns, thereby augmenting the separation between normal and abnormal patterns. In end-to-end Modeling, we employ a U-Net architecture in the proposed model to achieve more efficient feature matching and defect detection as shown in Fig. 1. Our model uses a pre-trained ResNet18 as an encoder for feature extraction during the training phase. In addition, to synthetic anomalies generated from normal images, our model incorporates the proposed skip-connection structure, spatial attention module, and multi-scale feature fusion module, with all convolutions replaced by depth-separable convolutions [27]. Ultimately, we adopt a comprehensive loss function, encompassing L_1 loss, focal loss (L_f), and our proposed L_{GWD} loss, to guide the training process and optimize the model's performance.

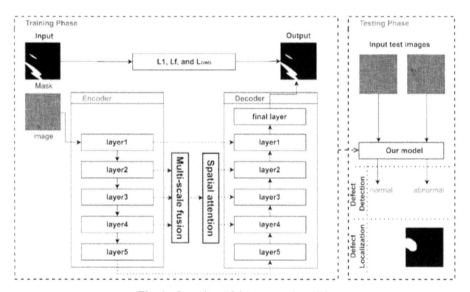

Fig. 1. Overview of the proposed model.

3.1 Anomaly Simulation Strategy

In some cases, gathering a sufficiently large dataset of real-world anomalies can lead to challenges in terms of difficulty or cost. To overcome this challenge, anomaly simulation can be used to generate synthetic anomalies that resemble real-world anomalies. These synthetic anomalies can be used to augment the dataset and train the deep learning model to detect anomalies more accurately. Synthetic anomalies can be generated using various techniques, including image inpainting or attribute restoration. In image inpainting, the idea is to use synthetic or simulated abnormal data to reconstruct the normal regions and repair abnormal regions. By generating defective training samples, the network can improve its ability to repair the anomalies. In attribute restoration, hidden attributes like color and orientation are used instead of masked areas [5]. There are several state-of-the-art anomaly simulation-based methods for visual industrial products, such as MemSeg

[11], CutPaste [13], RIAD[15], DRAEM [28], NSA [29], etc. However, it's important to note that while these methods can be effective, they are not foolproof. There can be a gap between real anomalies and simulated anomalies, and the relationship between accurate positioning results and more realistic defect synthesis methods is maybe not clear. To refine the ability of our model to detect anomalies, we propose a synthetic anomaly generation optimization algorithm. This algorithm incorporates an adaptive learning mechanism that adjusts the models sensitivity dynamically to different types and levels of anomalies. Adaptability allows the model to effectively handle various visual patterns and detect anomalies under different real-world conditions. Our optimized algorithm utilizes the set of generation masks (L_M) from Perlin noise [11], free-form noise [12], and CutPaste [13] with various techniques for generating noise images (L_N), including texture noise, structure noise, image blending noise, filling noise, and stain noise to create various synthetic anomalies (I_a).

These techniques are randomly selected and applied to normal images to generate an artificial anomaly image. These masks allow for greater diversification in size, shape, location, and appearance of anomalous patterns and can better generalize to real anomalous image samples. The process of synthetic anomaly generation (see Fig. 2) is described by Algorithm 1, and the generation is selected according to the importance factor as follows:

$$P_{WM} = \frac{e^{(\alpha * L_M)}}{\sum_i e^{(\alpha * L_{Mi})}} \tag{1}$$

$$P_{WN} = \frac{e^{(\alpha * L_N)}}{\sum_i e^{(\alpha * L_{Ni})}} \tag{2}$$

In these equations, P_{WM} and P_{WN} represent probability weights, computed as the exponential of α multiplied by L_M and L_N, respectively. The hyperparameter α is an importance factor, dynamically updated when optimal solutions are achieved. In case of performance decline, it is configured to explore new solutions. While L_M and L_N represent the list of the count of randomly selected make generations and noise image generators, respectively.

$$M = \text{random.choices}(L_M, P_{WM}) \tag{3}$$

$$N = \text{random.choices}(L_N, P_{WN}) \tag{4}$$

The mask M is randomly chosen and generated based on the calculated probabilities P_{WM} and the set L_M. Similarly, the noise image N is randomly selected and generated by the set L_N with the determined probabilities P_{WN}. The generating synthetic anomaly image equation is represented as:

$$I_a = I \odot \overline{M} + \beta \cdot (M \odot N) + (1 - \beta) \cdot (I \odot M) \tag{5}$$

where I_a is the resulting synthetic anomaly image after the blending process. I is the original normal image that has been resized to 256 x 256. The $\overline{M}$ is the complement of the mask image M. $\odot$ represents element-wise multiplication. N is the noise image. β

is the blending factor that controls the influence of the noise. The purpose of blending factor β is to balance the fusion of the I_p and N to simulate anomalies more realistically. The blending factor β is sampled randomly and uniformly from the range [0.40,1].

Algorithm 1. Synthetic anomaly generation algorithm

Input:
I: Normal image is resized to 256 × 256, α = 0.005: Importance factor, $L_M = [1,1,1]$: # of mask generations randomly selected, $L_N = [1,1,1,1,1]$: # of noise image generations randomly selected, β = [0.40,1]: blending factor.
Output:
I_a: Synthetic anomaly image, M: Generate binary mask.
Begin
1. P_{WM}, P_{WN} ← Calculate prob$_{weights}$ using Eq.(1), Eq.(2)
2. M ← Randomly select and generate a binary mask with L_M and P_{WM} using Eq.(3)
3. N ← Randomly choose and generate a noise image with L_N and P_{WN} using Eq.(4)
4. I_a ← Apply the selected N and to the I_p using the generated β Eq.(5)
5. return (I_a, M)

End

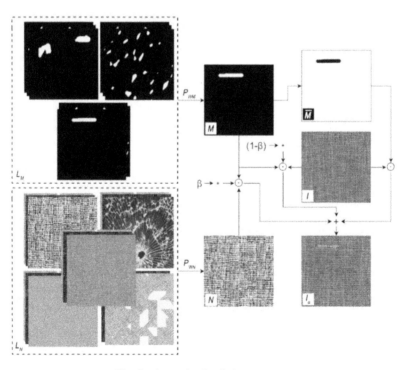

Fig. 2. Anomaly simulation strategy.

3.2 Optimization Objectives

We employ the Gamma-Weighted Discrepancy loss function (L_{GWD}), utilizing the hyperparameter gamma (γ), where $\gamma \geq 0$ in W_n for determining the weights associated with normal pixels, and W_a for abnormal pixels. As a result, we can be formulated as follows:

$$\mu_n = \frac{\sum_i^{N_n} D_{ni}}{N_n} \tag{6}$$

$$\mu_a = \frac{\sum_i^{N_a} D_{ai}}{N_a} \tag{7}$$

$$W_n = [\frac{D_n}{\mu_n}]^{\gamma} \tag{8}$$

$$W_a = [\frac{D_a}{\mu_a}]^{-\gamma} \tag{9}$$

where D_n and D_a are the most similar features between input and output samples for normal and artificial abnormal patterns, respectively. By adjusting the exponent, normal pixels with larger discrepancies are enhanced, while abnormal pixels with smaller discrepancies are assigned higher weights. The proposed loss function is defined as:

$$L_{GWD} = \frac{\sum_i^{N_n} w_{ni} * d_{ni} - \sum_i^{N_a} w_{ai} * d_{ai}}{\sum_i^{N_n} w_{ni} + \sum_i^{N_a} w_{ai}} \tag{10}$$

$$L = \lambda_{L_1} * L_1 + \lambda_{L_f} * L_f + \lambda_{L_{GWD}} * L_{GWD} \tag{11}$$

Here, $d_{ni} \in D_n : \{d_{n1}, d_{n2}, d_{n3}, \ldots, d_{Nn}\}$ represents the set of differences between normal feature descriptors, and $d_{ai} \in D_a : \{d_{a1}, d_{a2}, d_a, \ldots, d_{Na}\}$ represents the set of differences between artificial abnormal feature descriptors. The numbers of normal and synthetic abnormal pixels are denoted as N_n and N_a, respectively.

4 Experiments

4.1 Implementation Details

To reduce possible train-test discrepancies, we make use of the data pre-processing pipeline mentioned and built for all datasets. Data prep pipeline constituents per-pixel normalization using [0.48145466, 0.4578275, 0.40821073] as the mean and normalizing each RGB image to [0, 1] and standard deviation of [0.26862954, 0.26130258, 0.27577711]. In our experiments, we defaulted on the input size to 256 on the shorter edge for consistency with pre-trained models. For optimization detailed settings used in these experiments are listed in Table 1.

4.2 Evaluation Metrics

The Area Under the Receiver Operating Characteristic curve (ROC-AUC), is a widely utilized metric in defect detection. The datasets exhibit an imbalance in the quantity of normal and defective fabric images. ROC-AUC's virtue lies in its independence from the classification threshold, rendering it less susceptible to imbalanced datasets in comparison to alternative metrics. Furthermore, AUC gauges the area beneath the ROC curve, encapsulating the balance between the true positive rate (sensitivity) and false positive rate (1-specificity) across various decision thresholds [30]. This offers a comprehensive overview of the model's efficacy across all potential operating points, providing valuable insights into overall defect detection performance. The results for the Area Under the Receiver Operator Curve are documented in this work. We also supplement the image-level and pixel-level AUC results.

Table 1. Experiment configurations.

Parameter	Value	Parameter	Value
Input size	256 × 256	Hyper-parameters	$Lr = 0.003, \gamma = 4, \lambda l1 = 0.50, \lambda f = 0.40, \lambda GWD = 0.10$
Epochs	2700	Weight Decay	0.0005
Batch size	8	Optimizer	Adam
Pretrained Model	ResNet18	Frame	Pytorch
		Computer configuration	Graphics card: Tesla T4, RAM: 16GB

4.3 Dataset

MVTec-AD Dataset. The MVTec-AD dataset [31] is a widely used benchmark dataset for anomaly detection in industrial inspection tasks. This dataset contains various types of anomalies in 15 different categories, including 5 texture categories and 10 object categories. The dataset is composed of 5354 images with a resolution between 700x700 and 1024x1024 pixels, where 4096 normal and 1258 anomalous samples are represented. The anomalies are 73 distinct categories of defects. We focused exclusively on the carpet from the MVTec AD dataset as shown in Fig. 3.

AITEX Fabric Dataset. The AITEX Fabric dataset [32] comprises 245 images of 7 patterned fabrics, including 140 defect-free and 105 defective samples, with a resolution of 4096 × 256 pixels. The defective samples feature 12 different types of defects, such as broken ends, knots, weft cracks, etc. Access to the AITEX dataset is available online at www.aitex.es/afid. To overcome the challenge of small dataset size, the images are divided into 256 × 256 sizes. In conclusion, the size of the dataset increases from 245 to 3064 images (256 × 256). Within the AITEX dataset, there are 12 different categories of fabric defects identified by a functioning system in a factory during six months. These

Table 2. The public fabric datasets

Dataset	Category	# Train	# Test		# Defect types	Total
			# Good	# Defective		
MVTec-AD	Carpet	280	28	89	5	397
AITEX	00 (Plush)	241	39	7	2	287
	01 (Stripe)	357	58	26	5	441
	02 Sparse)	659	106	28	5	793
	03 (Dense)	404	66	77	6	547

marked defects are among the most common, though some may occur more sporadically, with variations observed from one factory to another [32]. Figure 4 illustrates examples of all 12 defects for at least one of the seven diverse fabric structures. We focused exclusively on four of the seven categories in the AITEX. Subsequently, the enlarged dataset is further split into training and test sets, as detailed in Table 2.

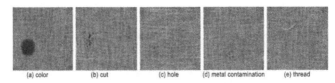

Fig. 3. ROI of 1024x1024 pixels of defective fabrics, with the names used in the database. (a) color, (b) cut, (c) hole, (d) metal contamination, and (e) thread [31].

4.4 Comparative Experiments

This section presents comparative experiments involving the proposed model and several state-of-the-art models, namely SPADE [20], STPM [25], RIAD [15], DFR [16], and HFFMM [23], which were talked about in the related work section. Based on the experiments with the selected models, Table 3 shows the results. The Image-level (Pixel-level) AUC (%) values, denoted as I(P)-AUC (%), represent the performance across the carpet category in the MVTecAD dataset and four categories of the AITEX dataset a. The superior performance of each dataset is indicated by boldfacing.

In Table 3, our proposed approach for image-level defect detection is performed the best compared to other methods regarding fabric categories over MVTecAD and AITEX datasets. The proposed method achieved an average performance of 98.4%, improving the HFFMM by 7.1%. In particular, our proposed method has resulted in increased performance compared to other methods on the AITEX data collection across 4 types of fabrics.

Table 3 also demonstrates that the average pixel-wise defect segmentation of our approach does significantly better than other methods. On the AITEX dataset, the proposed

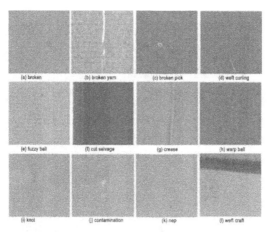

Fig. 4. ROI of 256x256 pixels of defective fabrics, with the names used in the database. (a) broken end, (b) broken yarn, (c) broken pick, (d) weft curling, (e) fuzzy ball, (f) cut selvage, (g) crease, (h) warp ball, (i) knot, (j) contamination, (k) nep, and (l) weft craft [32].

Table 3. Performance of the comparison models and our method in Image-level (Pixel-level)-AUC (%) in MVTecAD and AITEX datasets. Top performance is highlighted in bold.

Dataset	Category	SPADE	STPM	RIAD	DFR	HFFMM	Our
MVTecAD	Carpet	82.1(99.2)	99.0(99.0)	61.1(91.2)	97.4(98.4)	99.3(98.1)	**99.5(98.6)**
AITEX	00 (Plush)	82.6(99.2)	65.7(97.2)	73.5(91.5)	84.6(86.2)	94.1(93.7)	**100(99.5)**
	01 (Stripe)	59.3(99.5)	64.5(97.6)	57.9(96.0)	62.7(95.0)	82.4(**99.4**)	**98.7**(99.3)
	02 (Sparse)	70.2(86.7)	73.2(88.6)	78.6(72.6)	72.2(78.2)	96.2(95.5)	**97.1(98.5)**
	03 (Dense)	77.1(96.3)	63.1(88.7)	78.1(90.7)	83.3(89.7)	84.6(**94.9**)	**96.5**(94.5)
Average		74.3(96.2)	73.1(94.2)	69.9(88.4)	80.0(89.5)	91.3(96.3)	**98.4(98.1)**

method performed excellently for both plush and sparse fabric categories. The categories to experienced slightly reduced performance when compared to other products reviewed were stripes and dense fabrics. The reason could be that pixel-level segmentation requires an accurate prediction at a per-pixel level. The effectiveness of our method is further illustrated in Fig. 5, which shows a visual representation of the segmentation results from experiments performed on the test set of defect images. Our approach excels in identifying defective regions and effectively filters image noise when applied to regular fabric texture images like those in the AITEX dataset. Nevertheless, it is important to note that in some instances, the segmentation results may incorrectly classify negative samples on the boundary as positive or exhibit suboptimal performance in segmenting large target defects.

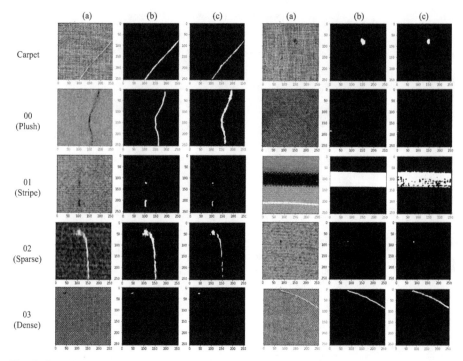

Fig. 5. Provides a visual representation of the segmentation results in the experiments, displaying (a) defective fabric images, (b) corresponding ground-truth images, and (c) the predicted mask generated by our method.

5 Conclusion

Our approach offers a solution to the overgeneralization challenges of visual anomaly detection by proposing a Gamma-Weighted Discrepancy loss function (L_{GWD}) and optimizing synthetic anomaly generation in the context of unsupervised deep learning-based defect detection. The proposed contribution aims to improve the current state of the art in anomaly detection, making it robust, interpretable, and applicable to a wide range of practical situations. However, it should be noted that in some cases the segmentation results may incorrectly classify negative samples as positive at the boundary or exhibit suboptimal performance in terms of segmentation error for the large target defect sizes. In the future, plans include enhancing the proposed approach by concentrating on the integration of more complex features and architecture. Additionally, exploring the capability of the approach to detect more complex types of anomalies will be a priority. Finally, testing the proposed approach in more challenging real-world scenarios is part of the agenda.

References

1. Zipfel, J., et al.: Anomaly detection for industrial quality assurance: a comparative evaluation of unsupervised deep learning models. Comput. Ind. Eng. **177**, 109045 (2023)
2. Liu, J., et al.: Deep industrial image anomaly detection: a survey. Machine Intelligence Res. **21**(1), 104–135 (2024)
3. Kahraman, Y., Durmuşoğlu, A.: Deep learning-based fabric defect detection: a review. Text. Res. J. **93**(5–6), 1485–1503 (2023)
4. Xia, X., et al.: GAN-based anomaly detection: a review. Neurocomputing **493**, 497–535 (2022)
5. Tao, X., et al.: Deep learning for unsupervised anomaly localization in industrial images: a survey. IEEE Transactions on Instrumentation and Measurement (2022)
6. Chen, C., et al.: Review of industry workpiece classification and defect detection using deep learning. Int. J.of Advanced Computer Science and Appl. **13**(4) (2022)
7. Cui, Y., Liu, Z., Lian, S.: A survey on unsupervised anomaly detection algorithms for industrial images. IEEE Access (2023)
8. Yang, J., et al.: Visual anomaly detection for images: a systematic survey. Procedia Computer Sci. **199**, 471–478 (2022)
9. Hojjati, H., Ho, T.K.K., Armanfard, N.: Self-supervised anomaly detection in computer vision and beyond: a survey and outlook. Neural Networks 106106 (2024)
10. Zhou, F., et al.: A comprehensive survey for deep-learning-based abnormality detection in smart grids with multimodal image data. Appl. Sci. **12**(11), 5336 (2022)
11. Yang, M., Wu, P., Feng, H.: MemSeg: a semi-supervised method for image surface defect detection using differences and commonalities. Eng. Appl. Artif. Intell. **119**, 105835 (2023)
12. Yang, J., Shi, Y., Qi, Z.: Learning deep feature correspondence for unsupervised anomaly detection and segmentation. Pattern Recogn. **132**, 108874 (2022)
13. Li, C.-L., et al.: Cutpaste: self-supervised learning for anomaly detection and localization. In: Proceedings of the IEEE/CVF Conference on Computer Vision and Pattern Recognition (2021)
14. Liu, J., et al.: Multistage GAN for fabric defect detection. IEEE Trans. Image Process. **29**, 3388–3400 (2019)
15. Zavrtanik, V., Kristan, M., Skočaj, D.: Reconstruction by inpainting for visual anomaly detection. Pattern Recogn. **112**, 107706 (2021)
16. Shi, Y., Yang, J., Qi, Z.: Unsupervised anomaly segmentation via deep feature reconstruction. Neurocomputing **424**, 9–22 (2021)
17. Deng, L.: The mnist database of handwritten digit images for machine learning research [best of the web]. IEEE Signal Process. Mag. **29**(6), 141–142 (2012)
18. Krizhevsky, A. and G. Hinton, Learning multiple layers of features from tiny images. 2009
19. Krizhevsky, A., Sutskever, I., Hinton, G.E.: Imagenet classification with deep convolutional neural networks. Advances in Neural Information Processing Syst. **25** (2012)
20. Cohen, N., Hoshen, Y.: Sub-image Anomaly Detection with Deep Pyramid Correspondences. arXiv (2020). arXiv preprint arXiv:2005.02357
21. Roth, K., et al.: Towards total recall in industrial anomaly detection. In: Proceedings of the IEEE/CVF Conference on Computer Vision and Pattern Recognition (2022)
22. Defard, T., et al.: Padim: a patch distribution modeling framework for anomaly detection and localization. In: International Conference on Pattern Recognition. Springer (2021)
23. Wan, D., et al.: Unsupervised fabric defect detection with high-frequency feature mapping. Multimedia Tools and Appl. **83**(7), 21615–21632 (2024)
24. Bergmann, P., et al.: Uninformed students: student-teacher anomaly detection with discriminative latent embeddings. In: 2020 IEEE/CVF Conference on Computer Vision and Pattern Recognition (CVPR), pp. 4182–4191 (2020)

25. Wang, G., et al.: Student-Teacher Feature Pyramid Matching for Anomaly Detection. arXiv preprint arXiv:2103.04257 (2021)
26. Cao, Y., et al.: Collaborative discrepancy optimization for reliable image anomaly localization. IEEE Transactions on Industrial Informatics (2023)
27. Cheng, L., et al.: Fabric defect detection based on separate convolutional UNet. Multimedia Tools and Appl. **82**(2), 3101–3122 (2023)
28. Zavrtanik, V., Kristan, M., Skočaj, D.: DRAEM-A discriminatively trained reconstruction embedding for surface anomaly detection. In: Proceedings of the IEEE/CVF International Conference on Computer Vision (2021)
29. Schlüter, H.M., et al.: Natural synthetic anomalies for self-supervised anomaly detection and localization. In: Computer Vision–ECCV 2022: 17th European Conference, Tel Aviv, Israel, October 23–27, 2022, Proceedings, Part XXXI. Springer (2022)
30. Wan, D., et al.: Unsupervised fabric defect detection with high-frequency feature mapping. Multimedia Tools and Applications, 1–18 (2023)
31. Bergmann, P., et al.: MVTec AD--a comprehensive real-world dataset for unsupervised anomaly detection. In: Proceedings of the IEEE/CVF Conference on Computer Vision and Pattern Recognition (2019)
32. Silvestre-Blanes, J., et al.: A public fabric database for defect detection methods and results. Autex Research J. **19**(4), 363–374 (2019)

Conservation of Wildlife in Hunting Tourism Using Artificial Intelligence and Image Processing in Smart Firearms

Ufuk Asil and Efendi Nasibov(✉)

Dokuz Eylul University, Izmir, Turkey
`ufukasil@hotmail.com, efendi.nasibov@deu.edu.tr`

Abstract. This paper presents an innovative technological approach to ensure the sustainability of state-controlled nature park hunting tourism and to protect wildlife. The developed smart fire control systems utilize artificial intelligence (AI) and advanced image processing techniques to prevent firing by activating the safety pin in various situations. Through an integrated camera module and a minicomputer from the Jetson family, animals can be detected in real-time, and various data analysis techniques, along with object recognition algorithms, allow for ethical intervention in many situations that are invisible to the human eye and could be exploited by hunters. Particularly in various national parks, the frequent reports of accidental or uninformed hunting of rare and endangered species, especially during their breeding seasons, underline the importance of this technology. Furthermore, hunters' preference for the most dominant animal in a herd as a trophy poses a significant threat in evolutionary terms. This technology contributes significantly to the conservation of wildlife while promoting ethical and sustainable practices in hunting tourism. The study addresses the development, testing, and implementation of this technology and adds a technological dimension to wildlife conservation strategies.

Keywords: Artificial Intelligence · Image Processing · Smart Firearms · Wildlife Conservation · Hunting Tourism

1 Introduction

Hunting tourism, known as "Trophy Hunting Tourism", is conducted in national parks worldwide, home to many region-specific rare animal species, as a significant economic and environmental activity. This activity is crucial for the conservation of natural habitats, wildlife management, and contributing to the economic well-being of local communities. However, the sustainability of hunting tourism relies on efforts to prevent overhunting and hunting during breeding seasons of targeted species. For instance, Tanzania's Serengeti National Park, which boasts rich biodiversity, hosts large mammals such as lions, elephants, leopards, and various antelope species.

Hunting dates and breeding periods, determined by the Tanzania Wildlife Research Institute and the Tanzania National Parks Authority (TANAPA), are published on the park's official website. Hunting of large mammals like lions, elephants, and leopards, as well as endangered species, is prohibited. This balance aims to ensure the preservation of these species while enabling controlled and sustainable hunting activities that contribute to the region's ecological and economic health [1].

Managed by the South African National Parks (SANParks), Kruger National Park ranks among the world's most popular destinations for sustainable hunting and wildlife observation. The park commits to balanced ecosystem management through strict conservation measures and hunting quotas. These efforts are crucial in maintaining the park's ecological integrity while allowing for regulated hunting and wildlife tourism, which contribute significantly to the region's environmental and economic sustainability. Kruger National Park's approach serves as a model for conservation and sustainable use of natural resources, highlighting the importance of responsible wildlife management in preserving biodiversity and supporting local communities [2].

In developing African countries, particularly those with low incomes, there is a clear policy direction towards protecting natural habitats outside of nature parks. This involves efforts to attract interest to the country and its national parks through sustainable hunting tourism [3].

To increase the long-term success of hunting tourism, there are numerous studies in the literature under the concept of "empowerment". These studies cover topics such as raising social awareness and enabling the local community to economically benefit from the revenue. Additionally, they address the importance of preventing any harm to the natural habitat to enhance sustainability [4].

In the context of trophy hunting tourism empowerment policies, there is a strong opposition to the claims of sustainability. There are numerous studies suggesting that trophy hunting can jeopardize the genetic integrity of populations [5, 6], and impede reproductive success [7, 8]. The largest and most dominant animals, often with the most impressive horns, pelts, or other physical features, are seen as symbols of prestige and success by hunters. This behaviour is counter to the evolutionary process of natural selection, where the fittest survive, and poses a clear and understandable threat to sustainability and long-term evolutionary processes.

Turkey stands among Europe's countries with the richest biodiversity, notably in its variety of wildlife. The wild goat (Capra egagrus), the ancestor of the domestic goat (Capra hircus), is capable of surviving and reproducing under challenging conditions. It can be found across various regions in Turkey, including the Aegean, Mediterranean, Southeastern Anatolia, Eastern Anatolia, and the Black Sea regions, inhabiting steep mountains that reach elevations of 4000–4500 m above sea level. The adult male wild goat typically measures 120–140 cm in length, has a shoulder height of 80–100 cm, and weighs between 50–85 kg. An adult female measures 60–80 cm in length, with a shoulder height of 80–90 cm, and weighs between 35–60 kg [9]. The distribution of the Capra aegagrus species has been identified across Afghanistan, Armenia, Azerbaijan, Georgia, Iran, Lebanon, Russia, Turkey, Cyprus, Greece, India, Iraq, Italy, Oman, Pakistan, Slovakia, Syria, and Turkmenistan. The wild goat is among the species considered endangered globally and within Turkey, classified under the "Vulnerable" (VU) category in the IUCN (International Union for Conservation of Nature) Red List of Threatened Species. It is protected under the Turkish Hunting Law.

Sexual dimorphism in adult wild goats (Capra aegagrus) is relatively easy to discern. As seen in Fig. 1, male wild goats possess significantly large and curved horns, with the rings on the horns providing morphological information about the animal's age.

Fig. 1. Age and Gender Discrimination of Capra aegagrus [10]

Anatolian wild goats are a popular destination in hunting tourism, hosting goats with horn sizes and masses that surpass many other regions. Hunters tend to measure and compare the horn length and weight of wild goats to showcase their success. This practice contradicts the natural ecosystem's principle, where the survival of the fittest leads to the elimination of the weaker individuals.

In this study, informed by the literature from the past and within the context of technological advancement, we propose that the use of artificial intelligence and image processing techniques can enable controlled selection on prey through morphological characteristics. We believe this selection can contribute to sustainable and ethical trophy hunting by ensuring that only specific individuals are targeted, aligning hunting practices with conservation goals.

2 Related Work

The literature review reveals that various details, such as age, sex, and race, can be accurately classified based on morphological characteristics. Such studies, primarily conducted on human populations, have achieved high accuracy rates [11, 12]. The results are promising, indicating that characteristics like age, race, and sex of different animal species can be correctly diagnosed through morphological analyses. For instance, a study by Surya T. et al. demonstrated that elephant tusk morphology could determine sex with a 98% success rate [13]. Similarly, Chen Y. and team developed CNN (Convolutional Neural Networks) based algorithms to identify the rare and protected Critically Endangered Myanmar snub-nosed monkey from similar species with a 96% success rate [14]. Pradana, Z. H. et al. used image processing techniques to analyse cattle size and weight [15]. In a further example, Bae, H. B. et al. successfully classified dogs based on their nose prints with a 98% accuracy rate [16].

These studies illustrate that advancements in image processing technologies and deep learning methods have enabled classifications beyond simple cat-dog distinctions, including intra-species classifications, gender identifications, size-weight comparisons, and even biometric analyses in animals.

Pose estimation is the process of determining the spatial locations and orientations of people or objects in an image. This technology is utilized to understand the posture and position of an object by identifying the body parts or specific points of individuals or objects in the image. Particularly, the use of pose estimation to determine the gender, intra-species size variations, and dominance ratios in various body parts of animals represents a significant application area in biological research and wildlife monitoring systems. In this context, key points placed on various organs and sections of animals enable detailed mapping of these creatures' anatomical structures. This method offers the possibility to automatically classify and analyse the gender, size within the species, and dominance ratios of individual body parts of animals through the processing of visual information with deep learning models or other algorithms. This technique, by providing valuable insights in areas such as ecological studies, species conservation efforts, and wildlife management, contributes to a better understanding and protection of animal populations [17, 18].

The literature clearly shows that integrating AI and modern technologies capable of deep learning algorithms into shooting control systems in hunting tourism and sports hunting is a potential solution for conserving wildlife habitats and addressing numerous ethical problems. This integration can help protect wildlife and play an effective role in solving ethical issues. These developments in hunting and sports hunting technologies are crucial steps towards preserving natural habitats and creating a sustainable ecosystem.

3 Used Devices and Features

We have developed a technological transformation kit, presented in Fig. 2. This kit offers a practical solution that does not add extra weight or occupy additional space for the hunters. The system is based on replacing the stock of a long-range rifle with a new one filled with electronic components. The typically empty spaces within the rifle's grip and stock provide areas for batteries and electronic components in this technological transformation. Users often employ electronic systems and cameras in gun sights for night vision or alternative views. This indicates that cameras and monitors are already commonly used electronic components, adding to the practicality and applicability of this technology.

The system utilizes the Xavier NX device from the Jetson family, which provides parallel image processing capabilities. Additionally, it includes a Raspberry Pi v2 camera and a 25-kilo servo motor to operate the safety pin. The system can operate for 6 h without recharging.

RF (Radio Frequency) based devices allow for remote control of the firearms, enabling them to be shut down or activated under primary administrative control. This feature is of great importance for safety and supervision. Furthermore, the integrated GPS module in the system enables the tracking of sportsmen's locations. This integration not only enhances participants' safety but also allows for more effective management of events. These advancements contribute significantly to making hunting sports and tourism technologically safer and more controllable. Studies targeting the enhancement of ecosystem safety through such transformation kits are available[19].

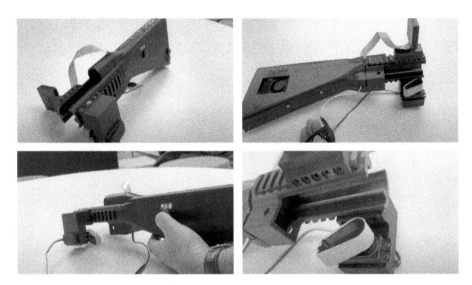

Fig. 2. Real Application of Smart Rifle

4 Software Algorithms and Principles

In this study, the Python programming language was employed, with the OpenCV library selected for image processing. Various object detection models were tested, and the YOLO V7 model achieved optimum success. The system processes 20–25 frames per second on tiny model. A key feature of the system is the prioritized identification of endangered species using a pre-trained model, ensuring the system starts in a passive safety mode.

The system includes features for detecting animals that are in their breeding season, are prohibited from hunting, or are identified as dominant individuals critical for maintaining a healthy genetic pool through various literature-based analysis methods. These detections are conducted in a way that facilitates beneficial and controlled selection, as determined by the relevant authority. This approach supports ethical hunting practices and wildlife conservation while integrating technological advancements into hunting sports. The flowchart of this system is illustrated in Fig. 3.

Our deep learning model has been trained on photographs of wild goats obtained from real footage published on free websites like YouTube, pertaining to hunting activities conducted in the rugged mountainous regions of Anatolia, such as Artvin, Kayseri, and Isparta. This system can recognize and distinguishing such animals as you can see Fig. 4. If the firearm is aimed at a species identified as endangered and in need of protection

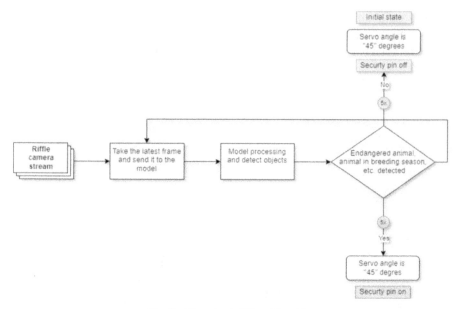

Fig. 3. Flowchart of Our Algorithm

by the authorities, a mechanism automatically disables the weapon. Additionally, the system can identify animals in their breeding season through date and species comparison. When such an animal is detected, the system automatically switches to safe mode, preventing the animal from being shot. This feature contributes to efforts to create a sustainable ecosystem and aids in the conservation of wildlife. These developments represent significant examples of how technology can be integrated into nature conservation and sustainable hunting practices.

Fig. 4. Different Situations' Screenshot from the Visor

The Yolo V7 model, trained with 300 distinct images, as illustrated in Fig. 5, can accurately predict the gender of mature individuals at a rate of 95% through the camera mounted on a firearm. This capability enables the firearm to lock itself when aimed at prey, acting as a deterrent to firing when the gender ratios within the species are disrupted,

Fig. 5. Gender Detection of Capra Aegagrus

thus maintaining balance. Additionally, statistical information related to gender, obtained through the firearm's scope during hunting, can further enlighten authorities.

5 Conclusion

This research investigates how artificial intelligence (AI) can be used in hunting tourism and how it might help protect animals and the environment. AI is getting better at processing information quickly and is changing a lot of areas. Because of this, it's important to investigate how AI can be used in ways that can really make a difference for the environment. This paper talks about encouraging more studies on using AI to help with wildlife conservation and making sure hunting tourism doesn't harm the ecosystem.

The advancements discussed enable comprehensive monitoring of hunting tourism systems, from the highest to the lowest levels. The ability to record all hunting actions on camera helps mitigate ethical issues and perform detailed data analyses for ecosystem conservation. Additionally, the system can prevent economic and human resource-related accidents. Our study demonstrates that the system operates effectively even on low-powered computers and can maintain stability with additional security measures, such as turning on or off after consecutively detecting the same object five times.

The computing capabilities of the Jetson family's minicomputers are noteworthy for running certain algorithms with sufficient accuracy levels at present. This suggests that, as these technologies evolve over time, analyses such as intra-species dominance could be performed more successfully, even when the prey is moving or partially obscured, by using algorithms like keypoints and pose estimation. Algorithms like object detection, which provide only the height and width of a detected object without considering its position, are not capable of analysing the dominance features of prey. However, pose estimation and keypoints can be successful in detecting dominance even if the prey is partially obscured or viewed from different angles. Our efforts to obtain ratios such as horn length/body size for intra-species dominance detection did not achieve visible success using object detection alone. In our study, we were able to make distinctions such as the type and sex of the prey, and the breeding season, without causing any additional hindrance to hunters during hunting, using only object detection. We also believe that GPS data and prey images can be used in many data analyses, such as inventory counts by authorities.

These features highlight the significance and efficacy of AI technologies in hunting tourism. Real images and experiences related to the transformation kit can be obtained from this news video [20].

References

1. Home |TANZANIA NATIONAL PARKS. Accessed: Jan. 28, 2024. (2024). https://www.tanzaniaparks.go.tz/
2. Conservation – SANParks. Accessed: Jan. 28, 2024. (2024). https://www.sanparks.org/conservation
3. Thomsen, J.M., Lendelvo, S., Coe, K., Rispel, M.: Community perspectives of empowerment from trophy hunting tourism in Namibia's Bwabwata National Park. J. Sustain. Tour. **30**(1), 223–239 (2022). https://doi.org/10.1080/09669582.2021.1874394
4. Aghazamani, Y., Hunt, C.A.: Empowerment in Tourism: A Review of Peer-reviewed Literature. Tour. Rev. Int. **21**(4), 333–346 (2017). https://doi.org/10.3727/154427217X15094520591321
5. Crosmary, W.G., et al.: Trophy hunting in Africa: long-term trends in antelope horn size. Anim. Conserv. **16**(6), 648–660 (2013). https://doi.org/10.1111/ACV.12043
6. Coltman, D.W., et al.: Undesirable evolutionary consequences of trophy hunting. Nature **426**(6967), 655–658 (2003). https://doi.org/10.1038/nature02177
7. Selier, S.A.J., Slotow, R., Di Minin, E.: The influence of socioeconomic factors on the densities of high-value cross-border species, the African elephant. PeerJ **2016**(10), e2581 (2016). https://doi.org/10.7717/PEERJ.2581/SUPP-1
8. Packer, C., et al.: Sport Hunting, Predator Control and Conservation of Large Carnivores. PLoS ONE **4**(6), e5941 (2009). https://doi.org/10.1371/JOURNAL.PONE.0005941
9. Paşalı, H., et al.: Türkiye'de Yaban Keçisi Capra aegagrus aegragrus, Animal Health Production and Hygiene, **3**(1), 245–247 (2014), Accessed: Feb. 01, 2024. (2024). https://dergipark.org.tr/tr/pub/aduveterinary/issue/58863/849026
10. Ekinci, H., Süel, H.: Population size and structure of Wild Goat (Capra aegagrus, Erxleben); Example of Lakes Region.
11. Ganorkar, S.S., et al.: Implementation on Real Time Age Estimation with Gender Detection with Image Processing, In: IJSDR1904077 International Journal of Scientific Development and Research, 2019, Accessed: Jan. 28, 2024. (2024). www.ijsdr.org
12. Ahmed, M.A., Choudhury, R.D., Kashyap, K.: Race estimation with deep networks. J. King Saud Univ. Comput. Inform. Sci. **34**(7), 4579–4591 (2022). https://doi.org/10.1016/J.JKSUCI.2020.11.029
13. Surya, T., Chitra Selvi, S., Selvaperumal, S.: The IoT-based real-time image processing for animal recognition and classification using deep convolutional neural network (DCNN). Microprocess. Microsyst. **95**, 104693 (2022). https://doi.org/10.1016/J.MICPRO.2022.104693

14. Chen, Y., Goorden, M.C., Beekman, F.J., Zeng, P.: Similar Animal Classification Based on CNN Algorithm. Peiyi Zeng. J. Phys, pp. 12001. (2021). https://doi.org/10.1088/1742-6596/2132/1/012001
15. Pradana, Z.H., Hidayat, B., Darana, S.: Beef cattle weight determine by using digital image processing. In: ICCEREC 2016 - International Conference on Control, Electronics, Renewable Energy, and Communications 2016, Conference Proceedings, pp. 179–184. (2017). https://doi.org/10.1109/ICCEREC.2016.7814955
16. Bae, H.B., Pak, D., Lee, S.: Dog nose-print identification using deep neural networks. IEEE Access **9**, 49141–49153 (2021). https://doi.org/10.1109/ACCESS.2021.3068517
17. Cao, J., et al.: Cross-Domain Adaptation for Animal Pose Estimation. Accessed: Feb. 02, 2024. (2024). www.jinkuncao.com/animalpose
18. Yu, H., et al.: AP-10K: A Benchmark for Animal Pose Estimation in the Wild. Accessed: Feb. 02, 2024. (2024). https://github.com/AlexTheBad/AP10K
19. Asil, U., Nasibov, E.: "Using Image Processing Techniques to Increase Safety in Shooting Ranges. Int. J. Comput. Sci. Eng. Appl. (IJCSEA) **11**(1), 1–10 (2021). https://doi.org/10.5121/ijcsea.2021.11103
20. Türk araştırmacı tabanca ve tüfeklere yapay zeka ekledi - YouTube. Accessed: Jan. 30, 2024. (2024.) https://www.youtube.com/watch?v=3oMmCLkr-Xg&t=26s1

Optimizing Microarray Gene Selection in Colon Cancer: An Enhanced Metaheuristic Algorithm for Feature Selection

Salsabila Benghazouani[1]($\boxtimes$), Said Nouh[1], and Abdelali Zakrani[2]

[1] Department of Mathematics and Computer Science, Faculty of Sciences Ben M'Sik, Hassan II University, Casablanca, Morocco
benghazouani.salsabila239@gmail.com
[2] Department of Computer Science Engineering, ENSAM, II University, Casablanca, Hassan, Morocco

Abstract. Microarray data has evolved into an indispensable tool for scrutinizing gene expression, prompting researchers to strategically utilize a minimal set of pertinent gene expression profiles to refine the accuracy of tumor identification. The imperative task of identifying key differential genes in colon cancers, crucial for distinguishing patients from the normal population, has led to the development of various techniques and algorithms. Swarm and Evolutionary Algorithms (SEA), renowned for their efficiency as global search agents, have emerged as particularly effective in optimizing the selection of a pertinent subset of genes related to colon cancer. This paper introduces an innovative approach that integrates swarm and evolutionary algorithms to tackle the challenge of feature selection in datasets related to colon cancer. A thorough comparative analysis is conducted, highlighting the differences between the proposed method and an alternative feature selection approach based on swarm and evolutionary algorithms. The extensive experimental results convincingly demonstrate the favorability and effectiveness of the proposed model, showcasing superior accuracy and a notable reduction in the number of selected features.

Keywords: Feature selection · Swarm and Evolutionary Algorithms · Microarray data · Colon cancer · Classification · Machine learning

1 Introduction

Colon cancer is a prevalent malignancy affecting the digestive tract, predominantly occurring in the colon. It constitutes around 10% of new cancer cases and cancer-related fatalities globally each year, with approximately 2.2 million new instances and 1.1 million deaths attributed to colon cancer [1], Additionally, it stands as the second leading cause of cancer-related mortality in the United States [2]. In [3], the authors pioneered a universal cancer classification method utilizing DNA microarray gene expression monitoring. They suggested that microarrays could serve as a classification tool for cancer. The utilization of microarray-based gene expression has been extensively applied in diagnosing and analyzing colon cancer. The timely identification of colon cancer is crucial for accurate diagnosis and treatment, with the challenge lying in pinpointing relevant genes from microarray datasets characterized by a small number of samples and inadequate follow-up information for many genes. Currently, feature selection serves as the primary method for extracting significant genes in cancer classification based on gene expression data [4]. Recognizing that swarm intelligence and evolutionary algorithms represent computational paradigms, this understanding is based on their engagement in an intelligent search process to identify an optimal solution [5]. This paper aims to introduce an innovative feature selection method by employing both swarm and evolutionary algorithms. The specific objective is to identify an optimal subset of genes associated with colon cancer. The subsequent sections of the paper are structured as follows: Sect. 2 offers an overview of previous work and includes a literature survey. Section 3 details the methodology employed in the proposed system, while Sect. 4 focuses on the analysis of experimental results and subsequent discussion. The study concludes with a summary of the findings in Sect. 5.

2 Backgrounds

Feature selection and classification algorithms have demonstrated significant and efficient applications in machine learning, contributing to scientific research in the medical field [6]. Numerous methods for feature selection employ meta-heuristics to manage computational complexity in high-dimensional datasets. Swarm intelligence (SI) and evolutionary algorithms represent computational approaches that engage in an intelligent search process for optimal solutions, utilizing non-deterministic or randomized techniques to explore and exploit extensive search spaces [7]. These algorithms draw inspiration from the collective social behaviors observed in decentralized and self-organized animals, such as birds, fish, ants, wolves, etc. Particle Swarm Optimization (PSO), created by Kennedy and Eberhart in 1995, is a potent optimization method rooted in swarm intelligence (SI). Drawing inspiration from the collective behavior of birds and fish, it has been applied in recent research to optimize the selection of the final feature subset [8]. In their study [9], the authors propose a model for colon cancer prediction employing a feature selection algorithm based on particle swarm optimization and a Convolutional Neural Network (CNN). By utilizing the gradient boosting classifier, they achieved an impressive accuracy of 99.73%. Differential Evolution (DE), a swarm intelligence-based algorithm, has been designed to tackle complex optimization problems by addressing the deficiency of local search in genetic algorithms, and it finds widespread application in various machine learning tasks, including feature selection [10]. In [11], The researchers presented a pioneering method that combines Differential Evolution optimization and Condorcet's Jury Theorem to optimize and identify optimal solutions, applying it to both pulmonary and colorectal cancer models, achieving an impressive accuracy of 99.88%. Firefly Algorithm (FA), conceptualized by Yang in 2010 based on the optical communication among fireflies, serves as an exemplary illustration of swarm intelligence, demonstrating how agents with lower performance can collaborate to achieve highly effective results, particularly in optimization problems [12]. In [13], the authors suggest an enhanced multilayer binary firefly algorithm designed to optimize the process of feature selection and classification for microarray data. The Cuckoo Optimization Algorithm (COA) is a nature-inspired optimization algorithm based on cuckoo birds' breeding behavior, utilizing Levy flights for global exploration and local search for exploitation to find optimal solutions iteratively. COA replaces poorer solutions with offspring generated through Levy flight, aiming to strike a balance between exploration and exploitation in the search space [14]. In [15], the researchers propose a two-stage gene selection method employing a hybrid strategy with the Cuckoo Optimization Algorithm and Harmony Search for enhanced cancer classification, aiming to improve the efficiency of gene selection in cancer-related research. The Salp Swarm Algorithm (SSA), introduced by Seyedali Mirjalili in 2017, mimics the swarming behavior of marine salps to optimize solutions [16]. It balances exploration and exploitation in solving optimization problems through iterative movement and leader-guided adjustments. In [17], the authors introduce a novel approach for multi-threshold image segmentation using an enhanced Salp Swarm Algorithm. The method is applied to segment breast cancer pathology images, demonstrating its efficacy in optimizing the segmentation process. The Jaya algorithm (JA), conceived by Venkata Rao in 2016, guides problem-solving by directing solutions towards the optimal while steering away from the least favorable alternatives [18]. In the referenced study [19], the researchers introduce a multi-objective feature selection approach for microarray data employing the Quasi-Oppositional based Jaya algorithm. This method is designed to improve feature selection in the analysis of microarray data. The Flower Pollination Algorithm (FPA), devised by Xin-She Yang in 2012, mimics the pollination process of flowering plants to iteratively enhance solutions for optimization problems through the transfer of information among flowers [20]. In the cited work [21], the authors propose a novel hybrid gene selection method for tumor identification. This approach combines multifilter integration with a recursive Flower Pollination Search algorithm, aiming to reduce the number of genes while improving classification accuracy.

A genetic algorithm (GA) is a search and optimization technique inspired by the principles of natural selection and genetics. It operates by iteratively evolving a population of potential solutions through processes like selection, crossover, and mutation to find optimal or near-optimal solutions to a given problem [22]. In the cited work [23], the authors present two memetic micro-genetic algorithms with diverse hybridization approaches for feature selection. The proposed methods demonstrate an enhancement in the feature selection process for cancer microarray data.

3 Proposed Survey Method

3.1 Methodology

Our hybrid approach, illustrated in Fig. 1, integrates outcomes from various swarm and evolutionary feature selection (FS) algorithms, including Genetic Algorithm (GA), FS_PSO, FS_DE, FS_COA, FS_FA, FS_SSA, FS_JA, and FS_FPA, as detailed in the preceding section. The primary objective of the proposed feature selection method, based on swarm and evolutionary algorithms (FS_SEA), is to identify subsets of informative features that demonstrate reduced size and/or enhanced classification performance compared to individual algorithms. In the initial phase, we employ feature selection methods based on swarm and evolutionary algorithms on a colon dataset. The resulting feature subsets are then integrated into a feature pool, forming the initial population for the Genetic Algorithm. The GA excels at navigating populations of hypotheses, where each hypothesis comprises intricately interacting components. Evaluation of each individual in the current population is conducted based on a specific fitness function. Subsequently, a new population is generated through genetic operations, including selection, crossover, and mutation. Our genetic algorithm is expressly designed to maximize classification accuracy while minimizing the size of feature subsets.

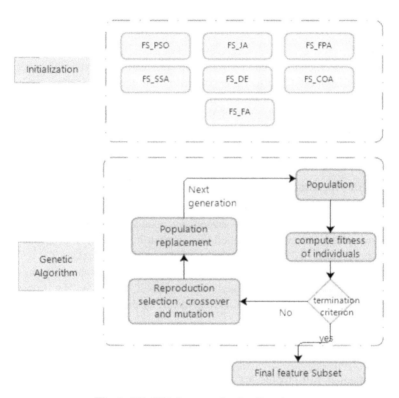

Fig. 1. FS_SEA feature selection flowchart.

3.1.1 Hypothesis Representation

In a genetic algorithm, the population consists of candidates referred to as chromosomes, typically encoded as binary sequences where each digit represents a gene. Each individual within the population signifies a feature subset and is represented by an n-bit binary vector. A binary value of 1 in the vector denotes the selection of the corresponding feature, while a value of 0 indicates the opposite.

3.1.2 Initialization of the Population

In the context of population initialization, the feature pool serves as an assemblage of potential candidate features, aimed at being chosen by the genetic algorithm to identify an optimal or nearly optimal subset of features. Instead of utilizing all features directly from the original dataset, we curate sets of features through the application of various feature selection techniques based on swarm and evolutionary algorithms. Consequently, the feature pool comprises valuable insights derived from diverse selection criteria.

3.1.3 Evaluation of Fitness

During the fitness evaluation process, the genetic algorithm is carefully crafted to achieve optimization with respect to two interconnected objectives. Firstly, the algorithm aims to enhance the classification accuracy of the selected feature subset, striving for a more robust and efficient model in discerning patterns within the data. Secondly, the genetic algorithm works to minimize the number of selected features, acknowledging the importance of simplicity in the resulting model. By striking a balance between these dual objectives, the genetic algorithm aims to generate a feature subset that improves classification performance and represents the relevant genes for diagnosing colon cancer.

3.1.4 Genetic Operators

In every iteration, a series of operations is implemented on candidates to improve the quality of individuals composing the subsequent population (Generation). The subsequent operators are successively applied during each iteration. The proposed method employs a population size of 30 and runs for a total of 100 generations.

Selection: Utilizing roulette wheel selection, individuals are chosen from the population for subsequent breeding through a probabilistic approach. The probability of selecting an individual, denoted as h_i, is determined by the formula (1):

$$P(h_i) = \frac{F(h_i)}{\sum_{i=1}^{N} F(h_i)} \qquad (1)$$

In this equation, $F(h_i)$ represents the fitness value of h_i. The probability of selection is directly tied to the individual's fitness, being proportionally higher for individuals with greater fitness and inversely proportional to the fitness of other competing hypotheses in the current population.

Crossover: The crossover operator simulates reproduction, where each pair represents parents giving birth to two children. Each child inherits a portion of its sequence from one of its parents. This sequence is derived by dividing the parental sequences at a specified position, referred to as the crossover point (with Pc=0.8).

Mutation: Following the creation of new individuals, they undergo an additional transformation involving the modification of their genes based on a predefined probability. In this context, the proposed method employs uniform mutation, wherein each gene has an independent probability of being modified (with pm=0.01). This process introduces randomness to the genetic material, contributing to the exploration of diverse genetic configurations within the population.

3.2 Dataset

Gene expression information related to colon cancer was acquired from[24] during the data acquisition stage. The datasets comprise 62 cases (tests) and 2000 genes (attributes) sourced from individuals with colon cancer. This includes 40 tumor biopsies (identified as abnormal) and 22 biopsies classified as normal.

3.3 The Employed Classifier

AdaBoost, short for Adaptive Boosting, is an ensemble learning algorithm designed to enhance the performance of weak classifiers and create a strong classifier. Introduced by Yoav Freund and Robert Schapire in 1996 [25], AdaBoost works iteratively by assigning higher weights to misclassified instances in each round of training. This emphasis on challenging samples allows the algorithm to sequentially improve its accuracy. The final model is a weighted combination of these weak classifiers, providing a robust and accurate prediction on diverse datasets [26]. Logistic regression, a model within generalized linear regression analysis, serves as a supervised learning algorithm. Its typical applications encompass regression, binary classification, and multi-classification tasks. The logistic regression process involves three main stages: defining a predictive function, formulating a loss function, and determining regression parameters that minimize the loss function. In adapting logistic regression for regression or classification challenges, it initially establishes a cost function. Subsequently, an iterative optimization approach is employed to ascertain the optimal model parameters [27].

3.4 Performance Metrics

In each experiment, every feature selection method is subjected to ten iterations, and the mean across these iterations is utilized for comparative analysis. Furthermore, normalization is applied to the colon dataset, and it is randomly divided into a training set (comprising 70% of the dataset) and a testing set (comprising 30% of the dataset).

The performance of the introduced feature selection method based on swarm and evolutionary algorithms (FS_SEA) was assessed based on the mean accuracy of the classifier, the minimum count and the proportion of retained features. Subsequently, this approach was juxtaposed with alternative metaheuristic algorithms using these evaluation criteria. Accuracy in this context represents the measure of correctness in predicting outcomes by the classifier, calculated as the ratio of correctly predicted instances to the total instances. The rates of retained features refer to the proportion of initially considered features that are preserved after the feature selection process, indicating the method's efficiency in maintaining relevant information while reducing dimensionality.

4 Results and Discussion

Within this section, we assess and compare the effectiveness of the proposed method FS_SEA against other metaheuristic feature selection techniques. The results are presented in terms of the rates of maintained features and classification accuracy.

Table 1. The rate of maintained features and the average classification accuracy of various feature selection techniques using the AdaBoost classifier and Logistic Regression

Feature Selection Technique	Logistic Regression		AdaBoost Classifier	
	Accuracy	Rate_maint_feat	Accuracy	Rate_maint_feat
FS_SEA	98.94	27.46	100	37.54
FS_PSO	95.25	38.25	95.26	44.36
FS_SSA	97.89	47.54	95.25	47.90
FS_DE	94.73	38.98	93.15	46.93
FS_FA	94.73	47.06	95.25	47.79
FS_FPA	94.37	48.05	92.62	48.33
FS_JA	96.31	40.76	90.52	45.57
FS_COA	95.25	43.21	99.47	47.24

Table 1 offers a comprehensive comparison of various feature selection techniques, such as FS_FPA, FS_JA, FS_SSA, FS_DE, FS_COA, FS_FA, and FS_PSO, alongside the proposed swarm and evolutionary algorithm-based feature selection method (FS_SEA). This evaluation is conducted in combination with both Logistic Regression and AdaBoost classifiers, focusing on accuracy and the proportion of maintained features. The results presented in Table 1 highlight the exceptional efficacy of FS_SEA in contrast to various other feature selection methods based on swarm and evolutionary algorithms. Notably, FS_SEA demonstrates exceptional accuracy, achieving 98.94% with Logistic Regression and a perfect 100% with AdaBoost. Additionally, it maintains relatively low rates of remaining features at 27.46% and 37.54%, respectively, when utilizing RL and AdaBoost. Following closely, FS_COA excels with an accuracy of 99.47% with AdaBoost, while FS_SSA achieves an accuracy of 97.89% with Logistic Regression. Moreover, FS_JA attains a 96.31% accuracy with Logistic Regression, and FS_PSO demonstrates an accuracy of 95.26 with AdaBoost. The outcomes uniformly reveal a substantial reduction in dimensionality across all methods, as each approach selectively retains only a fraction of the initial set of features.

Figure 2 displays the mean classification accuracy for different feature selection methods, encompassing FS_FPA, FS_JA, FS_SSA, FS_DE, FS_COA, FS_FA, FS_PSO, and the proposed FS_SEA. The evaluation employs both logistic regression and AdaBoost classifiers. The findings highlight the superior performance of the proposed method, FS_SEA, as it achieves remarkable accuracy scores of 98.94% with logistic regression and a perfect 100% with AdaBoost. This consistent outperformance positions FS_SEA as an effective feature selection technique, demonstrating its ability to enhance classification accuracy compared to the other evaluated methods across both regression and AdaBoost classifiers.

Figure 3 presents a comprehensive comparison of the rate of maintained features among various feature selection methods, namely FS_SEA, FS_FPA, FS_JA, FS_SSA, FS_DE, FS_COA, FS_FA, and FS_PSO, employed with both logistic regression and

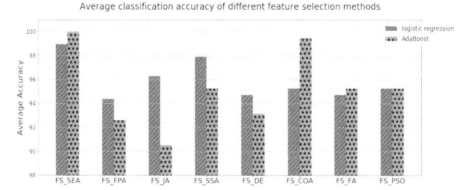

Fig. 2. Comparative mean Classification Accuracy of Feature Selection Methods with Logistic Regression and AdaBoost Classifiers in Colon Cancer Dataset.

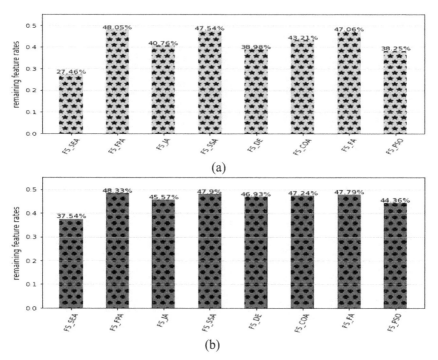

Fig. 3. The mean rate of maintained features using diverse feature selection techniques with LR (a) and AdaBoost (b) in the colon cancer dataset.

AdaBoost classifiers. The overall findings underscore the superior performance of the proposed method FS_SEA, which consistently maintains a reduced rate of selected features. Specifically, in Fig. 3(a), the proposed method achieves the lowest remaining feature rate at 27.46%, surpassing the efficiency of other methods. Moreover, in Fig. 3(b),

FS_SEA continues to exhibit remarkable performance with the lowest remaining feature rate at 37.54%. This consistent outperformance emphasizes the efficacy of the suggested method in minimizing dimensionality compared to alternative feature selection approaches.

Table 2. Table Comparative Analysis of Colon Cancer Studies by Various Authors.

Authors	Year	Methods	Accuracy Score (%)
J. Isuwa et al. [28]	2023	Cuckoo search algorithm (CSA) + Chi-Square filter method	99.47
M. Al-Rajab, J. Lu, and Q. Xu [29]	2017	PSO + SVM	94.00
A. S. M. Shafi et al. [30]	2020	Random forest + Mean Decrease Accuracy + Mean Decrease Gini	95.16
N. Alrefai and O. Ibrahim [31]	2020	particle swarm optimization + ensemble learning method	92.86
Our work		Proposed method FS_SEA Adaboost	100
		FS_SEA + LR	98.94

Table 2 provides a comparative analysis of colon cancer studies, featuring diverse models and their associated accuracy scores. The findings indicate that our proposed method, FS_SEA with Adaboost, attains a significantly higher accuracy score of 100%. In the second position is the Cuckoo Search Algorithm (CSA) combined with the Chi-Square filter method by J. Isuwa et al., achieving an accuracy score of 99.47%.

5 Conclusion

In conclusion, the integration of swarm and evolutionary algorithms, particularly in conjunction with genetic algorithms, provides a robust and innovative strategy for optimizing feature selection in colon cancer datasets. Through a comprehensive comparative analysis, our proposed model consistently demonstrated superior accuracy and a substantial reduction in selected features, showcasing its effectiveness in refining the precision of tumor identification. Future directions in the field involve exploring hybrid methodologies, incorporating deep learning, and addressing various challenges to advance the practical implementation of swarm intelligence-based gene selection methods in cancer research.

References

1. Ning, J., Ma, B., Huang, J., Han, L., Shao, Y., Wang, F.: Integrated Network Pharmacology and Metabolomics Reveal the Action Mechanisms of Vincristine Combined with Celastrol Against Colon Cancer. J. Pharm. Biomed. Anal. pp. 115883, (2023). https://doi.org/10.1016/j.jpba.2023.115883
2. Siegel, R.L., Wagle, N.S., Cercek, A., Smith, R.A., Jemal, A.: Colorectal cancer statistics. CA Cancer J. Clin. **73**(3), 233–254 (2023). https://doi.org/10.3322/caac.21772
3. Golub, T.R., et al.: Molecular Classification of Cancer: Class Discovery and Class Prediction by Gene Expression Monitoring. Science **286**(5439), 531–537 (1999). https://doi.org/10.1126/science.286.5439.531
4. Xi, M., Sun, J., Liu, L., Fan, F., Wu, X.: Cancer feature selection and classification using a binary quantum-behaved particle swarm optimization and support vector machine. Comput Math Methods Med **2016**, 1–5 (2016). https://doi.org/10.1155/2016/3572705
5. Thaher, T., Chantar, H., Too, J., Mafarja, M., Turabieh, H., Houssein, E.H.: Boolean Particle Swarm Optimization with various Evolutionary Population Dynamics approaches for feature selection problems. Expert Syst. Appl. **195**, 116550 (2022). https://doi.org/10.1016/j.eswa.2022.116550
6. Hassan, A.R., Subasi, A.: Automatic identification of epileptic seizures from EEG signals using linear programming boosting. Comput. Methods Programs Biomed. **136**, 65–77 (2016). https://doi.org/10.1016/j.cmpb.2016.08.013
7. Rostami, M., Berahmand, K., Nasiri, E., Forouzandeh, S.: Review of swarm intelligence-based feature selection methods. Eng. Appl. Artif. Intell. **100**, 104210 (2021). https://doi.org/10.1016/j.engappai.2021.104210
8. Kennedy, J., Eberhart, R.: Particle swarm optimization. In: Proceedings of ICNN'95-international conference on neural networks, IEEE, pp. 1942–1948. (1995). https://doi.org/10.1109/ICNN.1995.488968
9. Raihan, M.J., Nahid, A.-A.: Classification of histopathological colon cancer images using particle swarm optimization-based feature selection algorithm. In: Diagnostic Biomedical Signal and Image Processing Applications with Deep Learning Methods, Elsevier, pp. 61–82. (2023). https://doi.org/10.1016/B978-0-323-96129-5.00012-3
10. Pant, M., Zaheer, H., Garcia-Hernandez, L., Abraham, A.: Differential Evolution: A review of more than two decades of research. Eng. Appl. Artif. Intell. **90**, 103479 (2020). https://doi.org/10.1016/j.engappai.2020.103479
11. Srivastava, G., Chauhan, A., Pradhan, N.: Cjt-deo: Condorcet's jury theorem and differential evolution optimization based ensemble of deep neural networks for pulmonary and colorectal cancer classification. Appl. Soft Comput. **132**, 109872 (2023). https://doi.org/10.1016/j.asoc.2022.109872
12. Fister, I., Fister, I., Jr., Yang, X.-S., Brest, J.: A comprehensive review of firefly algorithms. Swarm Evol. Comput. **13**, 34–46 (2013). https://doi.org/10.1016/j.swevo.2013.06.001
13. Xie, W., Wang, L., Yu, K., Shi, T., Li, W.: Improved multi-layer binary firefly algorithm for optimizing feature selection and classification of microarray data. Biomed. Signal Process. Control **79**, 104080 (2023). https://doi.org/10.1016/j.bspc.2022.104080
14. Rajabioun, R.: Cuckoo optimization algorithm. Appl. Soft Comput. **11**(8), 5508–5518 (2011). https://doi.org/10.1016/j.asoc.2011.05.008
15. Elyasigomari, V., Lee, D.A., Screen, H.R., Shaheed, M.H.: Development of a two-stage gene selection method that incorporates a novel hybrid approach using the cuckoo optimization algorithm and harmony search for cancer classification. J. Biomed. Inform. **67**, 11–20 (2017). https://doi.org/10.1016/j.jbi.2017.01.016

16. Mirjalili, S., Gandomi, A.H., Mirjalili, S.Z., Saremi, S., Faris, H., Mirjalili, S.M.: Salp Swarm Algorithm: A bio-inspired optimizer for engineering design problems. Adv. Eng. Softw. **114**, 163–191 (2017). https://doi.org/10.1016/j.advengsoft.2017.07.002
17. Guo, H., et al.: Multi-threshold Image Segmentation based on an improved Salp Swarm Algorithm: Case study of breast cancer pathology images. Comput. Biol. Med. pp. 107769, (2023). https://doi.org/10.1016/j.compbiomed.2023.107769
18. Rao, R.: Jaya: A simple and new optimization algorithm for solving constrained and unconstrained optimization problems. Int. J. Ind. Eng. Comput. **7**(1), 19–34 (2016). https://doi.org/10.5267/j.ijiec.2015.8.004
19. Chaudhuri, A., Sahu, T.P.: Multi-objective feature selection based on quasi-oppositional based Jaya algorithm for microarray data. Knowl. Based Syst. **236**, 107804 (2022). https://doi.org/10.1016/j.knosys.2021.107804
20. Yang, X.-S.: Flower Pollination Algorithm for Global Optimization. In: Unconventional Computation and Natural Computation, vol. 7445, Springer, Berlin Heidelberg, pp. 240–249. (2012). https://doi.org/10.1007/978-3-642-32894-7_27
21. Li, M., Ke, L., Wang, L., Deng, S., Yu, X.: A novel hybrid gene selection for tumor identification by combining multifilter integration and a recursive flower pollination search algorithm. Knowl. Based Syst. **262**, 110250 (2023). https://doi.org/10.1016/j.knosys.2022.110250
22. Katoch, S., Chauhan, S.S., Kumar, V.: A review on genetic algorithm: past, present, and future. Multimed. Tools. Appl. **80**(5), 8091–8126 (2021). https://doi.org/10.1007/s11042-020-10139-6
23. Rojas, M.G., Olivera, A.C., Carballido, J.A., Vidal, P.J.: Memetic micro-genetic algorithms for cancer data classification. Intell. Syst. Appl. **17**, 200173 (2023). https://doi.org/10.1016/j.iswa.2022.200173
24. Alon, U., et al.: Broad patterns of gene expression revealed by clustering analysis of tumor and normal colon tissues probed by oligonucleotide arrays. Proc. Natl. Acad. Sci. U.S.A. **96**(12), 6745–6750 (1999). https://doi.org/10.1073/pnas.96.12.6745
25. Freund, Y., Schapire, R.E.: A decision-theoretic generalization of on-line learning and an application to boosting. J. Comput. Syst. Sci. **55**(1), 119–139 (1997). https://doi.org/10.1006/jcss.1997.1504
26. Wang, W., Sun, D.: The improved AdaBoost algorithms for imbalanced data classification. Inf. Sci. **563**, 358–374 (2021). https://doi.org/10.1016/j.ins.2021.03.042
27. Fan, Y., et al.: Privacy preserving based logistic regression on big data. JJ. Netw. Comput. Appl. **171**, 102769 (2020). https://doi.org/10.1016/j.jnca.2020.102769
28. Isuwa, J., et al.: Optimizing microarray cancer gene selection using swarm intelligence: Recent developments and an exploratory study. Egypt. Inform. J. **24**(4), 100416 (2023). https://doi.org/10.1016/j.eij.2023.100416
29. Al-Rajab, M., Lu, J., Xu, Q.: Examining applying high performance genetic data feature selection and classification algorithms for colon cancer diagnosis. Comput. Methods Programs Biomed. **146**, 11–24 (2017). https://doi.org/10.1016/j.cmpb.2017.05.001
30. Shafi, A.S.M., Molla, M.M.I., Jui, J.J., Rahman, M.M.: Detection of colon cancer based on microarray dataset using machine learning as a feature selection and classification techniques. SN Appl. Sci. **2**(7), 1243 (2020). https://doi.org/10.1007/s42452-020-3051-2
31. Alrefai, N., Ibrahim, O.: Optimized feature selection method using particle swarm intelligence with ensemble learning for cancer classification based on microarray datasets. Neural Comput. Appl. **34**(16), 13513–13528 (2022). https://doi.org/10.1007/s00521-022-07147-y

Securing Vehicular Ad-Hoc Networks: A Blockchain and Software Defined Network Approach for Enhanced Efficiency, Safety, and Security

F. Ajesh[1], Felix M. Philip[2], T. Triwiyanto[3], Danyalov Shafi[4], Mammadov Sabir[4], Vusala Abuzarova[4], Samira Aliyeva[4], Dursun Khurshudov[4(✉)], Vugar Hacimahmud Abdullayev[5], Latafat Mikailzade[5], and Taleh Asgarov[6]

[1] Department of Computer Science and Engineering, Sree Buddha College of Engineering, Alappuzha, Kerala, India
[2] Department of Computer Science and Information Technology, Jain (Deemed-to-Be University), Bangalore, India
[3] Department of Medical Electronics Technology, Poltekkes Kemenkes Surabaya, Surabaya, Indonesia
[4] Azerbaijan University of Architecture and Construction, Baku, Azerbaijan
dursuncosqun@gmail.com
[5] Department of Computer Engineering, Azerbaijan State Oil and Industry University, Baku, Azerbaijan
[6] Department of Aerospace Information Systems, Azerbaijan National Aviation Academy, Baku, Azerbaijan

Abstract. The Vehicular Ad-Hoc Network (VANET) is a vital component of the Internet of Things and smart buildings since it connects millions of autos to improve traffic efficiency. A VANET's peers (mostly automobiles) connect with one another by exchanging, informations on road and circumstances, improving passenger and street safety, and routing traffic through congested areas. Because VANET is so sensitive, it's critical to keep a safe, secure, and attack-free environment that allows for uninterrupted data transmission. However, in this paper, we employ blockchain and software defined networks (SDN) to more effectively and efficiently administer and control VANETs. It helps to lessen the strain on the controller by distributing management work between the blockchain and the SDN due to the ubiquitous processing that occurs. We have evaluated our model with some of deep learning models such as LSTM, DenseNet169, VGG16 and 3D-CNN under performance measures Accuracy, Specificity, Detection Rate, Sensitivity and mostly Security in which our model gives higher satisfaction in terms of accuracy and security.

Keywords: Block Chain · 5G · LSTM · Software Defined Networks(SDN) · Vehicular Ad-Hoc Network (VANET)

1 Introduction

The intelligent transportation system has become increasingly popular as the Internet of Vehicles (IoV) and artificial intelligence (AI) technologies continue to advance [1]. IoV plays an important role in the content collection and service delivery thanks to the deployment of powerful and diversified sensors, as well as AI technologies. Traditional IoV, on the other hand, frequently uses distributed control, making global optimization impossible. Furthermore, as the number of cars on the road grows, it becomes increasingly tough to satisfy the different service needs of a wide range of consumers. The Cognitive Internet of Vehicles (CIoV) has been found as a new IoV architecture to address these difficulties [2, 3]. Users may benefit from enhanced service quality by combining this architecture.

The traditional technique of relying on information providers to deliver data to vehicles or consumers is no longer practical, given the growing need for customised information and reduced latency in automobiles. Owing to the developmental scale of automobile networks and the automobile's diversity communication modalities, such as automobile-to-roadside units (RSUs) (V2R) [4], vehicle-to-vehicle (V2V) [5], and so on, it is no longer the only option for the users to obtain content from cloud-based data providers. They can also obtain content from network edge RSUs and other vehicles that are capable of communicating with one another. In reality, the majority of content acquisition research relies on customers providing their own content requirements. The content is transmitted to the car that requested it using RSUs or neighbouring vehicles that have it. This technique of content collection can quickly reveal a vehicle's content requirements, exposing the vehicle's privacy. For example, if a car repeatedly requests a movie, it may pique the user's curiosity, which could include personal information that the user does not want others to know. As a result, this way of obtaining content may result in the revealing of personal information [6].

To address the issue, some research has been conducted on encryption technology [7, 8], anonymization [9], and differential privacy [10]. To protect data secrecy, encryption technology, for example, can encrypt content that requires vehicle data. Most encryption algorithms, on the other hand, are computationally intensive, and most consumers have minimal processing and memory capacity. Traditional encryption methods cannot be easily applied to the CIoV since these complex calculations are impracticable, and K-anonymity and other anonymization approaches may result in redundancy. However, as the number of cars grows, it becomes more difficult to create a privacy mechanism that meets a variety of privacy needs. Furthermore, the content interaction contact time in CIoV is relatively short due to the fast-moving speed of the automobiles. As a result, the amount of time it takes to gather material should be considered. Delay will have an impact on how automobiles or users get material, especially when it comes to vehicle privacy protection. To address the issue of automotive privacy while also providing better information, we suppose in this work that automobiles no more transmit content requests to nearby RSUs or other automobiles, then instead broadcast contents to RSUs or neighbouring vehicles. Each vehicle in search of fuel must simply accept the chemical that meets its requirements. However, because it is hard to precisely estimate the real characteristics of automobiles, the data storage of neighbouring automobiles or RSUs may not match user expectations, and frequent broadcasting wastes bandwidth, this technique is prone to wasting computed and stored resources.This is due to the cognitive engine's ability to identify network vehicles' content demands, allowing them to obtain material more quickly [11, 12].Following the capture of this data, the cognitive engine can use ideas from machine learning and deep learning to identify vehicle care needs, which it can then send to RSUs or vehicles that make a request. As a result, content caching will be faster and more efficient.. The content requirements of users in diverse contexts are different. The cognitive engine can recognise underlying vehicle and intermediate RSU content requirements, cache content on RSUs ahead of time, and recommend certain cached contents to users. Users just need to wait for RSU or neighbouring car broadcast content to download selectively during the content request phase in order to obtain content. Users' direct request content will be reduced as a result of this active caching mechanism, and their privacy will be protected. The symbolic representation of IoV is shown in Fig. 1.

Fig. 1. Internet of Vehicles

The keyhighlights of this paper is mentioned below:

It consist of a well defined blockchain based security framework. Evaluation of this model is depicted using various performance measures such as accuracy, sensitivity, specificity, precision, recall, f1-measure, TPR, FPR and detection rate. Our proposed model is evaluated with some deep learning models such as LSTM, DenseNet169, VGG16 and 3D-CNN. Our proposed model outperforms well with highest average security of 95%.

Organization of Paper. As we already gone through introductory part in Sect. 1, then Sect. 2 depicts related works that is been so far done by researches, Sect. 3 brings system architecture of proposed model, Sect. 4 discuss about Blockchain based security framework, Sect. 5 gives the performance analysis of proposed model and finally the paper concludes with Sect. 6. Reference is given in the end part.

2 Literature Review

IoV offers Internet services to all aspects of vehicular communication, including cars, drivers, and passengers. When SDN is incorporated into the IoV paradigm, Cheng et al. [5] demonstrates the issues in the IoV environment, such as rising capacity, scalable administration and control, QoS requirements, and resource efficiency.The development of LTE mobile communication technologies to assist vehicular communication was triggered by several basic concerns with vehicle communication, such as limited capacity, poor flexibility, and connection problems. [19]. Furthermore, some open LTE concerns were aggressively investigated to ensure that future vehicle communications solutions would be available. Karagiannis et al. [21] give an outline of VANET applications, as well as the requirements and challenges that come with them.Zhang et al. [14] also developed a multi-hop relay-enabled analytical model for. The model was used to investigate the likelihood of uplink and downlink connectivity, with simulated results demonstrating how system parameters and performance indicators could be traded off. The performance of LTE vehicle networks is lowered when LTE communication technologies are integrated into VANETs [11]. As a result, millimeter-wave transmission technology was introduced as a means of connecting users inside automobiles.

Sharma et al. suggested a blockchain-based distributed architecture for vehicle networks in [26]. They looked at the architecture and how it worked, as well as discussing service scenarios and design concepts. Their work centred on vehicle data and networking in order to efficiently share resources and provide value-added services. Their research, on the other hand, did not contain any SDN components. In [27], Using blockchain technology, Xie et al. developed a credibile management system. To conduct periodic network elections, they used Proof of Work (PoW) and Proof of Stake (PoS) consensus-methods. These protocols, on the other hand, have substantial energy costs and significant network latency. Furthermore, issues with vehicle network handover were not addressed.

3 System Architecture

Figure 2 depicts our suggested architecture,End users are vehicles having SDN-enabled on-board equipment (OBUs). These OBUs can execute packet forwarding, collect vehicular data like speed, direction, vehicle type, and so on, as well as acquire environmental data. Their functions also include power control methods, transmission types (V2V or V2I), and channel selection.

The fog zone devices are also SDN-enabled, which means they can be managed by an SDN controller. The RSU hubs maintain control over the autos in fog zones (RSUHs). Unless expressly requested, the overhead is not conveyed to the SDN controller. As a result, RSUHs play a critical role in reducing network overhead. They use fronthaul

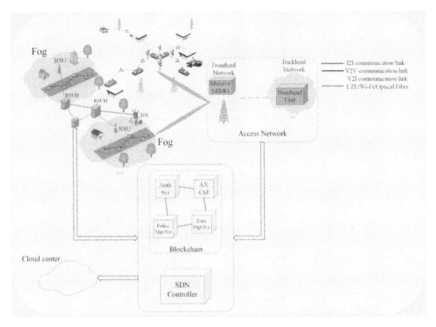

Fig. 2. Blockchain-SDN IoV architecture.

lines to connect various fog zones. As a result, RSUHs serve as a connection between the VANET architecture and the SDN controller. They gather data from the fog cells connected to it and use their own local intelligence to make forwarding judgments. As a result, the overhead of the centralised controller is decreased. Between the RSUHs and the BBUs, an inter-zonal link is built. As a result, RSUHs can be used for data as well as control. The SDN controller will be limited in its ability to manage the entire network. On the other hand, RSUHs and blockchain nodes are held accountable to one another. Based upon the data that we get from the data plane, the RSUHs will then provide a global viewmap of the network. The network monitoring module keeps track of the architecture's connections, while the resource allocation module allocates services based upon the data it collects.The application-plane is in charge of creating rules based upon various vehicle and user application requirements, as well as keeping the network running smoothly. The control plane is then informed of the application requirements (Fig. 3).

The SDN controller sends out the packet-handling instructions. The SDN can accept a range of topological changes in automotive networks at a cheaper cost in terms of software, hardware, and management. In the context of a VANET, this dramatically improves IoT service delivery.Due to capacity, scalability, and connection limitations, gNodeBs are also employed in VANETs to provide broadband wireless connectivity. The OpenFlow Protocol and high-capacity fibre optic back haul connections can also be used by the SDN controller to control RSUs and gNodeBs. On a worldwide scale, this helps with policy consistency and network administration.

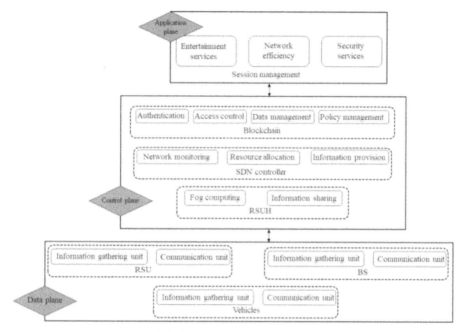

Fig. 3. Network's logical view.

The environment and/or coverage dictate the architecture of the 5G access network. The location may be either outside or indoors, and the coverage can be modest, medium, or huge. 5G will be implemented in densely populated urban regions in smart cities.Large number of cells, uninterrupted coverage, and quotidian coverage are common in this area.However, there will be substantial traffic and poor outdoor-to-indoor coverage in more densely populated regions.As a result, the networking will grow more intricate, with a slew of new features aimed at improving network performance..Cloud and fog computing, SDN, and other technologies will improve efficiency, while blockchain will be employed to establish a safe and trustworthy communication environment.

To ensure security and anonymity, it keeps track of all network transactions and maintains a distributed ledger. Transactions include resource sharing, energy transfer, and other activities. A consortium blockchain is employed because mining on the public blockchain incurs higher expenses and longer delays.If the transactions are signed, then they will be stored in a cryptographically secure structure known as a block, which is then linked chronologically by hash references to develop a chain, so the name block-chain.Then, to generate agreement on the block order as well as for validation purposes, a consensus technique is created.

The consensus algorithm is Byzantine Fault Tolerance (PBFT). In this scenario, the RSUHs take on the role of block miners. After a leader is chosen from among them, a block is formed. Following the receipt of the block, a small number of pre-selected nodes votes until a consensus is reached. The PBFT is used to determine the accuracy of a block. Because the system is decentralised, any RSUH can become a leader and have access to the entire transaction record. The consortium blockchain's leader is selected prior to the block generation process and remains in place until the consensus process is completed. 4th Figure displays a block's structure (Fig. 4).

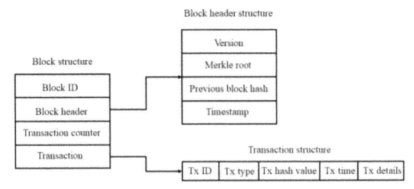

Fig. 4. Block structure

The transaction field defines the chain's transactions, whereas the transaction counter counts the number of transactions completed. In contrast, the transaction field contains the transaction ID, which distinctively identifies an exact transaction in the chain, the transaction type, which specifies the type of transaction being introduced, the transaction hash, which is the hash value generated for that specific transaction, the time of transaction, which specifies when the transaction occurred, and lastly, the details of transaction, which provides a brief explanation of the transaction. To provide efficient and effective network management both blockchain and software-defined networking (SDN) work together.

Authentication Server. Authentication services are provided by the authentication server to the network's entities. An entity must be registered before becoming a member of the network, and cryptographic keys must be issued to that entity. The entity must be authenticated and before establishing any type of network connection. The server authenticates network users to avoid malicious activity, as a network hack would result in serious security and privacy-concerns.

Data Management Server. The SDN controller is relieved of some management obligations by the data management server.The peers in the network will have access to and process multiple requests in such a large, broad environment. Because the SDN is centralised, there is a risk of a single-point of failure (SPoF), which will severely impede the network's performance. The blockchain's decentralised structure assures that the problem of SPoF is eliminated.When the SDN controller can't handle all of the requests, it sends some to the data management server, which handles the rest.

Access Control. Each network has an access control list in order to ensure that the data created is only accessed by the relevant entities.. The access controller is in charge of this. Before data is sent to a network peer, it must first be checked to see if the entity is part of the data owner's access list.If the check is successfully completed,then the accessibility controller allows the individual-access to the data. They are dropped if the request is unsuccessful.

Policy Management Server. Finally, the network's defaulting members are penalised by a policy management server. If a vehicle delivers inaccurate information and causes a network accident, the policy management server penalises the vehicle, which might take the form of toll tickets. Such erroneous information drove the network's adoption of the trust model. This system offers secure and efficient resource management as well as flexible networking. To help in the secure sharing of resources, the blockchain enables tamper-resistant block generation. SDN can recognise complex wireless networks as well as the changing demands of developing services, enabling it to give the best resource allocation strategy SDN may assess the network's architecture, channel assignment, and packet forwarding. Finally, the use of blockchain allows for decentralised synchronisation and replication of network configurations, which will make network orchestration and diagnosis easier. Furthermore, computations for blockchain-enabled systems become simpler.

3.1 Block Chain Based Security Framework

The detailed explanation on the Block Chain Based Security Framework of proposed method is discussed in following sub-sections.

Real -Time Video Report and Message-Sharing Services.
Vehicle Registration. Each automobile in the 5G-VANET has a 5G-subscriber identity module (SIM) number. In this situation, the SIM number is signified by ID. Each ID is required to be unique to a single vehicle. It's saved in the operator's database alongside the device number for the same vehicle. The device number is denoted by Dnum in this case. To demonstrate the registration process, consider the vehicle V_m. When V_m first enters the system, the operator must confirm that V_m'sDnum and ID are identical and that neither of them has been modified. Then, for the registration procedure, At random, a unique symmetric key SKE is generated by V_m. Only the Department of Motor Vehicles (DMV) is assumed to save all licence plate numbers and SIM numbers in this article. The number num represents the licence plate number of a car. A registration demand is sent with the DMV by TA. As a result, while TA possesses V_m's ID, it lacks V_m's genuine identity.

Road Condition Report. Video files of the road are recorded as V_m, status with its internal camera after registration and the message digest can be calculated [21]. To use a term, the technique is repeated every minute. In the middle of each term, the road condition is noted as $tag_{m,j}$, and $place_m$ is the marked position, where j is a timestamp. V_m encrypts the video's hash value and signs the data created inside that minute with SKE. The message is subsequently sent to all nodes in Area X. (which includes RSU and various other cars). If any node near V_m gets the details of message, it will validate V_m's public-key certificate to confirm V_m's legal identity for the message's lifespan. The message will be transmitted to further nodes after being verified before being uploaded to the gNBs via RSU. Step-by-step forwarding will be used to upload the encrypted data to the server.

Because the data transmission is encrypted, the video material is only known by the source vehicle and the cloud server. In addition to providing video recordings relating to itself, a vehicle is expected to send essential information to automobiles nearby or situated in a specified region, such as its own driving operations and road conditions. Although these communications are extremely beneficial for other cars' driving judgments, faked messages will dramatically reduce the mechanism's efficiency, creating a safety hazard. As a result, we regarded vehicle message sharing as a blockchain transaction, which implies that all messages can be identified and traced back to their origin vehicle, and all records are immutable. This ensures that the service's message source nodes are held accountable, and the strategy penalises rogue vehicles. Because one message equals one transaction, the amount of transactions on the blockchain is determined by the messagerate of each vehicle. Because hostile nodes and false messages are inevitable, a temporarily centered node, such as the miner, must be chosen to spread the message and gain consensus. The following is a proposed miner election scheme:

$$\text{Hash(nonce, prehash, bits, time} - \text{stamp)} \leq \text{Br} \qquad (1)$$

Br is the RSUr hash-threshold, and r is the number of RSU. The suggested system can use a variety of hash algorithms, the most prevalent of which is the SHA-256 algorithm. The hash threshold is calculated using the SHA-256 algorithm as follows:

$$\text{Br} = 2^{256-x} - 1 \qquad (2)$$

where Br satisfies the criterion that the first x-bit be exactly 0. Each RSU node adjusts the value of its nonce on a regular basis and computes the hash-value of the block header, which contains the nonce. If an RSU's computed hash value is less than the hash threshold.

Trust-management of Vehicular Messages.
Traffic information Collection. Automobiles evaluate the trustworthiness of road condition tags and assign a value to them during message delivery. The only way to rate a score is to give it a $+ 1$ or a -1. Assume that vehicle Vn accepts a communication from vehicle Vm. Vn verifies Vm's legal identity and retrieves the message's traffic tags tagm,j and the location of placem. Vidm.j's precise content is unknown to Vn. It just assesses tagm,dependability. j's Vn assesses the dependability of tagm,j. When Vn agrees with tagm,description j's of the road state, the tag receives a $+ 1$. When Vn detects that tagm,reported j's road condition is wrong, the tag earns a -1. Vn then signs its location and tagm,score j's and uploads these, along with its public key certificate, to the nearest RSU.

RSU receives the message at time j, examines the source, and categorises the obtained road condition information as Ej,1,Ej,2,Ej,3,... Ejp. Ej,p is composed of the following components: e 1j,p, e 2j,p, e 3j,p,... e mj,p, where e m j,p represents the traffic state of road segment p at the moment. The message division is the same at subsequent times as it is at time j.

Trust-Value Computation. Sj,1,Sj,2,Sj,3,Sj,p are the scores received by RSU from forwarding vehicles, where p is the section number of road, Sj,p is the score for the road condition of section p at time j, and Vn is the score Vn for the road condition of section p transmitted by Vm at time j. Because not all scores in set Sj,p are equally reliable, cars closer to the location where the tag was created frequently outperform vehicles farther away. As a result, the score of a forwarding vehicle's trust value is defined as:

$$o_{n,m}^{j,p} = e^{-\frac{|p_n - p_m|}{a}} \qquad (3)$$

where $o_{n,m}^{j,p}$ represents the trust value score for $s_{n,m}^{j,p}$, p_m denotes the road section of vehicle V_m and p_n denotes the road section of the video forwarding vehicle V_n, $|p_n - p_m|$ denotes

Considering that the hacker cannot hold the majority of the cars, weighted aggregation can increase the trust value's dependability. The trust-value of the road condition at section p transmitted by Vm at time j is defined as follows: (6).

$$o_m^{j,p} = \frac{\sum_{i=1}^{n} s_{i,m}^{j,p} * o_{i,m}^{j,p}}{n}, \left(s_{i,m}^{j,p} \neq 0, o_{i,m}^{j,p} \neq 0\right) \qquad (4)$$

where $d_m^{j,p}$ represents the trust value of the road condit section p broadeast by V_m at time j, V_m represents the recording vehicle, $s_{n,m}^{f,p}$ and $\alpha_{i,m}^{f,p}$ can only be calculated and $o_{i,m}^{f,p}$ are both equal to zero in the Formula (4). If the scoring vehicle is further from the broadcast address of the tag, the reliability of the scoring will be reduced. Therefore, when the trust value $o_{n,m}^{j,p}$ is greater than 0.5, $s_{n,m}^{j,p}$ is scored credible. The score with a trust value of no more than 0.5 will be discarded. Therefore, the trust value of score $S_{j,p}$ for the road condition of section p at time j is calculated as:

$$o^{j,p} = \frac{\sum_{i=1}^{n} \sum_{l=1}^{m} s_{i,l}^{j,p} * o_{i,l}^{j,p}}{n}, \left(s_{i,l}^{j,p} \neq 0, o_{i,l}^{j,p} > 0.5\right) \qquad (5)$$

Miner Election. The summation of the RSU's trust value o j,p for automobiles is used to determine the value of x, which is computed as follows:

$$x = int(e^{-((Ft*Gr/\alpha) - e)}) \qquad (6)$$

where is a variable parameter with a non-zero value. It was used to manage the generating time taken by a new block in Formula (3). In this investigation, the value was fixed to 100. Gr is the total of the trust values determined by RSUr. The bigger the value of Gr, the less significant the value of x. As x decreases, so does the hash threshold, suggesting that the higher the hash threshold, the more probable it is that an RSU will be voted as the miner and contribute a new block to the network. The formula for calculating Gr is as follows:

$$G_r = \sum_{o^{j,p} \in O_r} o^{j,p} \qquad (7)$$

The trust value RSUr derived for Sj,p is represented by o j,p. Set Or is made up of distinct trust-values from various times and places. The more precise the road traffic data RSUris collects, the more likely it is that RSUris will discover the miners. The following is the definition of the function Ft:

$$F_t = \begin{cases} 1 & (t < t_1) \\ 1 - \beta t & (t_1 < t < 2t_1) \\ 0 & (\text{if } F_t > 0 \text{ and } t = 2t_1) \end{cases} \qquad (8)$$

where t is the average time spent by RSUr for a successful miner election over a given period, since it last cleaned up elements in set Or, and t1 is the average time spent by RSUr for a successful miner election, over a given time. Ft always equals 1 when t is less than t1. When t exceeds t1 but less than 2t1, the value of Ft decreases monotonically. is a non-zero coefficient that controls the rate at which Ft decreases. When the value of Ftt is decreased to zero, the decrease stops. The [0, 1] range is used to control Ftis. At the same time, the items in the collection Or will be cleared, and t will be reset to zero to restart the timer. When other RSUs get a block given by a chosen miner, they verify the nonce value.

Vehicle Credibility Assessment. Tracing is only feasible with SIM numbers on 5G networks. Because the DMV is the only one that knows the licence plate numbers and will not release them, a trustworthy third party will be required in addition to a decentralised blockchain architecture.

$$MDA = (TP + TN)/(TP + FN + FP + TN) \tag{9}$$

Using all of the appropriate entries in the blocks, a vehicle's trust rating may be calculated, and the vehicle can subsequently be awarded or punished based on the trust-policies. In old techniques, individual effort is necessary for analysis and assess messages utilising data from a few close or connected vehicles. The unchangeable datas in the blocks that is required for validation on the blockchain, on the other hand, provides a consensus mechanism. All of the ratings for connected cars, as well as the information contained in the blocks, can be used to detect messages.

4 Conclusion

Because VANETs are more or less sensitive, it is vital to maintain a secure, dependable, and attack-free environment that allows for easy information dissemination across the network. Because every piece of information supplied is so important, it's crucial that it's genuine and generated by legitimate entities. In addition, the network must be able to handle and meet security requirements. In this study, we present our proposed model BC-SDN-CIoV, which surpasses the competition in terms of meeting the requirements. The proposed model is then compared to some deep learning models such as LSTM, DenseNet169, 3D-CNN, and VGG16 on a variety of metrics, and our model outperforms them by an average of 95 percent security and 94.4 percent accuracy.

References

1. Gao, J., et al.: A blockchain-SDN-enabled Internet of vehicles environment for fog computing and 5G networks. IEEE Internet Things J. **7**(5), 4278–4291 (2019)
2. Qian, Y., Jiang, Y., Hu, L., Hossain, M.S., Alrashoud, M., Al-Hammadi, M.: Blockchain-based privacy-aware content caching in cognitive internet of vehicles. IEEE Network **34**(2), 46–51 (2020)
3. Lin, H., Garg, S., Hu, J., Kaddoum, G., Peng, M., Hossain, M.S.: Blockchain and deep reinforcement learning empowered spatial crowdsourcing in software-defined internet of vehicles. IEEE Trans. Intell. Transp. Syst. **22**(6), 3755–3764 (2020)

4. Zhou, H., Wenchao, X., Chen, J., Wang, W.: Evolutionary V2X technologies toward the Internet of vehicles: Challenges and opportunities. Proc. IEEE **108**(2), 308–323 (2020)
5. Lv, Z., Lloret, J., Song, H.: Guest editorial software defined internet of vehicles. IEEE Trans. Intell. Transp. Syst. **22**(6), 3504–3510 (2021)
6. Dai, Y., Xu, D., Maharjan, S., Chen, Z., He, Q., Zhang, Y.: Blockchain and deep reinforcement learning empowered intelligent 5G beyond. IEEE Netw. **33**(3), 10–17 (2019)
7. Lu, Y., Huang, X., Zhang, K., Maharjan, S., Zhang, Y.: Blockchain empowered asynchronous federated learning for secure data sharing in internet of vehicles. IEEE Trans. Veh. Technol. **69**(4), 4298–4311 (2020)
8. Wan, S., Gu, R., Umer, T., Salah, K., Xu, X.: Toward offloading internet of vehicles applications in 5G networks. IEEE Trans. Intell. Transp. Syst. (2020)
9. Jaballah, W.B., Conti, M., Lal, C.: Security and design requirements for software-defined VANETs. Comput. Netw. **169** 107099 (2020)
10. Lu, Y., Wang, X., Yi, B., Huang, M.: The reliable routing for software-defined vehicular networks towards beyond 5G. Peer-to-Peer Netw. Appl. 1–15 (2021):
11. Yungaicela-Naula, N.M., Noe, M., Perez-Diaz, J.A., Zareei, M., Vargas-Rosales, C.: Towards the security automation in software defined networks. Comput. Commun. (2021)
12. Mollah, M.B., et al.: Blockchain for the internet of vehicles towards intelligent transportation systems: a survey. IEEE Internet Things J. **8**(6), 4157–4185 (2020)
13. Sharma, S., Ghanshala, K.K., Mohan, S.: Blockchain-based internet of vehicles (IoV): an efficient secure ad hoc vehicular networking architecture. In: 2019 IEEE 2nd 5G World Forum (5GWF), pp. 452–457. IEEE (2019)
14. Hou, X., et al.: Reliable computation offloading for edge-computing-enabled software-defined IoV. IEEE Internet Things J. **7**(8), 7097–7111 (2020)
15. Yao, W., et al.: A secured and efficient communication scheme for decentralized cognitive radio-based Internet of vehicles. IEEE Access **7**, 160889–160900 (2019)
16. Dai, Y., Xu, D., Maharjan, S., Qiao, G., Zhang, Y.: Artificial intelligence empowered edge computing and caching for internet of vehicles. IEEE Wireless Commun. **26**(3), 12–18 (2019)
17. Chattaraj, D., Bera, B., Das, A.K., Rodrigues, J.J., Park, Y.: Designing fine-grained access control for software defined networks using private blockchain. IEEE Internet Things J. (2021)
18. Rehmani, M.H., Davy, A., Jennings, B., Assi, C.: Software defined networks-based smart grid communication: a comprehensive survey. IEEE Commun. Surv. Tutorials **21**(3), 2637–2670 (2019)
19. Qian, Y., et al.: Towards decentralized IoT security enhancement: a blockchain approach. Comput. Electr. Eng. **72**, 266–273 (2018)
20. Siddiqui, S.A., Mahmood, A., Sheng, Q.Z., Suzuki, H., Ni, W.: A survey of trust management in the internet of vehicles. Electronics **10**(18), 2223 (2021)
21. Butt, T.A., Iqbal, R., Salah, K., Aloqaily, M., Jararweh, Y.: Privacy management in social internet of vehicles: review, challenges and blockchain based solutions. IEEE Access **7**, 79694–79713 (2019)
22. Xu, X., et al.: Secure service offloading for internet of vehicles in SDN-enabled mobile edge computing. IEEE Trans. Intell. Transp. Syst. **22**(6), 3720–3729 (2020)
23. Bhattacharya, P., Tanwar, S., Bodkhe, U., Kumar, A., Kumar, N. EVBlocks: a blockchain-based secure energy trading scheme for electric vehicles underlying 5G-V2X ecosystems. Wireless Pers. Commu. 1–41 (2021)
24. Garg, S., Guizani, M., Liang. Y.C., Granelli, F., Prasad, N., Prasad, R.R.: Guest editorial special issue on intent-based networking for 5G-envisioned internet of connected vehicles. IEEE Trans. Intell. Transp. Syst. **22**(8), 5009–5017 (2021)
25. Mershad, K.: SURFER: A secure SDN-based routing protocol for internet of vehicles. IEEE Internet Things J. **8**(9), 7407–7422 (2020)

26. Wu, Y., Dai, H.N., Wang, H., Choo, K.K.: Blockchain-based privacy preservation for 5g-enabled drone communications. IEEE Netw. **35**(1), 50–56 (2021)
27. Hasan, K.F., Overall, A., Ansari, K., Ramachandran, G., Jurdak, R.: Security, privacy and trust: cognitive internet of vehicles. arXiv preprint arXiv:2104.12878 (2021)
28. Chang, H., Ning, N.: An intelligent multimode clustering mechanism using driving pattern recognition in cognitive internet of vehicles. Sensors **21**(22), 7588 (2021)
29. Duo, R., Wu, C., Yoshinaga, T., Zhang, J., Ji, Y.: SDN-based handover scheme in cellular/IEEE 802.11 p hybrid vehicular networks. Sensors **20**(4), 1082 (2020)
30. Yang, Y., Hua, K.: Emerging technologies for 5G-enabled vehicular networks. IEEE Access **7**, 181117–181141 (2019)
31. Wu, J., Dong, M., Ota, K., Li, J., Yang, W.: Application-aware consensus management for software-defined intelligent blockchain in IoT. IEEE Netw. **34**(1), 69–75 (2020)
32. Rafique, W., Qi, L., Yaqoob, I., Imran, M., Rasool, R.U., Dou, W.: Complementing IoT services through software defined networking and edge computing: a comprehensive survey. IEEE Commun. Surv. Tutorials **22**(3), 1761–1804 (2020)
33. Garg, S., Aujla. G.S., Erbad, A., Rodrigues, J.J., Chen, M., Wang, X.: Guest editorial: blockchain envisioned drones: realizing 5G-enabled flying automation. IEEE Netw. **35**(1), 16–19 (2021)
34. Ayaz, F., Sheng, Z., Tian, D., Leung, V.C.M.: Blockchain-enabled security and privacy for internet-of-vehicles. In: Gupta, N., Prakash, A., Tripathi, R. (eds.) Internet of Vehicles and its Applications in Autonomous Driving. Unmanned System Technologies, pp. 123–148. Springer, Cham (2021). https://doi.org/10.1007/978-3-030-46335-9_9
35. Sharma, S., Agarwal, P., Mohan, S.: Security challenges and future aspects of fifth generation vehicular adhoc networking (5G-VANET) in connected vehicles. In: 2020 3rd International Conference on Intelligent Sustainable Systems (ICISS), pp. 1376–1380. IEEE (2020)
36. Alshamrani, S.S., Jha, N., Prashar, D.: B5G ultrareliable low latency networks for efficient secure autonomous and smart internet of vehicles. Math. Prob. Eng. (2021)
37. Zhang, K., Zhu, Y., Maharjan, S., Zhang, Y.: Edge intelligence and blockchain empowered 5G beyond for the industrial Internet of Things. IEEE Netw. **33**(5), 12–19 (2019)
38. Restuccia, F., D'Oro, S., Melodia, T.: Securing the internet of things in the age of machine learning and software-defined networking. IEEE Internet Things J. **5**(6), 4829–4842 (2018)
39. Jabbar, R., Kharbeche, M., Al-Khalifa, K., Krichen, M., Barkaoui, K.: Blockchain for the internet of vehicles: a decentralized IoT solution for vehicles communication using ethereum. Sensors **20**(14), 3928 (2020)
40. Mihailescu, M.I., Nita, S.L., Rogobete, M.G. Authentication protocol for intelligent cars using fog computing and software-defined networking. In: 2021 13th International Conference on Electronics, Computers and Artificial Intelligence (ECAI), pp. 1–6. IEEE (2021)

Influence of Process Noise Biases to Satellite Attitude Filters' Estimates

Chingiz Hajiyev(✉) 🄳 and Demet Cilden-Guler 🄳

Aeronautical Engineering Department, Faculty of Aeronautics and Astronautics, Istanbul Technical University, 34469, Maslak Istanbul, Türkiye
{cingiz,cilden}@itu.edu.tr

Abstract. In this study, an attitude filtering technique is proposed that adapts the bias type process noise uncertainties. To begin with, the Extended Kalman Filter (EKF) and Singular Value Decomposition (SVD) methods are combined to estimate a satellite's attitude.

Influence of the process noise bias type system changes to the innovation of EKF is investigated. It is proved that the bias type process noise change may be converted to the mean square of innovation of EKF and such type of changes can be compensated using the covariance scaling techniques.

For the aim of estimating a satellite's attitude, simulations are compared utilizing the adaptive and non-adaptive versions of the unconventional attitude filter in the presence of process noise bias. Three different forms of attitude estimation methods were investigated in the condition of process noise bias, and the findings results were compared. The simulation results show that, in the investigated cases the multiple fading factors based adaptive SVD-aided EKF can adapt to the changing environment better than the nonadaptive algorithms.

Keywords: Attitude Estimation · Process Noise · Extended Kalman Filter · Singular Value De-composition · Nanosatellite · Magnetometer · Sun Sensor

1 Introduction

An Extended Kalman Filter (EKF) can be used to determine satellite attitude and attitude rate. One approach is to add nonlinear measurements straight into the filter, like a magnetometer, solar sensor, etc. Conventional approaches that use nonlinear measurements of reference directions can be used to produce EKF for satellite attitude and rate estimates [1–5]. Nonlinear equations are utilized to connect the states and measurements.

In a non-traditional approach, vector measurements and the associated single-frame orientation determination method at each recursive step are used to first identify the orientation angles. After that, an attitude filter, such as the Extended Kalman filter (EKF) [6–8] or the Unscented Kalman filter (UKF) [9, 10], uses these attitude angles directly as measurement inputs.

The unconventional attitude filtering algorithm integrates the Singular Value Decomposition (SVD) and Extented Kalman Filter (EKF) techniques to estimate the attitude of a nanosatellite. The SVD method uses measurements from the magnetometer and Sun sensor as the first step of the algorithm to determine the attitude of the nanosatellite. Then, these attitude terms along with their error covariances are supplied into the EKF.

This study demonstrates that the process noise bias type system changes will result in alterations to the statistical features of the EKF innovation sequence. It has been demonstrated that the bias type process noise change may be translated into changes in the mean square of innovation of the EKF and that these kinds of changes can be compensated for with covariance scaling methods. We construct and provide the theoretical foundations of the Q-adaptive SVD-aided EKF with uncertain process noise mean. For the aim of estimating a satellite's attitude, simulations are compared utilizing the adaptive and non-adaptive versions of the unconventional attitude filter in the presence of process noise bias. Three different forms of attitude estimation methods were investigated in the condition of process noise bias, and the findings results were compared.

In this study, a covariance tuning is implemented for the process noise covariance matrix (Q matrix). The rest of the paper reads as follows. In Sect. 2, satellite rotational motion and attitude measurement models are given. Extended Kalman Based Estimator Aided by SVD is presented in Sect. 3. Section 4 gives the results of influence of process noise bias to the innovation of EKF. The theoretical basics of the Q-adaptive SVD-aided EKF with uncertain process noise mean are developed and presented in this Section. Section 5 includes simulation results of adaptive and non-adaptive attitude estimation algorithms. Discussion of the obtained results and conclusions are presented in Sect. 6.

2 SVD-Aided EKF for Attitude Estimation

This section presents the nontraditional attitude estimation process, which consists of two stages: extended Kalman filter and singular value decomposition.

2.1 Attitude Estimation via Singular Value Decomposition Method

Single-frame attitude estimate techniques include the Singular Value Decomposition (SVD, q-method, QUEST, FOAM, and others. Due to its greater robustness, the SVD approach is chosen as the single-frame method in this instance [8].

Given a set of $n \geq 2$ vector measurements, $\hat{u}_B^i$, in the body system, choosing to minimize the loss function given as for an ideal attitude matrix, A, is one possibility,

$$J(\mathbf{A}) = \sum_{i=1}^{n} w_i \left| \hat{\mathbf{u}}_B^i - \mathbf{A}\hat{\mathbf{u}}_R^i \right|^2 \tag{1}$$

where w_i is the weight of the ith vector measurement, $\hat{u}_R^i$ is the vector in the reference coordinate system.

A is the transformation matrix from reference frame to body frame. This optimization problem can be solved by SVD method. SVD based attitude determination algorithm is given in [8].

2.2 EKF for Attitude and Rate Estimation

Given that the EKF is constructed using discrete-time nonlinear equations, the process model is represented by the following equations:

$$x(k+1) = f(x(k), k) + w(k), \qquad (2)$$

$$y(k) = \mathbf{H}x(k) + v(k). \qquad (3)$$

Here, $x(k)$ is the state vector and $y(k)$ is the measurement vector. Moreover $w(k)$ and $v(k)$ are the process and measurement error noises, which, according to the assumption, are processes with Gaussian white noise with zero mean and a covariance of $Q(k)$ and $R(k)$ respectively, H is the measurement matrix of system.

The Extended Kalman filter for attitude and rate estimation is given in Fig. 1.

Predict	Update
Prediction : $\hat{x}(k+1/k) = \mathbf{f}[\hat{x}(k), k]$ Prediction covariance : $P(k+1/k) = \dfrac{\partial \mathbf{f}[\hat{x}(k),k]}{\partial \hat{x}(k)} P(k/k) \dfrac{\partial \mathbf{f}^T[\hat{x}(k),k]}{\partial \hat{x}(k)} + Q(k)$ Measurement Vector : $z(k) = [z_\varphi(k) \; z_\omega(k)]^T$	Innovation : $\tilde{e}(k) = z(k) - \mathbf{H}\,\hat{x}(k+1/k)$ Innovation covariance : $P_z(k+1/k) = \mathbf{H}\, P(k+1/k)\mathbf{H}^T + R(k)$ Kalman gain : $K(k+1) = P(k+1/k)\mathbf{H}^T \times P_z(k+1/k)^{-1}$ Estimation covariance : $P(k+1/k+1) = [\mathbf{I} - K(k+1)\mathbf{H}] P(k+1/k)$ Estimation : $\hat{x}(k+1) = \hat{x}(k+1/k) + K(k+1) \times \{z(k) - \mathbf{H}\,\hat{x}(k+1/k)\}$

Fig. 1. EKF structure for attitude and rate estimation.

2.3 EKF with Process Noise Bias Compensation

We try to make corrections in the Kalman filter equations in order to have the estimate that remains unbiased, since the known mean $E[\delta] = m_\delta$ must be incorporated into the estimation equations. EKF with biased process noise is presented by the following equation for the extrapolation value:

$$\hat{x}(k+1/k) = \mathbf{f}[\hat{x}(k), k] + m_\delta. \qquad (4)$$

The equations for the estimation value, the EKF gain, the covariance matrix of the extrapolation error and the covariance matrix of the estimation error are the same as for the conventional EKF.

3 Influence of Process Noise Bias to the Innovation of EKF

Assumption. In this section, the process noise is assumed to be biased and can be represented as sum of random ($w(k)$) and constant ($\delta(k)$) components.

The process and measurement model equations in this case can be written as,

$$\mathbf{x}_b(k+1) = \mathbf{f}(\mathbf{x}(k), k) + \boldsymbol{\delta}(k) + \mathbf{w}(k), \tag{5}$$

$$\begin{aligned}
\mathbf{z}_b(k) &= \mathbf{H}\mathbf{x}_b(k) + \boldsymbol{v}(k) = \\
&= \mathbf{H}\big[f(x(k-1), k-1) + \boldsymbol{\delta}(k-1) + \mathbf{w}(k-1)\big] + \boldsymbol{v}(k) = \\
&= \mathbf{H}\big[f(x(k-1), k-1) + \mathbf{w}(k-1)\big] + \mathbf{H}\boldsymbol{\delta}(k-1) + \boldsymbol{v}(k) = \\
&= \mathbf{H}\mathbf{x}(k) + \mathbf{H}\boldsymbol{\delta}(k-1) + \boldsymbol{v}(k) = \mathbf{z}(k) + \mathbf{H}\boldsymbol{\delta}(k-1).
\end{aligned} \tag{6}$$

Here b in the index shows that the value is biased. It is clear that in this case the system state estimates will be biased.

Theorem 1: If the measurements are processed by the EKF (Fig. 1), a process noise bias occurs at the iteration step $k = \tau$, then at the all $k > \tau$ steps the estimation and innovation of EKF are biased and innovation bias is equal to the observed difference between the process bias and prediction bias.

Proof. At the first step after the bias occurring at iteration $k = \tau$ we have:
Extrapolation value

$$\hat{\mathbf{x}}_b(k+1/k) = \mathbf{f}(\hat{\mathbf{x}}_b(k/k), k) = \hat{\mathbf{x}}(k+1/k) + \Delta\hat{\mathbf{x}}(k+1/k), \tag{7}$$

where $\Delta\hat{\mathbf{x}}(k+1/k)$ is the bias in the extrapolation value.
Estimation value

$$\begin{aligned}
\hat{\mathbf{x}}_b(k+1|k+1) &= \hat{\mathbf{x}}_b(k+1|k) + K(k+1)\big[\mathbf{z}_b(k+1) - \mathbf{H}\hat{\mathbf{x}}_b(k+1/k)\big] = \\
&= \hat{\mathbf{x}}(k+1/k) + \Delta\hat{\mathbf{x}}(k+1/k) + \\
&+ K(k+1)\big[\mathbf{z}(k+1) + \mathbf{H}\boldsymbol{\delta}(k) - \mathbf{H}\hat{\mathbf{x}}(k+1/k) - \mathbf{H}\Delta\hat{\mathbf{x}}(k+1/k)\big] = \\
&= \hat{\mathbf{x}}(k+1/k) + K(k+1)\big[\mathbf{z}(k+1) - \mathbf{H}\hat{\mathbf{x}}(k+1/k)\big] + \Delta\hat{\mathbf{x}}(k+1/k) + \\
&+ K(k+1)\mathbf{H}\boldsymbol{\delta}(k) - K(k+1)\mathbf{H}\Delta\hat{\mathbf{x}}(k+1/k) = \\
&= \hat{\mathbf{x}}(k+1/k+1) + \Delta\hat{\mathbf{x}}(k+1/k+1)
\end{aligned} \tag{8}$$

where

$$\begin{aligned}
\Delta\hat{\mathbf{x}}(k+1/k+1) &= \Delta\hat{\mathbf{x}}(k+1/k) + K(k+1)\mathbf{H}\boldsymbol{\delta}(k) - \\
&\quad - K(k+1)\mathbf{H}\Delta\hat{\mathbf{x}}(k+1/k) = \\
&= [\mathbf{I} - K(k+1)\mathbf{H}(k+1)]\Delta\hat{\mathbf{x}}(k+1/k) + K(k+1)\mathbf{H}\boldsymbol{\delta}(k)
\end{aligned} \tag{9}$$

is the bias in the estimation value. Innovation is,

$$\begin{aligned}
\tilde{\mathbf{e}}_b(k+1) &= \mathbf{z}_b(k+1) - \mathbf{H}\hat{\mathbf{x}}_b(k+1/k) = \\
\mathbf{z}(k+1) &+ \mathbf{H}\boldsymbol{\delta}(k) - \mathbf{H}\hat{\mathbf{x}}(k+1/k) - \mathbf{H}\Delta\hat{\mathbf{x}}(k+1/k) \\
&= \tilde{\mathbf{e}}(k+1) + \mathbf{H}\boldsymbol{\delta}(k) - \mathbf{H}\Delta\hat{\mathbf{x}}(k+1/k) = \tilde{\mathbf{e}}(k+1) + \boldsymbol{\mu}(k+1)
\end{aligned} \tag{10}$$

where

$$\boldsymbol{\mu}(k+1) = \mathbf{H}\boldsymbol{\delta}(k) - \mathbf{H}\Delta\hat{\mathbf{x}}(k+1/k) = \mathbf{H}\big[\boldsymbol{\delta}(k) - \Delta\hat{\mathbf{x}}(k+1/k)\big] \tag{11}$$

is the innovation bias.

The innovation bias is equal to the observed difference between the process bias and prediction bias. This circumstance is true for the all $k > \tau$ steps. Thus, the *Theorem 2* is proved.

A sampling covariance matrix of innovation is presented as statistics for detecting and compensating for changes in process noise. The sample covariance matrix of the innovation may be expressed as follows [11]:

$$\hat{\mathbf{S}}_e(k) = \frac{1}{M} \sum_{j=k-M+1}^{k} \tilde{\mathbf{e}}(j) \tilde{\mathbf{e}}^T(j) \tag{12}$$

where M is the width of the "sliding window".

If there is a bias in the mean of the innovation at time τ, and the biased innovation is denoted by.

$\tilde{e}_b(k)$, then the biased innovation is defined as,

$$\tilde{e}_b(k) = \tilde{e}(k), k = 1, 2, ..., \tau - 1, \tag{13}$$

$$\tilde{e}_b(k) = \tilde{e}(k) + \boldsymbol{\mu}(k), k = \tau + 1, \tau + 2, ... \tag{14}$$

where $\boldsymbol{\mu}(.)$ is an unknown innovation bias vector.

When $k < \tau$, the mathematical expectation of the sample innovation covariance matrix (12) can be determined by the following equation

$$E\left[\hat{\mathbf{S}}_e(k)\right] = \mathbf{P}_e(k/k-1) = \mathbf{H}\mathbf{P}(k/k-1)\mathbf{H}^T + \mathbf{R}(k) \tag{15}$$

Here $\boldsymbol{P}(k/k-1)$ is the extrapolation error covariance matrix.

In the case of $k > \tau$, in the sample innovation covariance a biased innovation $\tilde{e}_b(k) = \tilde{e}(k) + \boldsymbol{\mu}(k)$ is used instead of an unbiased innovation $\tilde{e}(k)$, where $\boldsymbol{\mu}(k)$ is the innovation bias [11]

$$\hat{\mathbf{S}}_{eb}(k) = \frac{1}{M} \sum_{j=k-M+1}^{k} \tilde{\mathbf{e}}_b(j) \tilde{\mathbf{e}}_b^T(j). \tag{16}$$

Remark. Note that the mean of the innovation $\tilde{e}_b(k)$ in this case is not zero, therefore the formula (16) is not a sample covariance. In the "sliding window," this is the mean square of innovation (MSI). Bias type process noise change may be converted to the mean square of innovation and such type of changes can be compensated using the covarince scaling techniques.

Statement. The bias in innovation for the $k > \tau$ iteration steps leads to an increase in the mathematical expectation of the mean square of innovation.

Proof. The mathematical expectation of mean square of innovation (16) for $k > \tau$ can be written as

$$E[\hat{S}_{eb}(k)] = \frac{1}{M}E\left(\sum_{j=k-M+1}^{k} \tilde{e}_b(j)\tilde{e}_b^T(j)\right) =$$

$$= \frac{1}{M}E\left(\sum_{j=k-M+1}^{k} [\tilde{e}(j) + \mu(j)][\tilde{e}(j) + \mu(j)]^T\right) = \quad (17)$$

$$= \frac{1}{M}E\left(\sum_{j=k-M+1}^{k} \left[\tilde{e}(j)\tilde{e}^T(j) + \tilde{e}(j)\mu^T(j) + \mu(j)\tilde{e}^T(j) + \mu(j)\mu^T(j)\right]\right)$$

Taking into account $E\left[\tilde{e}(k)\right] = 0$, and the absence of correlation between the parameters $\tilde{e}(j)$ and $\mu(j)$, we have

$$E[\hat{S}_{eb}(k)] = E[\hat{S}_e(k)] + \frac{1}{M}E\left(\sum_{j=k-M+1}^{k} \mu(j)\mu^T(j)\right) \quad (18)$$

Comparison of expressions (17) and (18) proves the *Statement 1*. Consequently, the process noise bias will increase the mathematical expectation of the mean square of innovation.

It can be seen from the *Theorem 2* and the *Statement* above that the process noise bias is transferred to the innovation bias and changes the mean square of innovation (16). The innovation bias $\mu(k)$ leads to a change in the values of $\tilde{e}_b(k)$ and the elements of the mean square of innovation. As a result, the bias in the process noise is transferred to the MSI. Thus, he MSI can be chosen as a monitoring statistic in the problem of compensating for changes in process noise.

4 Satellite Attitude Estimation via Q-Adaptive SVD/EKF Algorithm

The SVD method is used as the first step in the algorithm and yields one estimate of the nanosatellite attitude per single frame. These estimated attitude terms are then sent into the EKF. Thus, the attitude of the satellite is computed. "SVD-aided EKF" is abbreviated as "SaEKF." For the sake of convenience, "Q-Adaptive SVD-aided EKF" is referred to as "ASaEKF" throughout.

In addition to the estimated quaternions, the estimation covariance from the SVD is also entered into the EKF method and used as the measurement noise covariance matrix of the filter, i.e. $R(k+1) = P_{svd}(k+1)$. This is why the unconventional filter is inherently resistant to increases in measurement noise. When a measurement error occurs, the SVD's estimate covariance—which is also the filter's measurement noise covariance—increases, allowing the filter to continue operating without suffering appreciable harm [8].

For unconventional attitude filters, one problem is adjusting the filter's process noise covariance matrix. Optimizing the process noise covariance is crucial as the environment changes. These modifications include tweaks to the disturbance torques and tweaks to the inertia parameters (such as when the satellite passes through or passes through an eclipse).

The adjustment of the filter's process noise covariance matrix, Q presents one challenge for nontraditional attitude filters. When the environment changes, it's very important to optimize the process noise covariance. These changes include adjustments to the inertia parameters (as when the satellite enters or exits an eclipse) and adjustments to the disturbance torques.

An adaptive technique is used to tweak the EKF in order to adapt it to the changing environment. The adaptation rule is simple to implement because the measurement model is linear.

In order to adapt the covariance, an adaptive scaling factor is put into the procedure. Substituting the adaptive scaling matrix $\mathbf{\Lambda}(k)$ into the formula for extrapolation error covariance $P(k+1/k)$ of EKF, for the iteration steps $k \geq \tau$ we have:

$$\mathbf{P}_e(k+1/k) = \mathbf{H}\left[\frac{\partial \mathbf{f}[\hat{\mathbf{x}}(k),k]}{\partial \hat{\mathbf{x}}(k)}\mathbf{P}(k/k)\frac{\partial \mathbf{f}^T[\hat{\mathbf{x}}(k),k]}{\partial \hat{\mathbf{x}}(k)}\right]\mathbf{H}^T + \mathbf{H}\mathbf{\Lambda}(k)\mathbf{Q}(k)\mathbf{H}^T + \mathbf{R}(k). \tag{19}$$

The scaling matrix $\mathbf{\Lambda}$ can be determined from equality (19) [12],

$$\frac{1}{M}\sum_{j=k-M+1}^{k} \tilde{\mathbf{e}}(j)\tilde{\mathbf{e}}^T(j) \approx \mathbf{H}\mathbf{P}_e(k+1/k)\mathbf{H}^T + \mathbf{R}(k). \tag{20}$$

Here, M is the width of the moving window.

The left side of equality (20) represents the real, and the right side represents the theoretical value of the covariance of the innovation sequence.

Then, the scale matrix $\mathbf{\Lambda}(k)$ can be determined from the Eqs. (19) and (20) as,

$$\mathbf{\Lambda}(k) = \left[\mathbf{H}^T\mathbf{H}\right]^{-1}\mathbf{H}^T \times$$

$$\times \left\{\frac{1}{M}\sum_{j=k-M+1}^{k}\tilde{\mathbf{e}}(j)\tilde{\mathbf{e}}^T(j) - \mathbf{H}\left[\frac{\partial \mathbf{f}[\hat{\mathbf{x}}(k),k]}{\partial \hat{\mathbf{x}}(k)}\mathbf{P}(k/k)\frac{\partial \mathbf{f}^T[\hat{\mathbf{x}}(k),k]}{\partial \hat{\mathbf{x}}(k)}\right]\mathbf{H}^T - \mathbf{R}(k)\right\} \times$$

$$\times \mathbf{H}[\mathbf{Q}(k)\mathbf{H}^T\mathbf{H}]^{(-1)}. \tag{21}$$

5 Analysis and Results

The analysis takes into account a small satellite in a tumbling state to evaluate the algorithms under cruical measurement and environmental conditions. The small satellite has the principal moment of inertia $J = \text{diag}[\,0.055\ 0.055\ 0.017\,]$ kg m^2. Using a 1 Hz sample rate for both the filter and the sensors, the algorithm operates for nearly an orbital cycle. The three-axis magnetometers and three-axis Sun sensors by their respective standard deviations $\sigma_B = 300$ nT and $\sigma_S = 0.002$ are the ones that were chosen.

The single-frame attitude estimate methods are only applicable when at least two measurements are available simultaneously and the vectors are not parallel to each other. Consequently, the single-frame approach fails when the satellite is shadowed and two vectors are parallel. An eclipse period is included between the 2000th and 3000th seconds to show how the filter performs when integrated with a single-frame technique in these intervals. This section compares SVD-aided EKF (SaEKF), SVD-aided EKF with Process Noise Bias Compensation (SaEKF-PNBC), and adaptive SaEKF (ASaEKF) in order to assess the adaptation for process noise bias between 4500th and 5500th seconds.

The algorithm is tested in order to examine the filter's capacity to adjust in response to process bias faults. By include the constant bias component in the process noise between the 4500th and 5500th s, biases in the process noise are simulated. The constant term used for the test case is $\delta = \begin{bmatrix} 0.02 & 0.02 & 0.02 & 0.002 & 0.002 & 0.002 \end{bmatrix}^T$.

Taking into account the constant bias term, the process model (2) can be represented as follows:

$$x(k+1) = f(x(k), k) + w(k) + \delta, \ (4500 \leq k \leq 5500). \tag{22}$$

The Root Mean Squares Errors (RMS) for SaEKF, SaEKF-PNBC and ASaEKF algorithms for process noise bias cases are shown in Table 1.

Table 1. RMS error for SaEKF, ASaEKF and SaEKF-PNBC for process noise bias case.

RMS Error	Process Noise Bias δ		
	SaEKF	ASaEKF	SaEKF-PNBC
ϕ	14.68333	0.28974	1.10105
θ	23.02956	0.20964	1.25742
ψ	18.50896	0.31995	1.21126

In Fig. 2, attitude angles estimations of ASaEKF, SaEKF-PNBC and SaEKF algorithms are given when having process noise biases. As can be seen, the results of the ASaEKF algorithm are excellent; using this algorithm, rotation angles can be estimated with high accuracy in the presence of process noise bias. The next best algorithm for estimating rotation angles is SaEKF-PNBC. The worst estimation results were obtained using the SaEKF algorithm.

The obtained results show that the bias type process noise changes can be compensated using the covarince scaling techniques or EKF with process noise bias compensation algorithm.

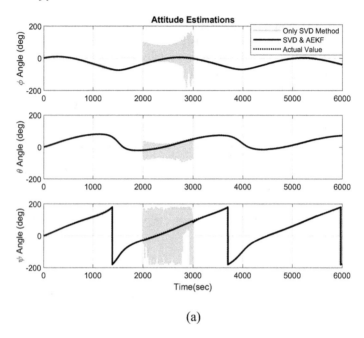

(a)

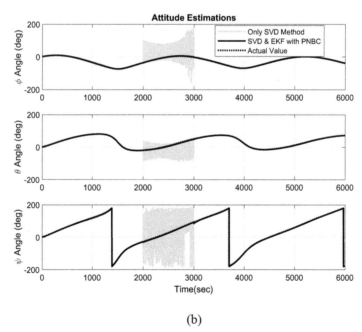

(b)

Fig. 2. Attitude angles estimates of ASaEKF (a), SaEKF-PNBC and SaEKF (c) algorithms when applying the process noise bias.

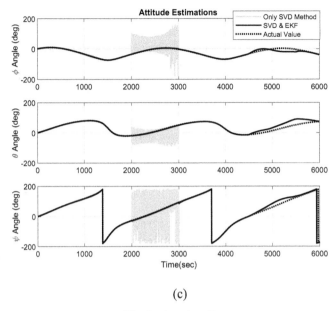

(c)

Fig. 2. (*continued*)

6 Conclusions

In this paper, an attitude filtering technique is proposed that adapts the bias type process noise uncertainties. To begin with, the Extended Kalman Filter and Singular Value Decomposition methods are combined to estimate a satellite's attitude.

Influence of the process noise bias type system changes to the innovation of EKF is investigated. It is proved that the bias type process noise change may be converted to the mean square of innovation of EKF and such type of changes can be compensatd using the covarince scaling techniques. In the case of process noise bias, three types of attitude estimation algorithms (SaEKF, SaEKF-PNBC and ASaEKF) were examined and the results were compared. Simulation results show that ASaEKF's estimation results are superior. The next best algorithm for estimating rotation angles is SaEKF-PNBC. The worst attitude estimation results were obtained using the SaEKF algorithm. The simulation results show that, in the investigated cases the multiple fading factors based adaptive SVD-aided EKF can adapt to the changing environment better than the nonadaptive algorithms.

References

1. Springmann, J.C., Cutler, J.W.: Flight results of a low-cost attitude determination systems. Acta Astronaut. **99**, 201–214 (2014). https://doi.org/10.1016/j.actaastro.2014.02.026
2. Cilden, D., Soken, H.E., Hajiyev, C.: Nanosatellite attitude estimation from vector measurements using SVD-AIDED UKF algorithm. Metrol. Meas. Syst. **24**(1), 113–125 (2017). https://doi.org/10.1515/mms-2017-0011

3. Hajiyev, C., Cilden-Guler, D.: Review on gyroless attitude determination methods for small satellites. Prog. Aerosp. Sci. **90**, 54–66 (2017). https://doi.org/10.1016/j.paerosci.2017.03.003
4. Ivanov, D., Ovchinnikov, M., Roldugin, D.: Three-axis attitude determination using magnetorquers. J. Guid. Control. Dyn. **41**(11), 2455–2462 (2018). https://doi.org/10.2514/1.G003698
5. Mashtakov, Y., Ovchinnikov, M., Wöske, F., Rievers, B., List, M.: Attitude determination & control system design for gravity recovery missions like GRACE. Acta Astronaut. **173**, 172–182 (Aug.2020). https://doi.org/10.1016/j.actaastro.2020.04.019
6. Hajiyev, C., Bahar, M.: Attitude determination and control system design of the ITU-UUBF LEO1 satellite. Acta Astronaut. **52**(2–6), 493–499 (2003). https://doi.org/10.1016/S0094-5765(02)00192-3
7. Mimasu, B.Y., Van der Ha, J.C.: Attitude determination concept for QSAT. Trans. Jpn. Soc. Aeronaut. Space Sci. Aerosp. Technol. Jpn. **7**, 63–68 (2009). https://doi.org/10.2322/tstj.7.Pd_63
8. Hajiyev, C., Cilden, D., Somov, Y.: Gyro-free attitude and rate estimation for a small satellite using SVD and EKF. Aerosp. Sci. Technol. **55**, 324–331 (2016). https://doi.org/10.1016/j.ast.2016.06.004
9. Cilden, D., Soken, H.E., Hajiyev, C.: Nanosatellite attitude estimation from vector measurements using SVD-aided UKF algorithm. Metrol. Meas. Syst. **24**(1), 113–125 (2017)
10. Hajiyev, C., Soken, H.E., Cilden-Guler, D.: Nontraditional attitude filtering with simultaneous process and measurement covariance adaptation. J. Aerosp. Eng. **32**(5), 04019054 (2019)
11. Hajiyev, C. An innovation-based actuator/surface fault detection, isolation and filter tuning. Aircr. Eng. Aerosp. Technol. **95**(3), 464–473 (2023)
12. Hajiyev C, Cilden-Guler D.: Satellite attitude estimation using SVD-Aided EKF with simultaneous process and measurement covariance adaptation. Adv. Sp. Res. Pergamon **68**(9), 3875–3890 (2021)

Information-Retrieval System of Rock Pictures of Different Countries and Continents

Aydin Kazim-zada[(✉)] [iD] and Hayat Huseynova [iD]

Azerbaijan University of Architecture and Construction, Ayna Sultanova Street, 5, Baku AZ1073, Azerbaijan
`aydin91@gmail.com, ibrahimlihayat@yahoo.com.tr`

Abstract. This article narrates about petroglyphs of the Azerbaijan Republic, countries of Central and Middle Asia, and other countries of the world, representing the cultural heritage of ancient generations. The article touches on the issues concerning the necessity of preserving the rock pictures for future generations using all modern available means (printed, editions, electronic media of information, articles etc.). The article contains examples of rock pictures on which our ancestors depicted people, animals, hunting scenes, the history of their daily life, various kinds of festivities, sacred ceremonies etc. with the use of tools they had at hand. As the said petroglyphs are of different geometrical dimensions (ranging from some centimeters to some meters) one of the methods of normalizing the rock pictures through the placement of the pictures within the circle area is considered. The coordinates of the coincidence of drawing points with the circle points complement a list of informative indicators. The presence of symmetry and asymmetry in the rock drawings is treated as other informative indicators. A number of rock pictures of deer and goats of different countries were studied and their graphs of distribution density along both the OX and OY axes are given. These graphs also are one of the additional informative features of the rock pictures. Correlation analysis of the pictures against both coordinate axes has been made.

A structure of the developed information retrieval system of petroglyphs of the world countries is demonstrated.

Keywords: informative features · petroglyphs · rock pictures · information retrieval system · distribution density · correlation analysis · recognition · identification

1 Introduction

Ancient people gouged rock pictures on stones and painted them in caves. The oldest rock pictures or petroglyphs belong to the epoch of the Paleolithic. They are a unique source of information on the life of our ancestors in the remote past and attract the attention of scientists: historians, archeologists, art experts, zoologists, ethnographers, linguists, folklore specialists as well as explorers from other fields of science. These monuments ancient culture have been found practically all over the globe. More than 35 million petroglyphs have been discovered in 120 countries of the world for example,

© The Author(s), under exclusive license to Springer Nature Switzerland AG 2025
G. Mammadova et al. (Eds.): ITTA 2024, CCIS 2226, pp. 111–122, 2025.
https://doi.org/10.1007/978-3-031-73420-5_10

petroglyphs have been discovered in Azerbaijan (Gobustan) [1]. Middle Asia (Saymaly Tash) [2], Russia (Tomsk writing, Elangash, Jalghiztobe, Kalbak-Tash) [3], and other countries in all continents of the world. Like everything that exists, rock pictures tend to disappear in time. As the majority of petroglyphs are located in the open air, they are exposed both to natural phenomena (sun, air, weather, wind, rains, landslides, earthquakes u.a.) and subjective factors (human factors). Cave petroglyphs have their intrinsic ecological peculiarities. Conservation of rock pictures for successive generations is the task of the living generation, the importance of which it is impossible to overestimate. Modern computer and information technologies play a special role in the conservation and exploration of petroglyphs.

2 Statement of the Problem

All modern available means (printed edition, electronic media of information, archives etc.) are employed for conserving rock drawings. Performance as procedures for database creation of information retrieval systems of petroglyphs from individual countries, reserves, as well as their unification by continents, will allow:

a. to conserve images of petroglyphs for successive generations.
b. to explore and analyze by different parameters the works of ancient art left to us by our ancestors.

Due to the huge number of rock pictures, it is necessary to determine the informative features of petroglyphs and resolve the problem of recognition and identification of rock pictures with the aim of creating information retrieval systems of petroglyphs.

To resolve the problem of recognition and identification of petroglyphs, the use of up-to-date computer and information technologies, the development of methods and ways of recognition and identification of rock drawings.

3 Solution

Rock pictures of people (Fig. 1a), animals (Fig. 1b), birds, fishes (Fig. 1c), and vegetation, boats, hunting scenes (Fig. 1d), festivities, etc. [1–3] discovered on different continents and in different countries demonstrate that despite huge distances and separation by oceans and seas, the likeness of life, daily and domestic lives, fauna etc. can be observed. There are also differences showing unique natives of this locality as compared to other ones. Also, there are likenesses and differences in the technical designs of some structures. For example, drawn pictures of boats in various reserves look different (Fig. 2) [1]. It is natural that pictures of boats are typical only of places located close to rivers, seas, and oceans.

Practically all locations of rock pictures represent state historical artistic reserves. The most popular of the world historical artistic reserves studied by us are the following:

1. Gobustan.
2. The Pamirs-Altai.
3. Western Tien-Shan.

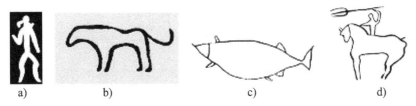

Fig. 1. Examples of rock pictures

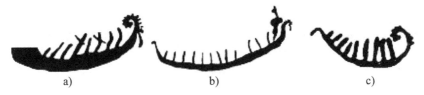

Fig. 2. Examples of boat pictures

4. Central Tien-Shan.
5. Northern region close to Tien-Shan.
6. Zungar massif.
7. Altamira – cave in Spain (epoch of the upper paleolithic).
8. Fond-de-Gaume – cave in France with wall pictures of animals of the epoch of late Paleolithic,
9. Lascaux cave – cave in France with rock pictures of the late Paleolithic period,
10. Combarelles – cave in France of the Upper Paleolithic,
11. Trois Freres – cave in France with rock pictures of the Paleolithic period,
12. Rouffignac – cave in France (the Upper Paleolithic).
13. Examples of coloured rock pictures discovered in the Altamira cave are given in Fig. 3.

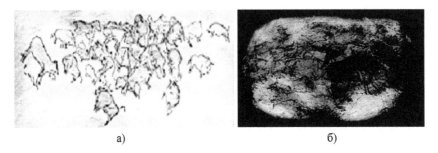

Fig. 3. Altamira-cave in Spain with a polychrome stone painting of the Upper Paleolithic epoch

Many petroglyphs are also present in the Sayan-Altai territory. The special interest of petroglyph explorers of this region is seen in the study of subject compositions [2]. Similar petroglyphs are also observed on the territories below:

1. Tarbagatai, Chingiz-tau, the upper reaches of the Irtish;
2. The Gorno-Altai;
3. The Mongolian and Gobi Altai
4. The Western Sayan u.a.
5. By their content and shape petroglyphs can be divided into the following main groups:
6. Pictures of living beings (people, wild animals, domestic animals, birds, fishes etc.)
7. Pictures of objects for everyday life (household articles, work tools, objects intended for hunting, etc.)
8. Vegetation elements (trees, bushes, flowers etc.)
9. Scenes of work, hunting
10. Elements of diverse shapes etc.
11. Rather complicated drawings requiring scientific analysis.

To determine a drawn object the work of specialists engaged in different fields of science and technology is employed.

Let's study several informative features of rock pictures. Petroglyphs relate to a continent, country, certain reserve, or mountainous locality, they are of various geometrical dimensions (ranging from some centimeters to some meters). Many of them, naturally, are asymmetrical. However, drawings that are symmetrical or closely resemble mirror symmetry, are also observed. Figure 4a,b,c,d,e,f display examples of symmetrical drawings of Russian petroglyphs [2, 3].

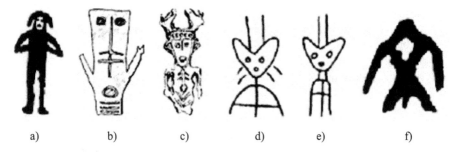

Fig. 4. Examples of symmetrical petroglyphs of Russia

In Fig. 5a,b,c,d,e show examples of Azerbaijani petroglyphs having incomplete vertical symmetry. The disposition of the drawings is the same as on the rocks, but with the use of graphic editors- such as, for example, ADOBE PHOTOSHOP, CORELDRAW etc., the pictures were turned into a vertical position, symmetries were studied, graphs of distribution functions along OX and OY axes were plotted.

In Fig. 5 a1,b1,c1,d1,e1 we have shown the graphs of distribution functions for a case when a petroglyph drawing would be ideally symmetrical. The right side would be ideally a mirror relative to the left side. In this case, the graphs demonstrate noticeably state character. But if the graphs of distribution functions of the drawings along the OX axis (Figs. 5 a2,b2,c2,d2,e2) and OY axis (Figs. 5 a3, b3, c3, d3, e3) were plotted as they are in reality, then dynamics and rhythm of the drawings can be felt. This is just what unknown painters of that time wanted to convey to us.

Information-Retrieval System of Rock Pictures 115

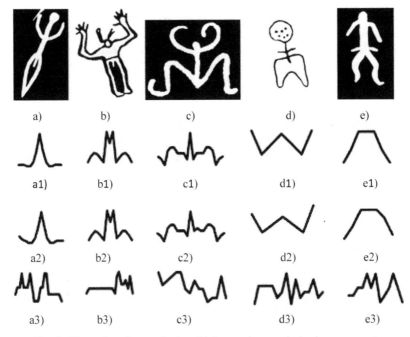

Fig. 5. Examples of petroglyphs with incomplete vertical mirror symmetry

The range of geometrical dimensions of the drawings showing people, domestic and wild animals, birds, fishes, work tools, vegetation elements, etc. is large enough: it is between some centimeters and some meters. On the basis of this, it can be concluded that even in antiquity people drawing on stones were able to create both miniatures and big-sized pictures.

The normalization of petroglyphs located as on the Azerbaijani territory, so in the other countries and continents with further determination and usage of informative features allows to bring rock pictures to a single scale and provides the process of recognition with subsequent identification of these rock pictures. To conduct the normalization of pictures petroglyphs are placed in the circle area. We will take as an assumption that a rock picture of the petroglyph can be presented as a drawing on a net of NxM dimensions.

The majority of rock pictures represent simple one-contour figures and that makes the task of recognition easier.

The indispensable condition is that the circle area must be maximally filled with a picture. The process of normalization being the necessary stage of the process of recognition and identification of rock pictures of different countries and continents is important enough when they are added to the database. Further analysis of results obtained from the process of recognition and identification by specialists engaged in different fields of science and technology will allow to add of new knowledge of the life and culture of ancient generations and their interrelationships. The gist of normalization is in the determination of unknown parameter transformations to which source pictures are subjected and their subsequent bringing to the benchmark form.

We will take a picture of a man in wide clothes as an example of the performance of the normalization process with the placement of the picture in the circle (Fig. 6a). The picture is made proportionally enough, and after being placed in the circle area it comes in contact with the circle in four points (Fig. 6b).

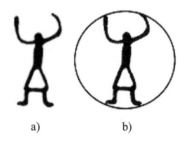

Fig. 6. Picture of a man in the circle area

The drawing fills maximally the circle area irrespective of its real size.

Figure 7 demonstrates rock pictures placed in the circle area. We obtain any additional opportunities for exploring the pictures which can be provided by the circumference. There is a possibility of dividing the picture into (n)-number of sectors, studying certain sectors individually, drawing the necessary radius, chords, parallels, and horizontal lines etc., and measuring, if needed, magnitudes of angles and distances among them (Fig. 8). in case of necessity all these characteristics are used as informative features in the process of recognition.

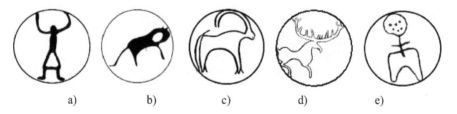

Fig. 7. Rock pictures in the circle area

Figure 8 shows an example of dividing the circle area into twelve sectors by six diagonal lines. If necessary, it is possible to conduct explorations and unifications of some sectors etc. This division makes it possible to investigate additional points of coordinates of coincidence with the circumference line.

To get a better quality of recognizing pictures inside the circle area one can turn them through a certain degree by means of graphic editors bringing the pictures into horizontal or vertical positions.

In the paper [4] informative features of the Gobustan reserve were analyzed. The list of the informative features being observed also comprises added informative features developed with the use of information technologies -such as, for example, the distribution

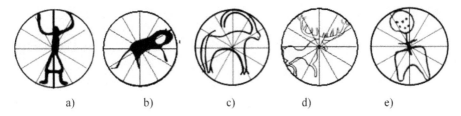

Fig. 8. Rock pictures in the circle area with diagonal divisions

density of petroglyph graphs. This parameter along with the rest of informative features is employed in performing the process of recognition and identification of rock pictures. In doing so it is obligatory to compare distribution density both along the OX and OY axes.

As an example, we demonstrate graphs of the distribution density of rock pictures of deer (Fig. 9) of the Gobustan reserve as well as rock pictures of mountain goats (Fig. 10) of various geographical regions of Europe and Asia.

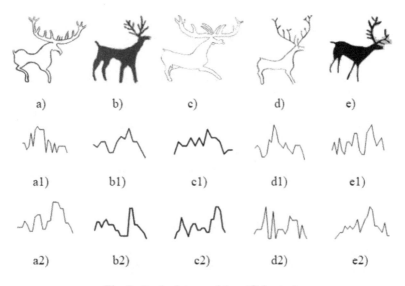

Fig. 9. Rock pictures of deer (Gobustan)

Figure 9a,b,c,d,e display rock pictures of deer. Figure 9a1,b1,c1,d1,e1 show graphs of the distribution density of the mentioned pictures along the OX-axis while Fig. 9a2,b2,c2,d2,e2 demonstrate graphs of the distribution density of these pictures along OY-axis. This designation is valid also for mountain goats (Fig. 10,11) of different world countries.

From Tables 1,2,3 it can be seen that for each pair of rock pictures correlation analysis was made, correlation coefficients were calculated, and their reliabilities were given. It should be noted that all parameters were calculated both along the OX and OY axes.

Table 1 [5] contains the result of the correlation analysis for pictures of deer. Table 2 shows the results of this analysis for goats of the Gobustan historical-cultural reserve of the Azerbaijan Republic and Table 3 presents also parameters related to pictures of goats of other reserves of the world.

Table 1.

№	Code of Figure 9	Name	Belonging to region	Sample n		Correlation coefficient		Reliability P	
				X	Y	X	Y	X	Y
1	a-b	Deer Deer	Gobustan Gobustan	24	24	-0,27	0,67	> 0,05	< 0,001
2	a-c	Deer Deer	Gobustan Gobustan	24	24	0,01	0,64	> 0,05	< 0,001
3	a-d	Deer Deer	Gobustan Gobustan	24	24	-0,54	0,22	< 0,01	> 0,05
4	a-e	Deer Deer	Gobustan Gobustan	24	24	0,16	0,54	> 0,05	< 0,01
5	b-c	Deer Deer	Gobustan Gobustan	24	24	0,44	0,36	< 0,05	> 0,05
6	b-d	Deer Deer	Gobustan Gobustan	24	24	-0,06	0,50	> 0,05	< 0,05
7	b-e	Deer Deer	Gobustan Gobustan	24	24	0,46	0,46	< 0,05	< 0,05
8	c-d	Deer Deer	Gobustan Gobustan	24	24	0,11	0,32	> 0,05	> 0,05
9	c-e	Deer Deer	Gobustan Gobustan	24	24	0,13	0,2	> 0,05	> 0,05
10	d-e	Deer Deer	Gobustan Gobustan	24	24	0,28	0,18	> 0,05	> 0,05

Having studied and analyzed the majority of typical rock and cave pictures we have revealed and determined informative features on the basis of which their classification takes place this list of the informative features:

- Epoch.
- Continent.
- Country.
- Name of territory where rock pictures are located.
- Name of cave, mountain.
- No. of picture.
- Display of picture.
- Name of picture.

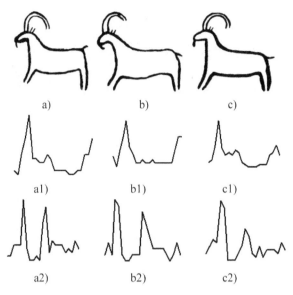

Fig. 10. Rock pictures of goats (Gobustan)

Table 2.

№	Code of Figure 9	Name	Belonging to region	Sample n		Correlation coefficient		Reliability P	
				X	Y	X	Y	X	Y
1	a-b	Goat Goat	Gobustan Gobustan	24	24	0.93	0.34	> 0,05	< 0,001
2	a-c	Goat Goat	Gobustan Gobustan	24	24	0,67	0.64	> 0,05	< 0,001
3	c-d	Goat Goat	Gobustan Gobustan	24	24	0.63	0.55	< 0,01	> 0,05

- Time of picture creation.
- Geometric dimensions of picture.
- Colour of picture.
- Additional strokes on the picture made later.
- Symmetry and asymmetry of picture.
- Graphs of distribution density of petroglyph pictures.

The above-mentioned list of informative indicators is utilized in the information retrieval system of petroglyphs.

Figure 12 demonstrates the structure of the developed information retrieval system of petroglyphs of world countries [6].

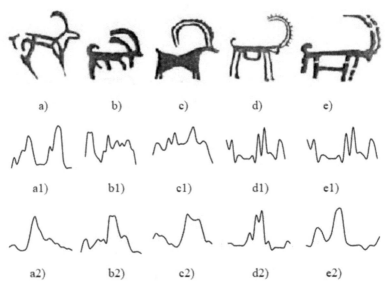

Fig. 11. Rock pictures of goats, a) Gobustan, b) Mongolia, c) Middle Asia, d) Gobustan, e) Tien-Shan

Table 3.

№	Code of Figure 9	Name	Belonging to region	Sample n		Correlation coefficient		Reliability P	
				X	Y	X	Y	X	Y
1	a-b	Goat Goat	Gobustan Mongolia	24	24	-0,03	0,53	> 0,05	< 0,01
2	a-c	Goat Goat	Gobustan Midd.Asia	24	24	0,05	0,09	> 0,05	> 0,05
3	a-d	Goat Goat	Gobustan Gobustan	24	24	-0,38	0,71	> 0,05	< 0,001
4	a-e	Goat Goat	Gobustan Tian-Shan	24	24	-0,00	0,58	> 0,05	< 0,001
5	b-c	Goat Goat	Mongolia Midd.Asia	24	24	0,25	0,47	> 0,05	< 0,05
6	b-d	Goat Goat	Mongolia Gobustan	24	24	0,11	0,48	> 0,05	< 0,05
7	b-e	Goat Goat	Mongolia Tian-Shan	24	24	0,07	0,23	> 0,05	> 0,05

(*continued*)

Table 3. (continued)

№	Code of Figure 9	Name	Belonging to region	Sample n		Correlation coefficient		Reliability P	
				X	Y	X	Y	X	Y
8	c-d	Goat Goat	Midd.Asia Gobustan	24	24	0,06	-0,06	> 0,05	> 0,05
9	c-e	Goat Goat	Midd.Asia Tian-Shan	24	24	0,12	-0,53	> 0,05	< 0,01
10	d-e	Goat Goat	Gobustan Tian-Shan	24	24	0,06	0,51	> 0,05	< 0,01

The system is designed on the hierarchical principal continents of the world countries are represented on the upper level of the system (I), countries of the continents are placed on the next level (II). Reserves of rock pictures located in world countries are included in the third level (III).

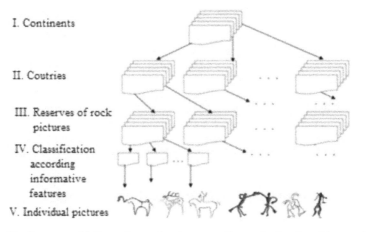

Fig. 12. Structure of information retrieval system of petroglyphs of world countries

Classification of petroglyphs according to informative features is made separately for each reserve because along with general features ones (IV) are observed for example, in some countries (Norway, France-Lascaux cave etc.) coloured rock pictures have been discovered. Themes of petroglyphs (V) vary depending on the geographical location of a reserve (on the seashore, ocean shore, in the heart of the continent, at the felt of mountains, on open plain territories). Differences in pictures of people, animals, objects of household, everyday life etc., are explained by the historical period to which a reserve belongs.

4 Conclusion

We have analyzed several hundreds of petroglyphs of Azerbaijan, Russia, Middle and Central Asia, and countries of Europe, Asia and Africa. Informative features of rock pictures of different countries have been studied. For example, nearby 90 percent of all petroglyphs under study can be related to the class of asymmetrical. One of the ways of normalization of rock pictures is presented. The way of placing pictures inside the circle area allows to bring pictures of various sizes (ranging from some centimetres to some metres) to a single scale. Informative features of rock and cave pictures of world countries of different continents have been determined and presented. The process of recognition and identification of petroglyphs takes place according to informative features.

At the next stage, rock pictures are included in the database of an information retrieval system.

On the basis of the informative features, an information retrieval system has been created. The created information retrieval system is of interest not only to specialists in computer and information technologies but also to specialists in diverse fields of science, technology, sphere of art etc.

References

1. Jafarzadeh, I.M.: Gobustan. Rock images. Baku, YNE-1999. 191p. (In Russian)
2. Sher, Y.A.: Petroglyphs of Middle and Central Asia. M.: 328 (1980) (In Russian)
3. Kovtun, I.V.: Petroglyphs of hanging stone and chronology of Tomsk scribe. Kuzbassvuzizdat, Kemerovo, 71 (1993) (In Russian)
4. Kazim-zada, A.K.: Determination of informative features of petroglyphs of Gobustan. In: 10th International Interdisciplinary Scientific and Practical School-Conference Modern Prolems of Science and Education Kharkov, pp. 258–259 (2010) (In Russian)
5. Kazim-zada, A.K., Kurbanova N.G.: Information retrieval system Rock paintings from various countries and continents. Int. Sci. J. Sci. World 2(114), 39–44 (2023) (In Russian)
6. Abdullaeva, G.G., Kazim-zada, A.K., Kurbanova, N.G.: Information system of petroglyphs of the countries of the world. In: 12th International Interdisciplinary Scientific and Practical School-Conference Modern Problems of Science and Education, Evpatoria, pp. 213–214 (2012) (In Russian)

Convergence of Sharpness-Aware Minimization with Momentum

Pham Duy Khanh[1], Hoang-Chau Luong[2], Boris S. Mordukhovich[3](✉), Dat Ba Tran[3], and Truc Vo[4]

[1] Department of Mathematics, HCM University of Education, Ho Chi Minh City, Vietnam
[2] Department of Computer Science, VNU-HCM University of Science,
Ho Chi Minh City, Vietnam
lhchau20@apcs.fitus.edu.vn
[3] Department of Mathematics, Wayne State University, Detroit, MI, USA
{aa1086,he9180}@wayne.edu
[4] Department of Economics, Wayne State University, Detroit, MI, USA
hi4482@wayne.edu

Abstract. This paper presents an analysis of Sharpness-Aware Minimization (SAM), a recently introduced efficient optimizer that has demonstrated remarkable improvements in the generalization of deep neural networks. A comprehensive analysis of the asymptotic convergence behaviors of the optimizer when integrated with momentum is proposed. We first show the convergence of the gradient sequence to zero and the topological properties of the set of accumulation points generated by the iterative sequence. Under the assumption of the isolation of stationary points, especially when the function is strongly convex, the convergence of the sequence of iterates is ensured. To validate the practical implications of our analysis, we conduct numerical experiments on classification tasks employing well-known deep learning models, including ResNet18 and ResNet34, with standard datasets CIFAR-10, CIFAR-100, MNIST, and Fashion-MNIST. The numerical results show that, in general, incorporating momentum improves both the training process and testing accuracy for SAM rather than just using standard SGD.

Keywords: Deep Learning · Gradient-based Optimizer · Momentum · Acceleration · Neural Networks · Sharpness-Aware Minimization

1 Introduction

This paper concentrates on optimization methods for solving the unconstrained optimization problem

$$\text{minimize} \quad f(x) \text{ subject to } x \in \mathbb{R}^n, \tag{1}$$

where $f : \mathbb{R}^n \to \mathbb{R}$ is a continuously differential (smooth) function with a Lipschitz continuous gradient. We study the convergence properties of the gradient-based optimizer called *Sharpness-aware minimization* together with its variants. Given an initial point $x^1 \in \mathbb{R}^n$, the original iterative procedure of SAM is designed as follows

$$x^{k+1} = x^k - t_k \nabla f\left(x^k + \rho_k \frac{\nabla f(x^k)}{\nabla f(x^k)}\right) \qquad (2)$$

for all $k \in \mathbb{N}$, where $t_k > 0$ and $\rho_k > 0$ are respectively the *stepsize* (in other words, the learning rate) and *perturbation radius*. The main motivation behind the construction algorithm is to solve a surrogate problem called *sharpness-aware optimization problem* of [intro problem] instead of solving it directly. The surrogate problem is defined as the following minimax problem

$$\min_{x \in \mathbb{R}^n} \max_{d \leq \rho_k} f(x+d).$$

To do so, the authors employed an approximation for the inner problem

$$d^k := \operatorname{argmax}_{d \leq \rho_k} \langle f(x) + d, \nabla f(x) \rangle = \rho_k \frac{\nabla f(x^k)}{\nabla f(x^k)},$$

and then consider the update $x^{k+1} = x^k - t_k \nabla f(x^k + d^k)$, which is exactly (2).

Due to its high performance in training deep neural networks, SAM is now widely used in various machine learning tasks within the vision and language domains [2, 3], federated learning [4], and meta-learning [5]. Several variants of SAM have been introduced, including Random Sharpness-Aware Minimization [6] and Variance Suppressed Sharpness-Aware Optimization [7]. The success of SAM is theoretically investigated in recent papers [8–10]. The fundamental convergence properties of SAM are also derived comprehensively in [11]. However, in practical implementations, SAM and its variants often deviate from the standard procedure [SAM intro] with two key features. Firstly, due to the prevalence of big data, the procedure [SAM intro] is typically considered in a stochastic form. This means that the gradients taken in both the inner and outer terms are only stochastic approximations of the full gradient. Consequently, investigations into SAM and its variants often involve stochastic settings, as seen in [9, 12]. Secondly, these methods are commonly incorporated with momentum to enhance performance. To the best of our knowledge, there is no existing work that considers the effect of momentum in the implementation of SAM or its variants. This gap in research motivates us to investigate the convergence behaviors of SAM and its variants when they are integrated with momentum. We adopt the momentum version proposed by Polyak [13, 14] that is in fact used in the original implementations of SAM [1] and its practical variants [6, 7]. The convergence properties under consideration include the convergence of the gradient sequence, the topological properties of the set of accumulation points, and the convergence of the iterative sequence. Subsequently, numerical experiments are conducted to validate the practical aspects of our analysis.

The organization for the rest of the paper is as follows. Section 2 considers preliminaries for the subsequent analysis. Our main results about the convergence properties of SAM and its variants when incorporating with momentum are presented in Sect. 3. Section 4 presents numerical experiments to confirm the practical aspects of our analysis. The conclusions are discussed in Sect. 5.

2 Preliminaries and Discussions

First, we recall some notions and notations frequently used in the paper. All our considerations are given in the space $\mathbb{R}^n$ with the Euclidean norm $\|\cdot\|$. As always, $\mathbb{N} := \{1, 2, \ldots\}$ signifies the collections of natural numbers. Let $f : \mathbb{R}^n \to \mathbb{R}$ be a $\mathcal{C}^1$-smooth (continuously differentiable) function. A point $\bar{x} \in \mathbb{R}^n$ is a *stationary point* of f if $\nabla f(\bar{x}) = 0$. The function is said to have a Lipschitz continuous gradient with the uniform constant $L > 0$ if

$$\langle \nabla f(x) - \nabla f(y) \rangle \leq L\|x - y\| \text{ for all } x, y \in \mathbb{R}^n. \tag{3}$$

The function f is said to be strongly convex with constant $\mu > 0$ if

$$\lambda f(x) + (1 - \lambda) f(y) \geq f(\lambda x + (1 - \lambda) y) + \tfrac{\lambda(1-\lambda)\mu}{2} \|x - y\|^2$$

for all $x, y \in \mathbb{R}^n$ and $\lambda \in [0, 1]$. This property has the following first-order characterization

$$f(y) \geq f(x) + \langle \nabla f(x), y - x \rangle + \tfrac{\mu}{2} \|x - y\|^2 \text{ for all } x, y \in \mathbb{R}^n. \tag{4}$$

The convergence analysis of the methods presented in the subsequent sections takes advantage of the following important results. The first result presents a crucial property of a function that has a Lipschitz continuous gradient [15], Lemma A.11.

Lemma 1. *For any $f : \mathbb{R}^n \to \mathbb{R}$ and any $x, y \in \mathbb{R}^n$, if f is differentiable on the line segment $[x, y]$ with its derivative being Lipschitz continuous on this segment with a constant $L > 0$, then*

$$|f(y) - f(x) - \langle \nabla f(x), y - x \rangle| \leq \tfrac{L}{2} \|y - x\|^2. \tag{5}$$

Next we recall the classical results from [16, 17] that describe important properties of the set of accumulation points generated by a sequence satisfying the Ostrowski condition; see (6) below.

Lemma 2. *Let $\{x^k\}$ be a sequence satisfying the Ostrowski condition*

$$\lim_{k \to \infty} \|x^{k+1} - x^k\| = 0. \tag{6}$$

Then the following assertions hold:

1. *If $\{x^k\}$ is bounded, then the set of accumulation points of $\{x^k\}$ is nonempty, compact, and connected in $\mathbb{R}^n$.*
2. *If $\{x^k\}$ has an isolated accumulation point, then this sequence converges to it.*

The following lemma from [18] is useful in deriving the convergence properties in subsequent sections.

Lemma 3. *Let $\{u_k\}, \{v_k\}, \{w_k\}$ be three sequences such that $\{w_k\}$ is nonnegative for all $k \in \mathbb{N}$. Assume that*

$$v_{k+1} \leq v_k - u_k + w_k \text{ for all } k \in \mathbb{N},$$

and that the series $\sum_{k=1}^{\infty} w_k < \infty$. Then either $v_k \to -\infty$ or else v_k converges to a finite value and $\sum_{k=1}^{\infty} u_k < \infty$.

3 Convergence Analysis of SAM with Momentum

Let us now formalize the algorithmic construction of SAM with momentum as follows.

Algorithm 1 (Sharpness-Aware Minimization with momentum).

Step 1 (initialization). Choose some initial points $x^0 = x^1 \in \mathbb{R}^n$, momentum constant $\beta \geq 0$, stepsize $\tau > 0$ and perturbation radii $\{\rho_k\} \subset [0, \infty)$. For $k = 1, 2, \ldots$, do the following.
Step 2 (momentum). Set $y^k = x^k + \beta(x^k - x^{k-1})$.
Step 3 (iteration update). Set $x^{k+1} := y^k - \tau \nabla f\left(x^k + \rho_k \frac{\nabla f(x^k)}{\|\nabla f(x^k)\|}\right)$.

Remark 4. When $\rho_k = 0$ for all $k \in \mathbb{N}$, Algorithm 1 reduces to the standard gradient descent method with momentum [13]. When the objective function f is strongly convex with constant $\mu > 0$ and the gradient is Lipschitz continuous with constant $L > 0$, the stepsize and the momentum constant are usually chosen as

$$\tau = \frac{4}{\left(\sqrt{L}+\sqrt{\mu}\right)^2}, \quad \beta = \left(\frac{\sqrt{L}-\sqrt{\mu}}{\sqrt{L}+\sqrt{\mu}}\right)^2. \tag{7}$$

However, in general, this selection only ensures local convergence properties of the method [13], Theorem 9(3). An example showing the divergence of the method using (7) when minimizing a smooth strongly convex function is presented in [19, Sec. 4.6]. Therefore, to validate the global convergence properties of Algorithm 1, we impose different conditions for stepsize, momentum constant, and perturbation radii in (8) instead of using the selection (7).

The convergence properties of SAM are given in the main result below.

Theorem 5. Let $f : \mathbb{R}^n \to \mathbb{R}$ be a smooth function whose gradient is Lipschitz continuous with constant $L > 0$ and let $\{x^k\}$ be generated by Algorithm 1 under the following conditions

$$\tau < \tfrac{1}{2L}, \quad \beta \leq \sqrt{\tfrac{1-2L\tau}{2(2L\tau+1)}}, \quad \sum_{k=1}^{\infty} \rho_k^2 < \infty. \tag{8}$$

Assume that $\nabla f(x^k) \neq 0$ for all $k \in \mathbb{N}$ and $\inf_{k \in \mathbb{N}} f(x^k) > -\infty$. Then the following convergence properties holds:

1. The gradient sequence $\{\nabla f(x^k)\}$ converges to the origin, $\{f(x^k)\}$ is convergent. As a consequence, if $\bar{x}$ is an accumulation point of $\{x^k\}$ then $\nabla f(\bar{x}) = 0$ and $f(x^k) \to f(\bar{x})$.
2. If $\{x^k\}$ is bounded, then the set of accumulation points of $\{x^k\}$ is compact, connected.
3. If $\{x^k\}$ has an isolated accumulation point $\bar{x}$ then $x^k \to \bar{x}$.

For each $k \in \mathbb{N}$, define $g^k := \nabla f\left(x^k + \rho_k \frac{\nabla f(x^k)}{\|\nabla f(x^k)\|}\right)$. It follows from the Lipschitz continuity of ∇f that

$$\left\|g^k - \nabla f(x^k)\right\| = \left\|\nabla f\left(x^k + \rho_k \frac{\nabla f(x^k)}{\nabla f(x^k)}\right) - \nabla f(x^k)\right\|$$
$$\leq L\rho_k \text{ for all } k \in \mathbb{N}.$$

By the Lipschitz continuity of ∇f with constant $L > 0$, we deduce from Lemma 1 that

$$-\tfrac{L}{2}\|y-x\|^2 \le f(y) - f(x) - \langle \nabla f(x), y-x\rangle \le \tfrac{L}{2}\|y-x\|^2 \text{ for all } x, y \in \mathbb{R}^n.$$

Take some $k \in \mathbb{N}$. Substituting $y = x^k$, $x = y^k$ in the left-hand side inequality and $y = x^{k+1}$, $x = y^k$ in the right-hand side inequality, we deduce that

$$f(x^k) - f(y^k) - \langle \nabla f(y^k), x^k - y^k\rangle \ge -\tfrac{L}{2}\|x^k - y^k\|^2,$$
$$f(x^{k+1}) - f(y^k) - \langle \nabla f(y^k), x^{k+1} - y^k\rangle \le \tfrac{L}{2}\|x^{k+1} - y^k\|^2.$$

Subtracting the first inequality from the second inequality gives us

$$f(x^{k+1}) - f(x^k) - \langle \nabla f(y^k), x^{k+1} - x^k\rangle \le \tfrac{L}{2}\|x^{k+1} - y^k\|^2 + \tfrac{L}{2}\|x^k - y^k\|^2,$$

which implies that

$$\begin{aligned}
f(x^{k+1}) - f(x^k) &\le \tfrac{L}{2}\|x^{k+1} - y^k\|^2 + \tfrac{L}{2}\|x^k - y^k\|^2 + \langle \nabla f(y^k), x^{k+1} - x^k\rangle \\
&= \tfrac{L}{2}\|x^{k+1} - y^k\|^2 + \tfrac{L\beta^2}{2}\|x^k - x^{k-1}\|^2 + \langle \nabla f(y^k) - \nabla f(x^k), x^{k+1} - x^k\rangle \\
&\quad + \langle \nabla f(x^k) - g^k, x^{k+1} - x^k\rangle + \langle g^k, x^{k+1} - x^k\rangle.
\end{aligned} \quad (9)$$

It follows from Cauchy-Schwarz, AM-GM inequalities, the Lipschitz continuity of ∇f and $y^k = x^k + \beta(x^k - x^{k-1})$ from Step 2 of Algorithm 1 that

$$\begin{aligned}
\langle \nabla f(y^k) - \nabla f(x^k), x^{k+1} - x^k\rangle &\le \|\nabla f(y^k) - \nabla f(x^k)\|\|x^{k+1} - x^k\| \\
&\le \|Ly^k - x^k\|\|x^{k+1} - x^k\| \\
&\le \tfrac{L}{2}\left(\|y^k - x^k\|^2 + \|x^{k+1} - x^k\|^2\right) \\
&= \tfrac{L}{2}\left(\beta^2\|x^k - x^{k-1}\|^2 + \|x^{k+1} - x^k\|^2\right).
\end{aligned} \quad (10)$$

It follows from Cauchy-Schwarz, AM-GM inequalities together with $x^{k+1} := y^k - \tau g^k$ and $\|g^k - \nabla f(x^k)\| \le L\rho_k$ from Step 2 of Algorithm 1 that

$$\begin{aligned}
\langle \nabla f(x^k) - g^k, x^{k+1} - x^k\rangle &\le \|\nabla f(x^k) - g^k\|\|x^{k+1} - x^k\| \\
&\le L\rho_k \|x^{k+1} - x^k\| \\
&\le \tau L^2 \rho_k^2 + \tfrac{1}{4\tau}\|x^{k+1} - x^k\|^2.
\end{aligned} \quad (11)$$

Using Steps 1, 2 of Algorithm 1, we also deduce that

$$\begin{aligned}
\langle g^k, x^{k+1} - x^k\rangle &= \tfrac{1}{\tau}\langle y^k - x^{k+1}, x^{k+1} - x^k\rangle \\
&= \tfrac{1}{2\tau}\left(\|y^k - x^k\|^2 - \|x^{k+1} - y^k\|^2 - \|x^k - x^{k+1}\|^2\right) \\
&= \tfrac{1}{2\tau}\left(\beta^2\|x^k - x^{k-1}\|^2 - \|x^{k+1} - y^k\|^2 - \|x^k - x^{k+1}\|^2\right).
\end{aligned} \quad (12)$$

Combining (9)–(12) while taking into account $2L\tau < 1$, we deduce that

$$\begin{aligned}
f(x^{k+1}) - f(x^k) &\le \tfrac{L\tau-1}{2\tau}\|x^{k+1} - y^k\|^2 + \tfrac{2L\tau-1}{4\tau}\|x^{k+1} - x^k\|^2 \\
&\quad + \tfrac{2L\tau\beta^2 + \beta^2}{2\tau}\|x^k - x^{k-1}\|^2 + \tau L^2 \rho_k^2 \\
&\le -\tfrac{1-2L\tau}{4\tau}\|x^{k+1} - x^k\|^2 + \tfrac{(2L\tau+1)\beta^2}{2\tau}\|x^k - x^{k-1}\|^2 + \tau L^2 \rho_k^2.
\end{aligned} \quad (13)$$

Define further $\alpha := \frac{1-2L\tau+2(2L\tau+1)\beta^2}{8\tau}$ and

$$v_k := f(x^k) + \alpha \|x^k - x^{k-1}\|^2 \text{ for all } k \in \mathbb{N}. \tag{14}$$

Combining this with estimate (13) and $2(2L\tau+1)\beta^2 + 2L\tau - 1 < 0$ from the selection of β gives us

$$\begin{aligned} v_{k+1} - v_k &= f(x^{k+1}) + \alpha \|x^{k+1} - x^k\|^2 - f(x^k) - \alpha \|x^k - x^{k-1}\|^2 \\ &\leq \left(\alpha - \tfrac{1-2L\tau}{4\tau}\right)\|x^{k+1} - x^k\|^2 + \left(\tfrac{(2L\tau+1)\beta^2}{2\tau} - \alpha\right)\|x^k - x^{k-1}\|^2 + \tau L^2 \rho_k^2 \\ &\leq \tfrac{2(2L\tau+1)\beta^2+2L\tau-1}{8\tau}\left(\|x^{k+1} - x^k\|^2 + \|x^k - x^{k-1}\|^2\right) + \tau L^2 \rho_k^2 \\ &\leq -\tfrac{1-2L\tau-2(2L\tau+1)\beta^2}{8\tau}\|x^{k+1} - x^k\|^2 + \tau L^2 \rho_k^2 \end{aligned} \tag{15}$$

Using $v_k \geq f(x^k)$ for all $k \in \mathbb{N}$ and $\inf_{k \in \mathbb{N}} f(x^k) > -\infty$, we deduce that $\inf_{k \in \mathbb{N}} v_k > -\infty$. Let

$$u_k := \tfrac{1-2L\tau-2(2L\tau+1)\beta^2}{8\tau}\|x^{k+1} - x^k\|^2 \text{ and } w_k := \tau L^2 \rho_k^2 \text{ for all } k \in \mathbb{N}.$$

We observe that $u_k \geq 0$ for all $k \in \mathbb{N}$ and $\sum_{k=1}^{\infty} w_k < \infty$. It follows from Lemma 3 that $\{v_k\}$ is convergent and $\sum_{k=1}^{\infty} u_k < \infty$. The latter yields $x^{k+1} - x^k \to 0$ as $k \to \infty$. Combining this with the update $y^k = x^k + \beta(x^k - x^{k-1})$ in Step 1 of Algorithm 1, we deduce that $y^k - x^k \to 0$, which in turn implies that $x^{k+1} - y^k \to 0$ as $k \to \infty$. Combining this with $x^{k+1} = y^k - \tau g^k$ and $\nabla f(x^k) - g^k \leq L\rho_k$ from Step 2 of Algorithm 1 together with $\rho_k \downarrow 0$, we arrive at $\nabla f(x^k) \to 0$. As $\{v_k\}$ and $\{x^k - x^{k-1}\}$ are convergent, we deduce that $\{f(x^k)\}$ is convergent as well. Therefore, (i) is verified.

Employing Lemma 2 with taking into account $x^{k+1} - x^k \to 0$ as $k \to \infty$ gives us (ii) and (iii). We now use the theorem above to derive the convergence of the sequence of iterates generated by Algorithm 1 for strongly convex functions.

Corollary 6. Let $f : \mathbb{R}^n \to \mathbb{R}$ be a smooth strongly convex function and let ∇f be Lipschitz continuous with constant $L > 0$. Let $\{x^k\}$ be generated by Algorithm 1 under the conditions in (8). Then $\{x^k\}$ converges to the unique global minimizer of f.

It follows from Theorem 5(i) that $f(x^k) \to f(x^*)$ with x^* being the unique global minimizer of f. By the first-order characterization of the strong convexity (4) and by $\nabla f(x^*) = 0$, we find some $\mu > 0$ such that

$$\tfrac{\mu}{2}\|x^k - x^*\|^2 \leq f(x^k) - f(x^*) \text{ for all } k \in \mathbb{N}.$$

which verifies the convergence of $\{x^k\}$ to x^*.

4 Numerical Experiments

To validate the practical aspect of our theory, this section compares the performance of SAM with and without the incorporating of momentum. All the experiments are conducted on a computer with NVIDIA RTX 3090 GPU. The algorithms are tested on two

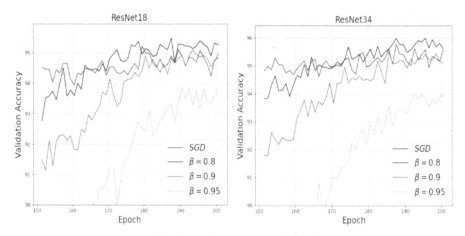

Fig. 1. Validation accuracy on CIFAR-10

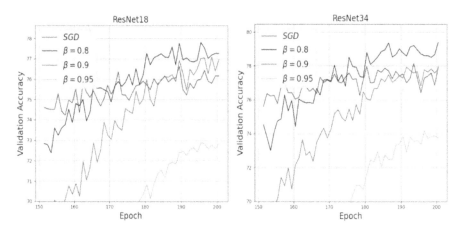

Fig. 2. Validation accuracy on CIFAR-100

widely used image datasets: MNIST, Fashion-MNIST, CIFAR-10 [20] and CIFAR-100 [20]. We train well-known deep neural networks including ResNet18 [21], ResNet34 [21] on this dataset by using 10% of the training set as a validation set. Basic transformations, including random crop, random horizontal flip, normalization, and cutout [22], are employed for data augmentation. All the models are trained by using SAM with SGD Momentum as the base optimizer for 200 epochs and a batch size of 128. This base optimizer is also used in the original paper [1] and in the recent works on SAM [7, 8]. Following the approach by [1], we set the initial stepsize to 0.1, the ℓ_2-regularization parameter to 0.001, and the perturbation radius ρ to 0.05. The algorithm with the highest accuracy, corresponding to the best performance, is highlighted in bold. The results in the test accuracy and validation accuracy in several tests are presented below. It is evident that incorporating momentum typically results in better performance in terms

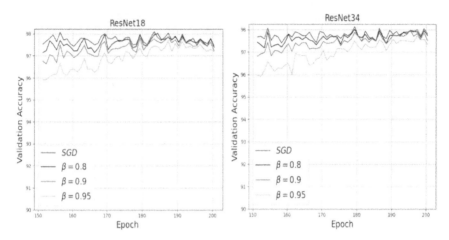

Fig. 3. Validation accuracy on MNIST

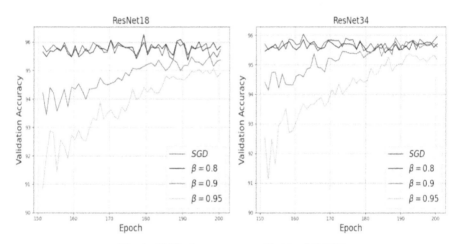

Fig. 4. Validation accuracy on Fashion-MNIST

of test accuracy. The method with a momentum constant of 0.8 stands out as the best, excelling in both test accuracy and validation accuracy (Figs. 1, 2, 3, and 4, Table 1).

Table 1. Test accuracy

Data	Architecture	SGD	$\beta = 0.8$	$\beta = 0.9$	$\beta = 0.95$
CIFAR-10	ResNet18	95.79	**96.39**	96.28	95.00
	ResNet34	96.22	**96.72**	96.53	95.25
CIFAR-100	ResNet18	78.32	**79.46**	79.24	76.31
	ResNet34	79.35	**80.95**	79.60	77.00
MNIST	ResNet18	99.31	**99.35**	99.32	99.27
	ResNet34	99.33	**99.34**	99.33	99.30
Fashion-MNIST	ResNet18	95.74	**95.95**	95.28	95.19
	ResNet34	**96.02**	95.99	96.00	95.39

5 Conclusions and Discussions

In this manuscript, we conduct a convergence analysis of SAM when it is combined with momentum. Our findings encompass the gradient sequence converging to zero, the topological characteristics of the set of accumulation points, and the convergence of the iterative sequence. Our analysis is carried out in deterministic settings, relying on standard assumptions that extend to various applications in nonconvex optimization. The numerical experiments affirm that our analysis aligns with the effective implementations of SAM commonly employed in practical scenarios.

References

1. Foret, P., Kleiner, A., Mobahi, H., Neyshabur, B.: Sharpness-aware minimization for efficiently improving generalization. In: Proceedings of International Conference on Learning Representations (2021)
2. Chen, X., Hsieh, C.-J., Gong, B.: When vision transformers outperform resnets without pre-training or strong data augmentations. In: Proceedings of International Conference on Learning Representations (2022)
3. Zhang, Z., Luo, R., Su, Q., Sun, X.: GA-SAM: gradient-strength based adaptive sharpness-aware minimization for improved generalization. In: Proceedings of Conference on Empirical Methods in Natural Language Processing (2022)
4. Qu, Z., Li, X., Duan, R., Liu, Y., Tang, B., Lu, Z.: Generalized federated learning via sharpness aware minimization. In: Proceedings of International Conference on Machine Learning (2022)
5. Abbas, M., Xiao, Q., Chen, L., Chen, P.-Y., Chen, T.: Sharp-MAML: Sharpness-aware model-agnostic meta learning. In: Proceedings of International Conference on Machine Learning (2022)
6. Liu, Y., Mai, S., Cheng, M., Chen, X., Hsieh, C.-J., You, Y.: Random sharpness-aware minimization. In: Advances in Neural Information Processing System (2022)
7. Li, B., Giannakis, G.B.: Enhancing sharpness-aware optimization through variance suppression. In: Advances in Neural Information Processing System (2023)
8. Ahn, K., Jadbabaie, A., Sra, S.: How to escape sharp minima (2023). arXiv:2305.15659

9. Andriushchenko, M., Flammarion, N.: Towards understanding sharpness-aware minimization. In: Proceedings of International Conference on Machine Learning (2022)
10. Barlett, P.L., Long, P.M., Bousquet, O.: The dynamics of sharpness-aware minimization: bouncing across ravines and drifting towards wide minima. J. Mach. Learn. Res. **24**, 1–36 (2023)
11. Khanh, P.D., Luong, H.-C., Mordukhovich, B.S., Tran, D.B.: Fundamental convergence analysis of sharpness-aware minimization (2024). arXiv:2401.08060
12. Si, D., Yun, C.: Practical sharpness-aware minimization cannot converge all the way to optima. In: Advances in Neural Information Processing System (2023)
13. Jiang, W., Yang, H., Zhang, Y., Kwok, J.: An adaptive policy to employ sharpness-aware minimization. Sov. Math. Dokl. (1983)
14. Polyak, B.: Introduction to Optimization. Optimization Software, New York (1987)
15. Izmailov, A.F., Solodov, M.V.: Newton-Type Methods for Optimization and Variational Problems. Springer (2014)
16. Facchinei, F., Pang, J.-S.: Finite-Dimensional Variational Inequalities and Complementarity Problems, vol. 2. Springer, New York (2003)
17. Ostrowski, A.: Solution of Equations and Systems of Equations, 2nd edn. Academic Press, New York (1996)
18. Bertsekas, D., Tsitsiklis, J.N.: Gradient convergence in gradient methods with errors. SIAM J. Optim. **10**, 627–642 (2000)
19. Lessard, L., Recht, B., Packard, A.: Analysis and design of optimization algorithms via integral quadratic constraints. SIAM J. Optim. **26**(1), 57–95 (2024)
20. Krizhevsky, A., Hinton, G., et al.: Learning Multiple Layers of Features From Tiny Images. Technical Report (2009)
21. He, K., Zhang, X., Ren, S., Sun, J.: Deep residual learning for image recognition. In: Proceedings of the IEEE Conference on Computer Vision and Pattern Recognition, pp. 770–778 (2016)
22. DeVries, T., Taylor, G.W.: Improved regularization of convolutional neural networks with cutout (2017). https://arxiv.org/abs/1708.04552

Vulnerability Assessment and Penetration Testing of University Network

Dursun Khurshudov[✉] [iD], Akif Imanov, Jamalladdin Nuraliyev, Malahat Nagiyeva, and Samira Aliyeva

Department of Information Technology and Systems, Azerbaijan University of Architecture and Construction, Baku, Azerbaijan
`dursuncosqun@gmail.com`

Abstract. Recent research in network management and the enhancement of network security efficiency has seen a widespread use of traffic classification and clustering The increasing use of the Internet and the development of protocols and applications have also made research in the field of traffic classification highly relevant.

Vulnerability assessments and penetration testing are determine the security posture of modern information networks. By assessing university network vulnerabilities and attempting to exploit found vulnerabilities through penetration testing security specialists are able to evaluate the effectiveness of their network defenses by identifying defense weaknesses, affirming the defense mechanisms in place. The purpose of this article is to determine the main tasks of penetration testing monitoring, investigate methods for vulnerability network traffic, and explore intellectual analysis methods used for penetration testing monitoring. Also, analyzing the of an penetration testing monitoring system capable of integrating components and methods to transmit information within a specified timeframe, detect interventions, control the development of events, identify, categorize, and effectively use modern security tools is envisaged.

Keywords: Vulnerability Clustering Methods · Machine learning · Network security · Algorithm · Penetration testing monitoring

As a result of this research, the creation of an penetration testing monitoring system capable of integrating components and methods to transmit information within a specified timeframe, detect interventions, control the development of events, identify, categorize, and effectively use modern security tools is envisaged.

1 Introduction

Assessing the security of systems is a complex problem. Most of the recent efforts in security assessment involve the detection of vulnerabilities in the security system. Detection of unknown vulnerabilities in the security system still remains a subjective procedure. The procedure knows how to improve by taking into account the characteristics of vulnerabilities in the known protection system. Therefore, the obtained information is included in the appropriate classification, which is then used as a systematic structure for the investigation of new systems in connection with the unknown weaknesses in the protection system. There have been several attempts to develop such classifications.

Vulnerability in network security is defined as flaw or weakness in the network or system from where any attacker can possess into our network or system to exploit. The network or system containing these vulnerabilities is called vulnerable network or vulnerable system.

In order to exploit a vulnerability, an attacker must have at least one tool or technique that allows them to connect to any weak point in the system. In this context, vulnerability detection is also known as the initial stage of an attack. Vulnerability management is the regular practice of identifying, organizing, correcting, and mitigating vulnerabilities.

Generally, such control is based on software vulnerabilities in computing systems.

A security vulnerability can be classified as a vulnerability. Risk is related to the possibility of a significant failure. Therefore, there are also vulnerabilities that do not have any threat [1]. Vulnerabilities resulting from business activities or fully executed attacks are classified as exploitable vulnerabilities. When a security hole in the software is detected, access is revoked, other methods of secure connection are offered, or the attack is stopped.

Non-software vulnerabilities also exist. The human factor can be cited as an example in systems without security gaps in technical means and software.

More the network system is vulnerable; more is the threat to exploit. Because of these vulnerabilities numbers of systems are exploit each year. Vulnerable network or system may be compromised by different attacks like, DDoS, DNS Spoofing, DHCP Snooping, ARP Poisoning, Man-in-the-Middle, Smurf attacks, Buffer overflow, SQL injection attack and other many cyber-attacks along with a number of malicious attacks including Viruses, Trojan horse, Worms, Malwares and root kits etc.

These vulnerabilities are due to week passwords, software bugs, nonpatching of software's and operating systems. Script code injection spaces etc.

Therefore, there is a need to enhance the penetration methods of existing computer networks of local university network (CN). For example, it has been shown in [2] that methods for identifying the requirements and threats of monitoring must be determined for the creation of secure and reliable intelligent systems.

The above considerations highlight the relevance of seamless and reliable monitoring for the effective operation of university computer networks and, more broadly, for the security of CNs. One of the current issues is the policy-based management of security – control must be exerted over the implementation of the security policy declared within the organization. [3] advocates for a proactive approach to the operation of security policies and mechanisms. Such Network Security Monitoring (NSM) contributes to the integrity, availability, confidentiality, and prevention of unauthorized access to the provided data.

For the realization of the above, NSM should combine the following functions: seamless monitoring, operability, availability, reliability, cost-effectiveness, forecasting, managed security, automatic execution, and intelligence.

2 The Main Functions of Vulnerability Assessment and Penetration Testing Monitoring (NSM)

NSM is a complex system comprising software and hardware components that gather information about the network, analyze it, and direct the results to relevant individuals or systems, meeting modern requirements.

The primary tasks of NSM can be explained as follows:

Monitoring and Recording Internet Traffic: This process begins with observing traffic, recording current sessions, measuring the volume of used traffic, controlling Internet traffic usage, and logging packets based on timestamps.

Security Control for Systems Operating in Real-Time: Security control for systems operating in real-time on the Internet varies based on the potential threats they may face. It is essential to classify these systems based on the level of risk they pose and regulate the characteristics of the control level accordingly.

Security, Confidentiality, and Protection of Internet Users: Certain types of malicious activities pose a threat to information security, including network attacks against services with vulnerabilities, attacks to gain elevated privileges, unauthorized access to confidential information, and the installation of malicious programs. These activities are highly dangerous for both the network and its users in terms of security and confidentiality.

In summary, NSM performs various functions crucial for enhancing the efficiency and security of networks, encompassing the monitoring and recording of Internet traffic, real-time security control for Internet-connected systems, and ensuring the security, confidentiality, and protection of Internet users.

Discovery of Vulnerabilities in Systems: Vulnerabilities in information security in systems may exist in the initial code of software, the database, system configurations, or management. These gaps can impact the state of information security in the network.

Seamless Operation: The primary goal is to eliminate factors that may degrade the quality of Internet traffic and ensure its use only for legitimate purposes.

Risk Assessment: The core of managing network security lies in the process of assessing and managing risks. The effectiveness of these processes is determined by the precision of monitoring, the accuracy of the conducted analysis, the assessment of risk factors, and monitoring the execution of decision-making mechanisms within any organization.

3 Penetration Testing

A penetration test, also known as a pen test, is a simulated cyber attack against your computer system to check for exploitable vulnerabilities. In the context of web application security, penetration testing is commonly used to augment a web application firewall (WAF).

Pen testing can involve the attempted breaching of any number of application systems, (e.g., application protocol interfaces (APIs), frontend/backend servers) to uncover vulnerabilities, such as unsanitized inputs that are susceptible to code injection attacks.

Insights provided by the penetration test can be used to fine-tune your WAF security policies and patch detected vulnerabilities.

Penetration testing stages

The pen testing process can be broken down into five stages (Fig. 1).

Fig. 1. Penetration testing stages

1. **Planning and Reconnaissance**

    The first stage involves:

- Defining the scope and goals of a test, including the systems to be addressed and the testing methods to be used.
- Gathering intelligence (e.g., network and domain names, mail server) to better understand how a target works and its potential vulnerabilities.

2. **Scanning**

    The next step is to understand how the target application will respond to various intrusion attempts. This is typically done using:

- Static analysis – Inspecting an application's code to estimate the way it behaves while running. These tools can scan the entirety of the code in a single pass.
- Dynamic analysis – Inspecting an application's code in a running state. This is a more practical way of scanning, as it provides a real-time view into an application's performance.

3. **Gaining Access**

    This stage uses web application attacks, such as cross-site scripting, SQL injection and backdoors, to uncover a target's vulnerabilities. Testers then try and exploit these vulnerabilities, typically by escalating privileges, stealing data, intercepting traffic, etc., to understand the damage they can cause.

4. **Maintaining Access**

The goal of this stage is to see if the vulnerability can be used to achieve a persistent presence in the exploited system— long enough for a bad actor to gain in-depth access. The idea is to imitate advanced persistent threats, which often remain in a system for months in order to steal an organization's most sensitive data.

5. **Analysis**

The results of the penetration test are then compiled into a report detailing:

- Specific vulnerabilities that were exploited
- Sensitive data that was accessed
- The amount of time the pen tester was able to remain in the system undetected

This information is analyzed by security personnel to help configure an enterprise's WAF settings and other application security solutions to patch vulnerabilities and protect against future attacks [4].

3.1 Penetration Testing Methods

In many cases, large and medium-sized computer networks (CNs) have complex infrastructures and heterogeneous structures, necessitating the need to work with multiple platforms. Security tools' analysis indicates that there is no fully integrated solution that can provide complete or seamless integration to ensure information security in heterogeneous networks. Figure 2 below shows the challenges of network security monitoring systems.

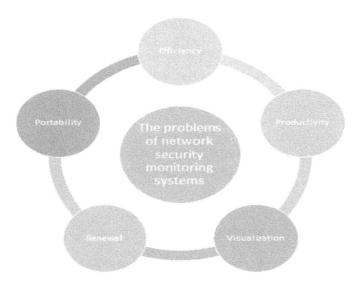

Fig. 2. Problems of network security monitoring systems.

1. Efficiency: Often, efforts are made to detect all known attacks in system intrusion detection, leading to a series of false positives. For example, these systems often create numerous rules for themselves, which can be resource-intensive. Moreover, the multitude of rules may only reveal a direct correlation between events.
2. Productivity: Assessing the productivity of Network Security Monitoring in real-world conditions is challenging. Additionally, there is no set of universal rules that allow for the assessment of NSM, meaning there is no standardized set of guidelines that can be applied to evaluate the system's productivity under specific conditions.
3. Visualization: Analyzing information is a complex task considering its diversity and the volume of data. It requires the use of appropriate visualization methods to make sense of the information effectively.
4. Renewal: Replacing existing systems with new NSM technologies is challenging. New subsystems must interact seamlessly with all systems, and achieving universal capabilities for mutual interaction is sometimes not possible. The installation of NSM often requires significant additional investments in security advancements that vary significantly. For instance, expertise is required to update numerous rules in incident detection systems. This idea also applies to statistical measurements of systems that identify vulnerabilities.
5. Portability: Many NSMs are created for specific devices, making it challenging to apply them to other systems requiring the realization of similar security policies. Network devices, such as switches, routers, multiplexors, high-capacity servers, network entry devices, modems, and network adapters, are used for establishing network infrastructure and traffic flow. The resources of many network devices are utilized for implementing monitoring functions, negatively impacting the reliability and data transmission capability of these devices. All these problems reduce the potential benefits of the obtained information. Typically, high-performance computing systems or a cluster of computers are used for data analysis. Such systems can be employed in the emulation process for analyzing security threats, examining data flow, detecting network anomalies, conducting Quality of Service (QoS) analysis, or ensuring total monitoring (complete interception of traffic and writing collected data to disk). Such a system must filter large volumes of data and write to the system memory. The main challenges in system analysis arise from the need to analyze the entire information flow at the lower levels of the OSI model. To address this problem, the computing power of hundreds or even thousands of computers may be required. The dynamic nature of internet traffic, the latest cyber threats, and software advancements underscore the necessity of employing efficient intellectual monitoring methods

3.2 Penetration Methods of Network Security

Many current software products rely on the use of signatures to detect information security threats.

Signature analysis is one of the methods applied to identify incidents. This method is based on matching the sample with program codes. The incoming information packet is carefully analyzed, and it is compared with the signature database, which is a characteristic line indicating the characteristics of malicious traffic. This signature may include keywords or commands related to the attack. If a match is found, an alert signal is triggered. The advantages of signature methods include effective identification of attacks, absence of numerous false positives, and reliable diagnostics of specific tools or attack technologies. This allows administrators, regardless of their expertise level, to initiate incident response procedures in the field of security and make corrections to ensure security.

The downside of the signature method is the necessity to update the information base for obtaining signatures of new attacks.

Like other complex systems, networks require seamless monitoring for various reasons. The monitoring system should have the ability to predict the state of the network, take preventive measures to manage disruptions and losses in the information flow. Therefore, real-time monitoring is crucial for ensuring operational security. It is noted in that there is a strong correlation between the human immune system and the computer network security system. Just as the immune system protects humans from pathogenic viruses, the information security system also protects computer networks from threats.

Collecting necessary data about the activity and security of computer networks, and employing intelligent analysis technologies during network security monitoring to make informed decisions for ensuring network security, have become crucial.

Ensuring reliable service in modern heterogeneous networks remains an essential challenge. Efficient utilization of network and computing resources is crucial for the reliable development and scalability of network services. Achieving this with shared local solutions using existing resources is possible. However, such solutions need to be integrated for general-purpose use. Generally, one of the fundamental blocks for systems is having the ability to monitor network parameters to obtain necessary and relevant information through measurements. The heterogeneous structure of modern networks, as mentioned earlier, complicates the local implementation of monitoring.

Many security tools, such as network monitors, intrusion detection systems (IDS), and comprehensive antivirus solutions, make effective management challenging due to the generation of numerous security events. The application of correlation mechanisms in the architecture of Security Information and Event Management (SIEM) systems, as well as the use of continuous intelligent analysis methods for the discovery process, is explained in. Correlating security events based on intelligent analysis of data creates automated rules, analyzes events, and identifies new threats.

While Security Information and Event Management (SIEM) systems have the authority to enforce precise security policies and intervene in real-time, they can also perform high-level audits. Through audits, it becomes easy to identify deviations from the security policy and take timely corrective actions. For instance, restricting communication channels (protocols and ports) limits the opportunities for malicious actors to execute the attack process, and SIEM controls restricted communication channels. This indicates that for the SIEM system to identify malicious traffic, the nature of normal traffic should be known.

The execution of traffic monitoring and system analysis mainly depends on the volume of information flow and the nature of incoming data packets. The hyper-security adaptive monitoring system, which combines several components used to collect information over time, performs operations such as identifying interventions, monitoring the development of events, characterizing security events, and determining them. Thus, appropriate security measures are adopted and implemented in a timely and effective manner. In special cases, such as [2], the algorithm of an analytically proven deterministic solution for the identification of interventions based on information from distributed sensors can be applied. Unlike conventional rules, systems proposed for comparison with security event samples are presented graphically. The correlation of networks emerges in the calculation of network compatibilities during the course of the journey, using the theory of random matrices. The identification of events through extensive modeling demonstrates the effectiveness of the proposed system.

The establishment of the monitoring and threat detection requirements fulfillment structure operates as an integral part of supporting secure and reliable systems. In this regard, the infrastructure model should be constructed based on templates. Templates generate rules for calculating events expressing security requirements for real-time monitoring. During this process, attack and threat signatures, or signatures of possible attacks on the system, are generated. In this theory, the probability of threats is assessed within a certain time frame based on Bayesian networks.

To ensure timely and reliable transmission of information, the use of multiple network additions and complex security solutions is required. To achieve this, it is necessary to control the network and protect it from increasingly damaging attacks. However, the management of hyper phases is significantly complex due to the lack of an effective mechanism to convert information from the lower level of the information state to the high level of human imagination for managing the situation, making decisions, and taking actions. Network systems suffer from the poor configuration of the used screens and monitoring functionality. Visual Firewall software and various levels of directed four synchronized views, time slices, and individual packets are used for configuring the inter-screen and ensuring the network monitoring process [3]. In the case of four synchronized views, real-time traffic, visual signature, statistics, and IDS are considered to implement active or passive monitoring processes in systems. The reports obtained from this process easily identify anomalies in traffic. In the monitoring of network security within institutions or universities, detecting attacks from internal networks to external networks becomes more important [3]. If such attacks are detected, determining the location of individual computers can sometimes be crucial. This work illustrates visualized security monitoring systems for large-scale networks. The system combines logical, temporal, and graphical information in a 3D view. Additionally, it provides mechanisms for filtering and mutual interaction of the system components. The adaptive monitoring system for cybersecurity gathers information about the situation over a certain time frame by combining various components, identifies interventions, monitors the development of events, identifies security events, and presents the results in visual form to assist in making informed decisions The visualization process is performed with the help of the Stemplot method to determine the location and scale of the event, and the results of modeling demonstrate the effectiveness of the proposed model. The SIEM system can also observe malicious activities, thereby strengthening control.

It is noted in that adaptive-resilient management of security provides self-defense and self-healing functions on critical business appliances platforms and networks. The foundation for decision-making to ensure adaptive security lies in the accumulation of authentic and sufficiently secure evidence within the system. The technical and economic analysis of the proposed mechanisms should also be conducted.

The widespread use of the Internet and the development of protocols and applications have brought research in the field of traffic classification to the forefront. Monitoring Internet traffic, managing QoS, identifying security events, etc., necessitate the study of the characteristics of the TCP/IP stream for classification.

The taxonomy of traffic classification methods is explored in for hierarchical and bidirectional identification of traffic. Four classification criteria are proposed to support bidirectional identification in such a systematic categorization: service, program, protocol, and function. Additionally, the suggested taxonomy supports the results of both upward and downward hierarchical organizations.

The use of the ARP sppofing can be employed for packet-level classification of network traffic [5]. Recently, "Machine learning" methods have gained widespread use in traffic classification. Classification based on the nearest neighbours method demonstrates excellent results. However, when the amount of training data is limited, there is a risk of reduced accuracy. To address such occurrences, the correlation of information should be performed during the classification process [5].

Web systems, due to their close connection with databases and browsers, face unique challenges and threats. Security threats in web systems are classified to identify their characteristics. Initially, the impact scale, type, and time of the incident are determined for analysis. Then, using "Machine learning" methods, security events are classified into two classes: scanning for vulnerabilities and attacks.

Various methods are available to enhance the reliability of Internet traffic classification results, but they are not considered reliable for lower-level classifications. Several intellectual analysis methods are proposed for online classification of Internet traffic. These methods specifically focus on detecting false events and prevent the evaluation process of false events.

Since collecting all Internet traffic would take a considerable amount of time during the processing, information is periodically collected from the flow of Internet traffic. This allows for an overall analysis of the environment during the processing, reducing computation time. When network management is ensured, the classification of HTTP traffic can be carried out based on the "parallel neural network classifier architecture." The effectiveness of this approach during traffic classification is assessed to be between 85–91% based on initial information [5].

Traffic classification and identification are generally considered one of the broadest research areas. The results of the analyzed studies above indicate that traffic classification is crucial for ensuring network security, proper interpretation of events, and bidirectional learning of Internet traffic. Based on these, intellectual, hierarchical, operational, and multifaceted methods should be developed and prepared for traffic classification.

4 Vulnerability Clustering Methods

The "Machine learning" approach is widely used for determining anomalous flows in network traffic based on unique statistical characteristics. Non-hierarchical clustering is faster and more suitable for detecting interventions and the natural emulation of data compared to traditional clustering.

Many clustering methods aim to separate normal and anomalous traffic for the detection of interventions. Clustering methods are applied to find the differences and similarities in traffic sessions, grouping them into respective clusters. These clusters are represented by assigned labels. Later, these labels are used to predict the type of incoming network traffic.

The quick and accurate identification of network traffic is one of the most important issues for functions such as QoS management, network security monitoring, etc. However, recently, the use of peers in P2P has increased, and they generate unnecessary information flow by hiding under any device, necessary data flow, or encrypted data flow by using various ports. In such cases, classical approaches, such as "port mapping" or "payload analysis," are not effective. An alternative approach is to classify by examining the behavior within the first few packets of TCP traffic in the network. This would simplify the identification process by clustering all the information in the future.

The increasing demand for wireless networks has drawn significant attention to their security issues. The methods used for the intelligent analysis of intrusions in wired networks are not directly applicable to wireless networks. In 802.11 wireless networks, measurements of the relevant network traffic should be conducted for the detection of intrusions and the modeling of wireless network security. An approach proposed in utilizes a distance-based heuristic metric to label clusters as normal or malicious.

Vulnerability assessment plays a crucial role in ensuring network penetration test. An adaptive clustering method is used to identify the characteristics of DDoS attacks using the steps in Fig. 3.

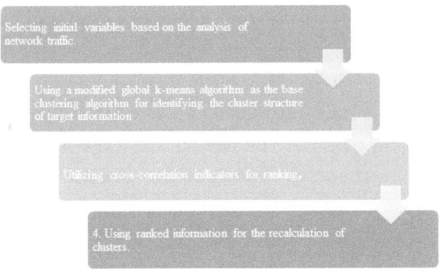

Fig. 3. Composition of the adaptive clustering method to identify the characteristics of DDoS attacks

This adaptive process can make necessary adjustments to the feature vector of DDoS attack samples, improving the quality of clusters and the efficiency of the algorithm.

A botnet is an infrastructure that leverages malicious software to remotely control infected computers without the user's knowledge and connects them in a network. Such malicious technology creates favorable conditions for creating armies of zombified computers. Hierarchical clustering is used for the discovery of control channels of botnets, determining the information exchange required for controlling botnets. This approach can be considered effective in preventing external control of a user's computer as part of a botnet network, given the differences in carrier protocols and encoding in normal traffic, which can be identified through traffic monitoring.

5 Discussion and Conclusion

This article has analyzed the possibility of applying intelligent analysis methods in vulnerability assessment of university network. Vulnerability so pervasive these days that we often don't even realise we use it anymore.

The application of intelligent analysis methods provides extensive opportunities for solving network penetration testing. Using modern penetration method methods, it is possible to automatically detect and identify problems, find solutions, and regulate operational management by eliminating vulnerabilities in existing university network.

References

1. Foreman, P: Vulnerability Management, p. 1. Taylor & Francis Group, Milton Park (2010). ISBN 978–1–4398–01505
2. Simpson, S.L.: The Vulnerability Assessment and Penetration Testing of Two Networks. Regis University (2011)
3. Vulnerability assessment and penetration testing in the military and ihl context Vojnotehnicki glasnik/Military Technical Courier, vol. 65, núm. 2, pp. 464–480. University of Defence (2017)
4. https://www.imperva.com/learn/application-security/penetration-testing/
5. . Umrao, S., Kaur, M., Gupta, G.: Vulnerability assessment and penetration testing. Comput. Sci. Eng. (2012)

Preliminary Orbit Estimation for Turksat 5B via Extended and Unscented Kalman Filtering

Hasan Kinatas[(✉)] [iD] and Chingiz Hajiyev [iD]

Istanbul Technical University, 34469 Maslak, Istanbul, Turkey
{kinatas16,cingiz}@itu.edu.tr

Abstract. Turksat 5B is a Turkish geostationary communications satellite operated by Turksat A.Ş which was launched on 19 December 2021 for commercial purposes. To determine its orbit, that is, its state vector in space, range, azimuth, and elevation angles are utilized in conjunction with Extended and Unscented Kalman filter (EKF and UKF) algorithms. Following the estimation results from these filters, their performance is compared in terms of both accuracy and computational burden. In this endeavor, several simulations are performed in which it is assumed that a single-station antenna exists in Istanbul at 41.10° latitude and 29.02° longitude. This single-station antenna provides range and angle measurements, which are then indirectly used in the filtering algorithm. Utilizing these measurements indirectly implies that they undergo pre-processing to obtain the satellite's position vector in the Earth Centered Inertial (ECI) frame. This pre-processing simplifies the EKF observation matrix and enhances the initial filter convergence.

Keywords: Orbit Estimation · Kalman Filtering · Satellite · Extended Kalman Filter · Unscented Kalman Filter

1 Introduction

Accurate orbit estimation is crucial for space operations and directly affects the quality of a satellite mission, whether it is communication, navigation, or Earth observation. Not knowing the satellite's position in orbit well enough may cause communication link disruption [1], poor positioning [2], or inaccurate data [3] depending on the mission objective. However, high-accuracy orbit estimation in satellites involves difficulties due to the natural perturbative forces in space. These perturbative forces inherently increase the uncertainty of the mathematical models used for orbit determination, necessitating advanced techniques.

Orbit estimation goes back a long way in history. Classical algorithms, including Gauss's and Laplace's methods, rely on astrodynamical principles to compute orbital elements from limited observational data [4]. Although, these methods proven to be effective in history, they generally lack the dynamic complexities of manmade satellite orbits. This led to the widespread adoption of least squares estimation in satellite orbit determination. Modern least squares techniques, which minimize the sum of the squares of residuals between observed and computed values, offer enhanced precision and adaptability in fitting a satellite's orbit to extensive and varied observational data sets [5, 6]. These advancements in statistical approaches have set the stage for even more sophisticated methods, like Kalman filtering in orbit estimation.

Kalman filtering, a groundbreaking advancement in statistical methods, has now been widely used in the field of orbit estimation [7–9]. Especially, the Extended Kalman Filter (EKF), a nonlinear adaptation of the original filter, handles the nonlinear dynamics characteristic of satellite orbits. It achieves this by linearizing the orbit dynamics around the current estimate, offering a more accurate estimation result [10]. Complementarily, the Unscented Kalman Filter (UKF) was developed to address some limitations of the EKF, particularly in scenarios with highly nonlinear dynamics. It employs a deterministic sampling strategy to better approximate the mean and covariance of the state estimate, enhancing the precision and reliability of orbit predictions [11].

Building upon the enhanced capabilities of advanced Kalman filtering techniques, this study presents a preliminary orbit estimation of the TURKSAT 5B satellite. TURKSAT 5B, a Turkish geostationary communications satellite operated by Turksat A.Ş, represents a significant asset in Turkey's space endeavors. The primary objective of this research is to utilize Extended and Unscented Kalman Filter algorithms (EKF and UKF) to estimate the satellite's orbit using range, azimuth, and elevation angle measurements from a single-station antenna located in Istanbul. This approach leverages the adaptability and precision of EKF and UKF in handling the non-linear dynamics of satellite orbits. Through methodical simulations and comparative analysis, this study aims to demonstrate the effectiveness of these filtering methods in accurately predicting the satellite's trajectory.

2 Coordinate Frames and Transformations

In the intricate field of satellite orbit estimation, the role of coordinate frames and transformations is pivotal. Precise orbit determination and subsequent predictions hinge crucially on the accurate representation and transformation of satellite positions and velocities across various coordinate systems. These systems, such as the Earth Centered Inertial (ECI) frame or the Earth-Centered Earth-Fixed (ECEF) frame, offer distinct perspectives for observing satellite motion. Each frame serves specific purposes, providing insights into the satellite's dynamics and environmental interactions. The meticulous transformation of data between these frames ensures fidelity in capturing the satellite's trajectory, enabling more accurate predictions and analyses. Subsequent subsections delve into the specifics of various coordinate systems, highlighting their essential roles in refining orbit estimation accuracy.

2.1 Earth Centered Inertial (ECI) Frame

The Earth Centered Inertial (ECI) frame, pivotal in satellite orbit determination, offers a non-rotating reference centered at Earth's mass center. Its axes, oriented fixedly relative to distant celestial objects, facilitate consistent tracking of orbital objects without the complexities introduced by Earth's rotation [12]. In the ECI frame, the x-axis points towards the vernal equinox, establishing a fundamental plane aligned with the Earth's equator, the z-axis aligns with the Earth's rotational axis, and the y-axis completes the right-handed coordinate system. The ECI frame's utility in satellite tracking is exemplified in its widespread application for long-term orbit prediction and analysis, bypassing the need for continuous adjustments for Earth's diurnal motion [13].

2.2 Earth Centered Earth Fixed (ECEF) Frame

The Earth-Centered Earth-Fixed (ECEF) frame, crucial in satellite navigation and geodesy, is distinguished by its fixed orientation relative to the Earth's surface. Centered at the Earth's center of mass, it aligns the z-axis with the Earth's rotational axis, pointing towards the North Pole, and the y-axis with the intersection at the equator and prime meridian, forming a right-handed system [14]. This frame's stability and alignment with the Earth's rotation are particularly advantageous for geostationary satellites, which require a constant positional reference for accurate tracking and seamless communication. The ECEF frame thus serves as an essential tool for both ground-based positioning and for monitoring satellites that appear stationary from an Earth-based viewpoint [15]. Transformations from ECEF to ECI, essential for certain operational purposes. This transformation is time-dependent and requires accounting for the Earth's rotation, which is characterized by the Greenwich Sidereal Time (GST). The GST represents the angle between the vernal equinox and the prime meridian. The transformation matrix $R_{\frac{ECEF}{ECI}}$ is defined as follows:

$$R_{\frac{ECEF}{ECI}} = \begin{bmatrix} \cos(GST) & \sin(GST) & 0 \\ -\sin(GST) & \cos(GST) & 0 \\ 0 & 0 & 1 \end{bmatrix} \quad (1)$$

2.3 Perifocal Frame

The perifocal frame, integral in celestial mechanics for satellite orbit analysis, is defined with respect to a satellite's orbit, having its origin at the center of mass of the orbiting body. In this frame, the unit vector $\hat{p}$ is defined along the x-axis, which points towards the periapsis, the point of closest approach to the primary body. Along the y-axis, which lies in the orbital plane and is perpendicular to the x-axis, the unit vector $\hat{q}$ is defined, indicating the direction of orbital motion. The unit vector $\hat{w}$ is defined along the z-axis, normal to the orbital plane and aligned with the orbital angular momentum vector. These unit vectors provide a clear and precise reference for describing the geometry and dynamics of the satellite's orbit, essential for accurate orbital element analysis.

The transformation matrix from the perifocal frame to the ECI frame, crucial for orbit determination, is presented as follows [4]:

$$R_{\frac{PF}{ECI}} = \begin{bmatrix} -\sin(\Omega)\cos(i)\sin(\omega) + \cos(\Omega)\cos(\omega) & -\sin(\Omega)\cos(i)\cos(\omega) - \cos(\Omega)\sin(\omega) & \sin(\Omega)\sin(i) \\ \cos(\Omega)\cos(i)\sin(\omega) + \sin(\Omega)\cos(\omega) & \cos(\Omega)\cos(i)\cos(\omega) - \sin(\Omega)\sin(\omega) & -\cos(\Omega)\sin(i) \\ \sin(i)\sin(\omega) & \sin(i)\cos(\omega) & \cos(i) \end{bmatrix} \quad (2)$$

where i is the inclination, Ω is the right ascension of the ascending node, and ω is the argument of periapsis. This transformation is essential as it bridges the gap between the satellite's orbital dynamics, expressed in the perifocal frame, and its position and velocity relative to the Earth's surface, as represented in the ECI frame.

2.4 Topocentric Horizon Coordinate Frame

The Topocentric Horizon Coordinate Frame is a local reference system essential for observation and communication with satellites from a specific Earth location. This frame is defined with its origin at the observer's position on the Earth's surface, providing a direct, localized perspective for tracking satellites. In this frame, the x-axis points towards the true north, aligning with the Earth's rotational axis, the y-axis points east, and the z-axis points towards the zenith, perpendicular to the Earth's surface at the observer's location [4]. This orientation makes the Topocentric Horizon Frame particularly useful for calculating angles of elevation and azimuth for satellites from a ground station's perspective. It allows for precise determination of a satellite's apparent position in the sky, which is critical for ground-based tracking, communication, and data acquisition systems.

The transformation from the Topocentric Horizon Coordinate Frame to the ECEF frame is crucial for integrating local observations with global positioning data. This transformation enables the conversion of coordinates from a local, observer-centric viewpoint to a global, Earth-centric system, essential for comprehensive satellite tracking and navigation. The matrix facilitating this transformation is outlined as follows [4]:

$$R_{\frac{TF}{ECEF}} = \begin{bmatrix} -\sin(\theta) & -\sin(\phi)\cos(\theta) & \cos(\phi)\cos(\theta) \\ \cos(\theta) & -\sin(\phi)\sin(\theta) & \cos(\phi)\sin(\theta) \\ 0 & \cos(\phi) & \sin(\phi) \end{bmatrix} \quad (3)$$

where θ is the local sidereal angle.

3 Orbital Dynamics and Mathematical Modeling of Measurements

3.1 Orbital Elements

Orbital elements are essential for describing and predicting satellite orbits. These elements include the semi-major axis (a), which defines the orbit's size and period; eccentricity (e), determining the orbit's shape; inclination (i), indicating the tilt of the orbit relative to a reference plane; right ascension of the ascending node (Ω), establishing the orbit's horizontal orientation; argument of periapsis (ω), describing the orientation within the orbital plane; and true anomaly (v), locating the satellite at a specific time [12]. The semi-major axis and eccentricity together dictate the size and shape of the orbit, while the inclination, right ascension of the ascending node, and argument of periapsis collectively determine its orientation in space. The true anomaly is crucial for pinpointing the satellite's position within this orbit. These elements are indispensable for accurately characterizing satellite orbits and are essential for effective orbit estimation.

In addition to these elements, the mean anomaly (M_e) and eccentric anomaly (E) are integral to understanding the satellite's orbit. The mean anomaly represents the satellite's position along its orbit as a fraction of the complete orbit period, providing a time-averaged measure of its location. The eccentric anomaly is an intermediate parameter that connects the mean anomaly with the true anomaly. Kepler's Equation, which is fundamental in astrodynamics, relates the mean anomaly to the eccentric anomaly as follows:

$$E - esin(E) = M_e \tag{4}$$

From the eccentric anomaly, the true anomaly (v), which directly indicates the satellite's actual position in its orbit, can be derived. These mathematical relationships are essential for precisely determining the satellite's orbit at any given time.

3.2 Obtaining State Vector from Orbital Elements

The state vector in the perifocal frame is obtained through a series of calculations that transform the orbital elements into position (r) and velocity (v) vectors [4]. This procedure begins with the solution of Kepler's Equation, given by Eq. (4). This equation is transcendental and cannot be directly solved due to the sine term. Instead, an iterative numerical method such as Newton's method is employed. Newton's method uses the function:

$$f(E) = E - esin(E) - M_e \tag{5}$$

and it's derivative with respect to E:

$$f(E) = 1 - ecos(E) \tag{6}$$

to iteratively converge to an accurate value of E. The iteration process follows the formula:

$$E_{n+1} = E_n - \frac{f_{E_n}}{f'_{E_n}} \tag{7}$$

Once the eccentric anomaly (E) is accurately determined, it is used to compute the position and velocity of the spacecraft in the perifocal frame. The position vector (r) in the perifocal frame has components calculated by:

$$r_p = a(\cos(E) - e) \tag{8a}$$

$$r_q = a\sqrt{1 - e^2} \sin(E) \tag{8b}$$

These components reflect the satellite's location in the orbital plane. The velocity vector (v) components, representing the satellite's motion, are derived as:

$$v_p = -\sqrt{\frac{\mu}{a}} \left(-\frac{\sin(E)}{\sqrt{1-ecos(E)}} \right) \tag{9a}$$

$$v_q = \sqrt{\frac{\mu(1-e^2)}{a}} \left(\frac{\cos(E)}{\sqrt{1-ecos(E)}} \right) \tag{9b}$$

After obtaining the state vector in perifocal frame, it can be transformed into geocentric equatorial inertial frame by using transformation matrices given by Eqs. (1) and (2).

3.3 Satellite Orbital Motion

The Kepler equations system is one of the systems that can define elliptical orbits of satellites. This equations system includes three equations in differential form as follows [4]:

$$\frac{d^2x}{dt^2} = -\frac{\gamma Mx}{r^3} \tag{10a}$$

$$\frac{d^2y}{dt^2} = -\frac{\gamma My}{r^3} \tag{10b}$$

$$\frac{d^2z}{dt^2} = -\frac{\gamma Mz}{r^3} \tag{10c}$$

where $\gamma = 6.67 \times \frac{10^{-11} m^3}{kgs^2}$ is the Kepler constant, r is the magnitude of the position vector in ECI frame, and $M = 5.976 \times 10^{24} kg$ is the mass of the Earth.

3.4 Radiolocation and Indirect Position Measurement Models

In this study, it is assumed that a ground-station exists in a specific location on Earth which provides range $\tilde{\rho}$, azimuth $\tilde{\alpha}$, and elevation $\tilde{\beta}$ via radiolocation measurements. Mathematical models for these measurements are given as follows:

$$\tilde{\rho} = \rho + \eta_\rho \tag{11}$$

$$\tilde{\alpha} = \alpha + \eta_\alpha \tag{12}$$

$$\tilde{\beta} = \beta + \eta_\beta \tag{13}$$

where η_ρ, η_α, and η_β are measurement noises which are all assumed to be zero-mean Gaussian white noises with the characteristics of:

$$E\left\{\eta_{\rho_k} \eta_{\rho_j}^T\right\} = \sigma_\rho^2 \delta_{kj} \tag{14a}$$

$$E\left\{\eta_{\alpha_k} \eta_{\alpha_j}^T\right\} = \sigma_\alpha^2 \delta_{kj} \tag{14b}$$

$$E\left\{\eta_{\beta_k} \eta_{\beta_j}^T\right\} = \sigma_\beta^2 \delta_{kj} \tag{14c}$$

where $E\{\cdot\}$ is the expected value operator and σ_ρ, σ_α, and σ_β are standard deviation of the corresponding sensor errors.

Using direct range, azimuth, and elevation measurements, indirect measurement of the position vector in ECI frame can be obtained. Components of this inertial position vector measurement are calculated as:

$$\begin{aligned} r_x = &\, \rho \cos(\theta) \cos(\lambda) \sin(\beta) - \rho \cos(\theta) \sin(\lambda) \cos(\beta) \cos(\alpha) - \\ &\, \rho \sin(\theta) \cos(\beta) \sin(\alpha) + R_E \cos(\lambda) \cos(\theta) \end{aligned} \tag{15a}$$

$$r_y = \rho \cos(\theta) \cos(\lambda) \sin(\beta) - \rho \sin(\theta) \sin(\lambda) \cos(\beta) \cos(\alpha) - \\ -\rho \cos(\theta) \cos(\beta) \sin(\alpha) + R_E \cos(\lambda) \sin(\theta) \quad (15b)$$

$$r_z = \rho \sin(\lambda) \sin(\beta) - \rho \cos(\lambda) \cos(\beta) \cos(\alpha) + R_E \sin(\lambda) \quad (15c)$$

where r_x, r_y, and r_z are the inertial position components of the satellite, R_E is the Earth's radius at the equator, λ is the latitude of the ground antenna location.

The coordinates of the satellite calculated by Eq. (15) are nonlinear functions of the radiolocation measurements measured by the single ground station antenna. By assuming that these measurements and measurement errors are independent and that errors of parameters R_E, λ, and θ are negligible, the variance of inertial position component measurement errors can be determined as [16]:

$$\sigma_x^2 = \left(\frac{\partial x}{\partial \rho}\right)^2 \sigma_\rho^2 + \left(\frac{\partial x}{\partial \alpha}\right)^2 \sigma_\alpha^2 + \left(\frac{\partial x}{\partial \beta}\right)^2 \sigma_\beta^2 \quad (16a)$$

$$\sigma_y^2 = \left(\frac{\partial y}{\partial \rho}\right)^2 \sigma_\rho^2 + \left(\frac{\partial y}{\partial \alpha}\right)^2 \sigma_\alpha^2 + \left(\frac{\partial y}{\partial \beta}\right)^2 \sigma_\beta^2 \quad (16b)$$

$$\sigma_z^2 = \left(\frac{\partial z}{\partial \rho}\right)^2 \sigma_\rho^2 + \left(\frac{\partial z}{\partial \alpha}\right)^2 \sigma_\alpha^2 + \left(\frac{\partial z}{\partial \beta}\right)^2 \sigma_\beta^2 \quad (16c)$$

where σ_x^2, σ_y^2, and σ_z^2 are variances of the calculated x, y, and z coordinate's errors, respectively.

4 Estimation of Spacecraft Position and Velocity via Kalman Filtering

4.1 Extended Kalman Filter (EKF) Formulation

The state vector which consists of spacecraft position and velocity components can be expressed as follows:

$$x = \begin{bmatrix} r_x \\ r_y \\ r_z \\ v_x \\ v_y \\ v_z \end{bmatrix} \quad (17)$$

Then, the following common nonlinear truth model with discrete-model measurements can be considered [10]:

$$\dot{x}_k = f(x_k, u_k, k) + G_k w_k \quad (18)$$

$$\tilde{y}_k = h(x_k, k) + \varepsilon_k \quad (19)$$

where $f(x_k, u_k, k)$ is a set of nonlinear functions of states, inputs, and iteration number k. Since no input is considered in this study, it can also be expressed as $f(x_k, k)$. w_k is a zero-mean Gaussian noise process with the characteristics of:

$$E\{w_k w_k^T\} = Q_k \delta_{kj} \tag{20}$$

On the other hand, $h(x_k, k)$ is the nonlinear measurement model consists of a set of nonlinear function of states and iteration number k. Similarly, ε_k is a zero-mean Gaussian noise process with the characteristics of:

$$E\{\varepsilon_k \varepsilon_k^T\} = R_k \delta_{kj} \tag{21}$$

The differential equations system given by Eq. (10) can be written in explicit form as:

$$\dot{r}_{x_k} = v_{x_k} \tag{22a}$$

$$\dot{r}_{y_k} = v_{y_k} \tag{22b}$$

$$\dot{r}_{z_k} = v_{z_k} \tag{22c}$$

$$\dot{v}_{x_k} = -\frac{\gamma M r_{x_k}}{r} + w_{v_{x_k}} \tag{22d}$$

$$\dot{v}_{y_k} = -\frac{\gamma M r_{y_k}}{r} + w_{v_{y_k}} \tag{22e}$$

$$\dot{v}_{z_k} = -\frac{\gamma M r_{z_k}}{r} + w_{v_{z_k}} \tag{22f}$$

And, considering that inertial position vector measurements are available, Eq. (19) can be written in explicit form as:

$$\begin{aligned} \tilde{r}_{x_k} &= r_{x_k} + \varepsilon_{r_{x_k}} \\ \tilde{r}_{y_k} &= r_{y_k} + \varepsilon_{r_{y_k}} \\ \tilde{r}_{z_k} &= r_{z_k} + \varepsilon_{r_{z_k}} \end{aligned} \tag{23a}$$

where error standard deviations are given by Eq. (16).

EKF algorithm requires the calculation of Jacobians of function sets $f(x_k, k)$ and $h(x_k, k)$ with respect to the state vector which can be achieved as follows:

$$F = \frac{\partial f}{\partial x}\Big|_{\hat{x}_{k-1}^+}, \quad H = \frac{\partial h}{\partial x}\Big|_{\hat{x}_k^-} \tag{24}$$

where $\hat{x}_{k-1}^+$ is the previously estimated state vector and $\hat{x}_k^-$ is the predicted state vector for the current iteration after the propagation which is performed as follows:

$$\hat{x}_k^- = f\left(\hat{x}_{k-1}^+, k-1\right) \tag{25}$$

The propagation of the estimation error covariance, on the other hand, is done as follows:

$$P_k^- = \Phi_k P_{k-1}^+ \Phi_k^T + Q_k \tag{26}$$

where Φ_k is known as the discrete state transition matrix calculated as:

$$\Phi_k = I + F_k \Delta t \tag{27}$$

After the propagation step is completed, the Kalman gain is calculated as follows:

$$K_k = P_k^- H^T \left[H_k P_k^- H_k^T + R_k \right]^{-1} \tag{28}$$

which then, followed by the update equations:

$$\hat{x}_k^+ = \hat{x}_k^- + K_k \left[\tilde{y}_k - h(\hat{x}_k^-, k) \right] \tag{29}$$

$$P_k^+ = \left[I_{6 \times 6} - K_k H_k \right] P_k^- \tag{30}$$

4.2 Unscented Kalman Filter (UKF) Formulation

Today, although the EKF is the most popular method for nonlinear estimation, there are other Kalman filtering methods that have been developed over time. Unscented Kalman filter (UKF) is another important method developed by Julier and Uhlmann [11] for use in nonlinear problems.

UKF is generally known to be computationally more demanding than the EKF, however, has several advantages. It can be applied to non-differentiable functions, does not require the calculation of Jacobian matrices, and generally offers a lower estimation error [17]. In the UKF algorithm, unlike the EKF, updating the error estimation covariance is done as follows:

$$P_k^+ = P_k^- - K P_k^{yy} K^T \tag{31}$$

where P_k^{yy} is the innovation covariance, that is the covariance of:

$$z_k = \tilde{y}_k - h(\hat{x}_k^-, k) \tag{32}$$

In addition to this, Kalman gain is also calculated differently which is given as:

$$K_k = P_k^{xy} (P_k^{yy})^- \tag{33}$$

where P_k^{xy} is the cross-correlation matrix. Like EKF, UKF starts with the propagation equations. First, for an n-dimensional state vector (6-dimension in this case), $2n + 1$ sigma points are determined as follows [17]:

$$x_{k0} = \hat{x}_k^+ \tag{34}$$

$$x_{k_\xi} = \hat{x}_k^+ + \left(\sqrt{(n+\kappa)P_k^+}\right)_\xi \tag{35}$$

$$x_{k_{\xi+n}} = \hat{x}_k^+ - \left(\sqrt{(n+\kappa)P_k^+}\right)_\xi \tag{36}$$

where x_{k_0}, x_{k_ξ}, and $x_{k_{\xi+n}}$ are sigma points and κ is a scaling factor for fine tuning. ξ indicated the columns number of the related matrix and defined as $\xi = 1, 2, \ldots, n$.

After determining the sigma points, these sigma points are propagated as follows:

$$\hat{x}_{k_l}^- = f\left(\hat{x}_{(k-1)_l}^+, k\right), \ l = 0, 1, \ldots, 2n \tag{37}$$

which is then, followed by the calculation of the predicted mean and covariance:

$$\hat{x}_k^- = \sum_{l=0}^{2n} \zeta_l \hat{x}_{k_l}^- \tag{38}$$

$$P_k^- = \left\{ \sum_{l=0}^{2n} \zeta_l \left[\hat{x}_{k_l}^- - \hat{x}_k^-\right]\left[\hat{x}_{k_l}^- - \hat{x}_k^-\right]^T \right\} + Q_k \tag{39}$$

where weights ζ_l are defined as:

$$\zeta_0 = \frac{\kappa}{n+\kappa}, \ \zeta_l = \frac{1}{2(n+\kappa)}, \ l = 1, 2, 3 \ldots, 2n \tag{40}$$

After the propagation part is completed, the update part starts with calculating the mean for the predicted observation vector and observation covariance matrix as:

$$\hat{y}_k^- = \sum_{l=0}^{2n} \zeta_l \hat{y}_{k_l}^- \tag{41}$$

$$P_k^{yy} = \left\{ \sum_{l=0}^{2n} \zeta_l \left[\hat{y}_{k_l}^- - \hat{y}_k^-\right]\left[\hat{y}_{k_l}^- - \hat{y}_k^-\right]^T \right\} + R_k \tag{42}$$

where $\hat{y}_{k_l}^-$ is given by:

$$\hat{y}_k^- = h\left(\hat{x}_{k_l}^-, k\right) \tag{43}$$

And finally, the cross-correlation matrix P_k^{xy} is calculated as:

$$P_k^{xy} = \left\{ \sum_{l=0}^{2n} \zeta_l \left[\hat{x}_{k_l}^- - \hat{x}_k^-\right]\left[\hat{y}_{k_l}^- - \hat{y}_k^-\right]^T \right\} \tag{44}$$

and then, states are updated as follows:

$$\hat{x}_k^+ = \hat{x}_k^- + K_k(\tilde{y}_k - \hat{y}_k^-) \tag{45}$$

5 Simulation Results

To assess the performance of both the EKF and UKF for TURKSAT 5B position and velocity estimation, several simulations are performed. The true state vector for the satellite is obtained using the process explained in Sect. 3.2 with the starting two-line element (TLE) set given by Table 1.

After obtaining the state vector in the perifocal frame, it is converted to ECI frame using the transformation matrices given by Eqs. (2) and (3). Figure 1 shows the true 3D orbit, and Figs. 2 and 3 show the true position and velocity components for the TURKSAT 5B for one orbital period. Note that although there are perturbations, in this study the true orbit is assumed to be in an environment where there are only Earth and the satellite.

Table 1. 11-Nov-2022 14:52:42 TLE set for TURKSAT 5B

1	50212U 21126A 22315.61993696 + .00000155 00000 − 0 + 00000 − 0 0 9995
2	50212 00.0718 336.3338 0003735 317.9539 021.6582 01.00272546000000

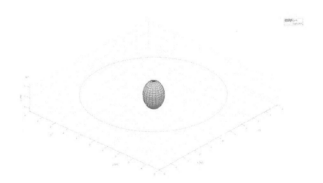

Fig. 1. True orbit for TURKSAT 5B in ECI frame

For simulations, it is assumed that there is a single-station antenna in Istanbul with 41.10° latitude and 29.02° longitude. Range, azimuth, and elevation measurement are simulated using Eqs. (11) through (13) with error standard deviations of $100\,m$, $0.01°$, and $0.01°$, respectively. Later, these measurements are converted to inertial position vector via Eq. (15) and corresponding error standard deviations are calculated via Eq. (16). Figure 4 shows the resulting absolute error of the converted measurements. In addition, the root mean square error (RMSE) values for these indirect measurements are calculated for the x-position, y-position, and z-position as 3.7232 km, 3.6637 km, and 6.5406 km, respectively.

Preliminary Orbit Estimation for Turksat 5B

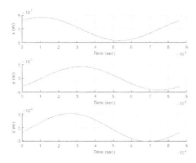

Fig. 2. True position components in ECI frame

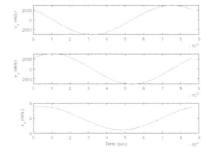

Fig. 3. True velocity components in ECI frame

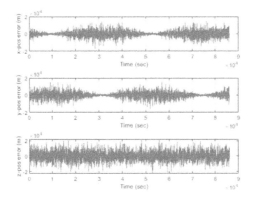

Fig. 4. Absolute error of indirect position vector measurements

The implementation of EKF and UKF algorithms are both initialized at the same state vector and state estimation error covariance which are given as:

$$x_0 = \begin{bmatrix} 2.8785 \times 10^7 m \\ -3.0765 \times 10^7 m \\ -2.2065 \times 10^4 m \\ 2.3522 \times 10^3 \frac{m}{s} \\ 2.1887 \times 10^3 \frac{m}{s} \\ 5.4196 \frac{m}{s} \end{bmatrix}, \quad P_0 = diag \left(\begin{bmatrix} 2.2953 \times 10^{12} m^2 \\ 2.1462 \times 10^{12} m^2 \\ 1.3523 \times 10^7 m^2 \\ 4.5725 \times 10^5 \frac{m^2}{s^2} \\ 4.8875 \times 10^2 \frac{m^2}{s^2} \\ 3.2635 \frac{m^2}{s^2} \end{bmatrix} \right) \quad (46)$$

and the process noise covariance matrix for the filters are determined to be:

$$Q = diag \left(\begin{bmatrix} 0 \\ 0 \\ 0 \\ 0.1\Delta t \\ 0.1\Delta t \\ 0.1\Delta t \end{bmatrix} \right) \quad (47)$$

According to these initial conditions, the filters are run for an orbital period with time steps of 23.9348 s. Figures 5 and 6 show the EKF and UKF estimation errors within ±3σ, respectively. In addition, Table 2 shows the RMSE values for the filters after convergence.

As can be seen from Figs. 5 and 6 both filters are working properly since their error are within ±3σ. Looking at Table 2, on the other hand, it is seen that the UKF algorithm gives slightly better estimation results than the EKF algorithm which is an expected result. However, the computational burden that an UKF algorithm brings is much higher than the EKF algorithm. While EKF can give estimates for an orbital period in an average of 0.1036 s, this time increases to 1.0166 s if UKF is used. It is thought that the use of indirect measurements also contributed to this difference, because the need to calculate the measurement Jacobian required for EKF is eliminated (Figs. 7 and 8).

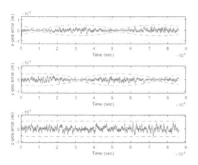

Fig. 5. EKF position estimation errors

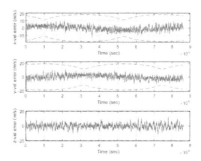

Fig. 6. EKF velocity estimation errors

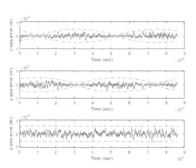

Fig. 7. UKF position estimation errors

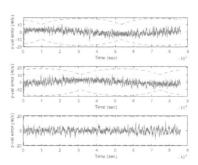

Fig. 8. UKF velocity estimation errors

Table 2. RMSE values for EKF and UKF estimation errors (avg. 100 Monte Carlo runs)

	RMSE EKF	RMSE UKF
x-pos error (m)	1077.41239900087	1077.39987449822
y-pos error (m)	1063.11607766108	1063.10723445325
z-pos error (m)	1655.27078287501	1655.26564920339
x-vel error (m/s)	3.30069148795181	3.30076115348366
y-pos error (m/s)	3.3047861509552	3.30494472770949
z-pos error (m/s)	3.22148430621486	3.22146395227791

6 Conclusion

In this study, a preliminary orbit estimation algorithm is developed for TURKSAT 5B satellite. The problem is considered as the preliminary orbit determination since no perturbations are considered throughout the work, that is the problem is fundamentally as a two-body problem. EKF and UKF estimation results show that position of the satellite can be estimated with approximately ±2.2 km error. The velocity estimation, on the other hand, can be obtained with approximately ±5.67 m/s error. Considering that TURKSAT 5B is a geostationary orbit (36786 km altitude) satellite, these estimation results are quite satisfactory. Comparing the estimation results of EKF and UKF, it is seen that UKF gives slightly better results, however, this improvement is negligible. In addition, it is observed that the computational burden of UKF is much higher than that of EKF. This makes EKF a much more suitable algorithm for this estimation problem.

References

1. Raines, R.A., Janoso, R.F., Stenger, D.K.: Performance studies of Low Earth Orbit Satellite (LEOS) communication networks for global communications. AIP Conf. Proc. **387**(1) (1997)
2. Lou, Y., et al.: The impact of orbital errors on the estimation of satellite clock errors and PPP. Adv. Space Res. **54**(8), 1571–1580 (2014). https://doi.org/10.1016/J.ASR.2014.06.012
3. Zhang, K., et al.: Precise orbit determination for LEO satellites with ambiguity resolution: improvement and comparison. J. Geophys. Res. Solid Earth **126**(9) (2021). https://doi.org/10.1029/2021JB022491
4. Curtis, H.D.: Orbital Mechanics for Engineering Students, 3rd edn. Butterworth-Heinemann, Oxford, UK (2013)
5. Schiemenz, F., Utzmann, J., Kayal, H.: Least squares orbit estimation including atmospheric density uncertainty consideration. Adv. Space Res. **63**(12), 3916–3935 (2019). https://doi.org/10.1016/j.asr.2019.02.039
6. Selvan, K., Siemuri, A., Prol, F.S., Välisuo, P., Bhuiyan, M.Z.H., Kuusniemi, H.: Precise orbit determination of LEO satellites: a systematic review. GPS Solut. **27** (2023). https://doi.org/10.1007/s10291-023-01520-7
7. Bierman, G., Thornton, C.: Numerical comparison of Kalman filter algorithms: orbit determination case study. Automatica **13**(5), 477–490 (1977)

8. Li, Q., Li, R., Ji, K., Dai, W.: Kalman filter and its application. In: Proceedings of the 2015 8th International Conference on Intelligent Networks and Intelligent Systems (ICINIS). IEEE (2015)
9. Roh, K.-M., Park, S.-Y., Choi, K.-H.: Orbit determination using the geomagnetic field measurement via the unscented Kalman filter. J. Spacecr. Rocket. **44**(1), 246–253 (2007). https://doi.org/10.2514/1.23693
10. Crassidis, J., Junkins, J.: Optimal Estimation of Dynamic Systems, 2nd edn. CRC Press Taylor & Francis Group, Boca Raton, FL (2012)
11. Julier, S., Uhlmann, J.: Unscented filtering and nonlinear estimation. Proc. IEEE **92**(3), 401–422 (2004)
12. Vallado, D.A.: Fundamentals of Astrodynamics and Applications, 3rd edn. Microcosm Press, El Segundo, CA (2007)
13. Wertz, J.R., Larson, W.J.: Space Mission Analysis and Design, 3rd edn. Microcosm Press and Kluwer Academic Publishers, Torrance, CA (1999)
14. Kaplan, M.H.: Understanding Space: An Introduction to Astronautics, 3rd edn. McGraw-Hill, New York, NY (2005)
15. Seeber, G.: Satellite Geodesy, 2nd edn. Walter de Gruyter, Berlin, Germany (2003)
16. Hajiyev, C.: Radio Navigation (in Turkish). Istanbul Technical University Press, Istanbul, Turkey (1999)
17. Hajiyev, C., Soken, H.: Fault Tolerant Attitude Estimation for Small Satellites, 1st edn. CRC Press Taylor & Francis Group, Boca Raton, FL (2020)

Three-Axis Attitude Determination of a Nanosatellite Using Optimized TRIAD Algorithms

Orhan Kirci(✉) and Chingiz Hajiyev

Faculty of Aeronautics and Astronautics, Istanbul Technical University,
34469, Maslak Istanbul, Turkey
kircio20@itu.edu.tr

Abstract. This paper aims to develop a comprehensive approach for determining the attitude of a nanosatellite. Accurate determination of these parameters is essential for the proper operation of the nanosatellite and its successful mission completion. To achieve this goal, a detailed simulation of the magnetometer and sun sensor measurements in the body and orbital frame of the nanosatellite is conducted. These measurements are critical for determining the attitude of the nanosatellite and ensuring that it is pointing in the right direction.

The attitude determination process is carried out using the Three-Axis Attitude Determination (TRIAD) algorithm, which is a popular and effective approach for estimating the attitude of a nanosatellite. The optimization of TRIAD with three different approaches is analyzed to ensure high accuracy and efficiency in determining the attitude of the nanosatellite. A covariance analysis is performed to evaluate the accuracy and reliability of the attitude determination process. The analysis provides valuable insights into the error sources and uncertainties associated with the measurements and estimation process. This information is used to improve the performance of the system and enhance the accuracy of the results. Overall, the results of this paper are expected to contribute significantly to the development of efficient and accurate methods for determining the attitude of a nanosatellite.

Keywords: nanosatellite · attitude determination · Euler angles · optimized TRIAD · data fusion · sun sensor · magnetometer

1 Introduction

Attitude determination refers to the process of determining the orientation of a nanosatellite relative to a reference frame. In a nanosatellite, attitude determination is crucial for its proper operation and the success of its mission. Knowing the attitude of a nanosatellite allows its instruments to point in the right direction, communicate with the ground station, and perform the desired tasks.

Attitude determination of a nanosatellite can be done using various methods which are using star trackers, sun sensors, magnetometers, and gyroscopes. These sensors as mentioned in Ref. [1] measure the direction and strength of the magnetic field, the position of the stars and the sun, and the angular velocity of the nanosatellite, respectively. The data from these sensors are then processed using algorithms to estimate the attitude of the nanosatellite in real-time.

The first approach to be published in the subject of the nanosatellite attitude determination is the algebraic method, which Harold Black established in 1964 [2]. In his 1981 publication, Malcolm Shuster also introduced this technique under the name TRIAD, which stands for "Three-Axis Attitude Determination" [3]. Finding the transformation matrix between the body frame of the nanosatellite and the reference frame is the goal of the TRIAD algorithm.

Optimized TRIAD that is studied in Refs. [4–6], combines the two transformation matrices held in two different TRIAD system which two different direction vectors act as anchor one by one. In this study, the optimization of TRIAD is performed with 3 different approaches to obtain lowest error for the attitude of a nanosatellite.

2 Mathematical Modeling of Attitude Motion and Reference Direction Sensors

The focus of this section is characterizing a nanosatellite's attitude motion and the sun and magnetic field sensors which are used in the TRIAD as reference direction vectors for the simulation environment.

2.1 Mathematical Model of Rotational Motion of the Nanosatellite

The nanosatellite's rotational motion is mathematically described in terms of Euler angles and angular velocities, and the problem is solved iteratively using the initial values of these parameters. To visualize the data, MATLAB, a popular piece of engineering software, will be utilized. The required programming algorithm is created using a mathematical model of the nanosatellite's rotating motion.

The following is the mathematical representation of the nanosatellite's rotational motion around its center of mass.

Expressions for Euler angles

$$\psi_{(i+1)} = \psi_{(i)} + \Delta t \left(-w_{x(i)} tan(\theta_{(i)}) * \cos(\psi_{(i)}) + w_{y(i)} \sin(\psi_{(i)}) * \tan(\theta_{(i)}) + w_{z(i)} \right) \tag{1}$$

$$\theta_{(i+1)} = \theta_{(i)} + \Delta t \left(w_{x(i)} sin(\psi_{(i)}) + w_{y(i)} cos(\psi_{(i)}) \right) \tag{2}$$

$$\phi_{(i+1)} = \phi_{(i)} + \Delta t \left(w_{x(i)} \cos(\psi_{(i)}) - \frac{w_{y(i)} \sin(\psi_{(i)})}{\cos(\theta_{(i)})} \right) \tag{3}$$

Expressions for angular velocities

$$\omega_{x(i+1)} = \omega_{x(i)} + \frac{\Delta t}{J_x} \left(\omega_{z(i)} \omega_{y(i)} + N_T \right) (J_y - J_z) \tag{4}$$

$$\omega_{y(i+1)} = \omega_{y(i)} + \frac{\Delta t}{J_y}(\omega_{x(i)}\omega_{z(i)} + N_T)(J_z - J_x) \tag{5}$$

$$\omega_{z(i+1)} = \omega_{z(i)} + \frac{\Delta t}{J_z}(\omega_{x(i)}\omega_{y(i)} + N_T)(J_x - J_y) \tag{6}$$

In the Eqs. (1)–(6) ϕ is the roll angle, θ is the pitch angle, ψ is the yaw angle, w_x, w_y, and w_z are the angular velocities, J_x, J_y, and J_z are the moments of inertias of the nanosatellite, w_{orbit} is the angular orbit velocity of the nanosatellite, N_T is the disturbance torque acting on the nanosatellite, Δt is the sample time and the N is the iteration number.

2.2 Mathematical Model of Earth Magnetic Field

The purpose of this section is to investigate how the Earth's magnetic field vector, one of the reference vectors used in the TRIAD approach, behaves when the orbital position of the nanosatellite varies. The magnetic field vector varies significantly with the orbital parameters as the nanosatellite travels along its path as described in Ref. [7]. The magnetic field tensor vector that affects satellites can be proven analytically as a function of time if those parameters are known as follows,

$$H_x(t) = \frac{M_e}{r_0^3}[\cos(\omega_0 t)(\cos(\varepsilon)\sin(i) - \sin(\varepsilon)\cos(i)\cos(\omega_e t)) - \sin(\omega_0 t)\sin(\varepsilon)\sin(\omega_e t)] \tag{7}$$

$$H_y(t) = -\frac{M_e}{r_0^3}[\cos(\varepsilon)\cos(i) + \sin(\varepsilon)\sin(i)\cos(\omega_e t)] \tag{8}$$

$$H_z(t) = \frac{2M_e}{r_0^3}[\sin(\omega_0 t)(\cos(\varepsilon)\sin(i) - \sin(\varepsilon)\cos(i)\cos(\omega_e t)) - 2\sin(\omega_0 t)\sin(\varepsilon)\sin(\omega_e t)] \tag{9}$$

In the Eqs. (7)–(9) M_e is the magnetic dipole moment of the Earth, μ is the Earth Gravitational constant, i is the orbit inclination, ω_e is the spin rate of the Earth, ε is the magnetic dipole tilt, r_0 is the distance between the center of mass of the nanosatellite and the Earth, ω_0 is the angular velocity of the orbit with respect to the inertial frame, found as $\omega_0 = (\mu/r_0^3)^{1/2}$.

To find the direction of the magnetic field vector, we can track its direction cosines which are computed as

$$H_0 = \begin{bmatrix} H_{x0} \\ H_{y0} \\ H_{z0} \end{bmatrix} = \frac{1}{\sqrt{H_x^2 + H_y^2 + H_z^2}} \begin{bmatrix} H_x \\ H_y \\ H_z \end{bmatrix} \tag{10}$$

where H_0 is the direction cosine elements of the magnetic field vector in the orbital frame. This vector measurement can be obtained in the body frame as follows.

$$H_B(k) = A(k)H_o(k) + \eta_H \tag{11}$$

In the Eq. (11), $A(k)$ is the direction cosines in terms of Euler angles which is given below in Eq. (12), η_H is assumed to be zero mean Gaussian white noise of the magnetic field sensor error.

$$A = \begin{bmatrix} c(\theta)c(\psi) & c(\theta)s(\psi) & -s(\theta) \\ -c(\varphi)s(\psi)+s(\varphi)s(\theta)c(\psi) & c(\varphi)c(\psi)+s(\varphi)s(\theta)s(\psi) & s(\varphi)c(\theta) \\ s(\varphi)s(\psi)+c(\varphi)s(\theta)c(\psi) & -s(\varphi)c(\psi)+c(\varphi)s(\theta)s(\psi) & c(\varphi)c(\theta) \end{bmatrix} \quad (12)$$

where c defines the cosine of the angle and s defines the sine of the angle.

2.3 Mathematical Model of Sun Direction

The sun model assesses whether the Earth is blocking the sun at any given time throughout the mission by computing the sun line in the inertial reference system. The inertial sun vector varies slightly as the nanosatellite moves in its orbit as a result of the Earth orbiting the nanosatellite and the Earth rotating around the sun.

The following standard algorithm in Eqs. (13)–(16) as stated in Ref. [1] describes the true motion of the Earth around the inertially fixed sun in the form of the sun orbiting the Earth being fixed in inertial space. The right ascension of the ascending node of this virtual orbit of the sun around the Earth is 0° by definition.

The unit sun vector is given in the inertial frame

$$\hat{s}_{ECI} = \begin{bmatrix} cos\lambda_{ecliptic} \\ sin\lambda_{ecliptic} \times cos\varepsilon \\ sin\lambda_{ecliptic} \times sin\varepsilon \end{bmatrix} \quad (13)$$

where ε is the linear model of the ecliptic of the sun, $\lambda_{ecliptic}$ is the ecliptic longitude of the sun.

This vector measurement can be obtained in the orbital frame by using transformation matrix C in Eq. (14) which is calculated in each iteration by using $\lambda_{ecliptic}$, i which is the inclination and u as the yaw angle of the nanosatellite in the inertial frame

$$C = \begin{bmatrix} c(u)c(i)c(\lambda) - s(u)s(\lambda) & c(u)s(i) & -c(u)c(i)s(\lambda) - s(u)c(\lambda) \\ -s(i)c(\lambda) & c(i) & s(i)s(\lambda) \\ s(u)c(i)c(\lambda) - c(u)s(\lambda) & s(u)s(i) & -s(u)c(i)s(\lambda) - c(u)c(\lambda) \end{bmatrix} \quad (14)$$

Then, unit sun vector in the orbital frame is

$$S_o(k) = C(k)\hat{s}_{ECI}(k) \quad (15)$$

Finally, obtained vector in the orbital frame can be converted to body frame as follows.

$$S_B(k) = A(k)S_o(k) + \eta_S \quad (16)$$

In the Eq. (16), $A(k)$ is the direction cosines, η_S is assumed to be zero mean Gaussian white noise of the sun direction sensor error.

3 TRIAD Algorithm

TRIAD (TRI-axial Attitude Determination) as described in Ref. [3] is an attitude determination algorithm that uses a vector-based approach to estimate the nanosatellite's attitude based on measurements from two sets of known vectors.

3.1 Classic TRIAD

TRIAD uses two unparalleled unit vectors to construct a new coordinate. The attitude matrix was calculated by the algebraic method as stated in Ref. [3]. In the classic TRIAD, first vector which is called as anchor vector is the sun direction vector. The equation is given as,

$$\begin{aligned} r_1 &= v_1 \\ r_2 &= (v_1 \times v_2)/|v_1 \times v_2| \\ r_3 &= (v_1 \times r_2)/|v_1 \times r_2| \end{aligned} \quad (17)$$

The equation includes two vectors, v_1 and v_2, that are in the orbital frame where first vector v_1 is the sun direction vector.

$$\begin{aligned} s_1 &= w_1 \\ s_2 &= (w_1 \times w_2)/|w_1 \times w_2| \\ s_3 &= (w_1 \times s_2)/|w_1 \times s_2| \end{aligned} \quad (18)$$

and w_1 and w_2 are in the body frame.

$$M_o = [r_1 r_2 r_3] \quad (19)$$

After orthogonalizing the reference vectors, the base vectors of the orbital frame matrix M_o are represented by r_1, r_2, and r_3.

$$M_b = [s_1 s_2 s_3] \quad (20)$$

The base vectors of body frame matrix, M_b, are represented by s_1, s_2, and s_3. Then the attitude matrix is obtained in the Eq. (21).

$$A = M_b * M'_o \quad (21)$$

For the application of TRIAD algorithm, two unit vectors in two reference frames is needed. H_0 is the Earth magnetic field vector in the orbital frame, H_B is the magnetometer measurement vector in the body frame, S_o is the sun direction vector in the orbital frame and S_B is the sun sensor measurement in the body frame.

After the application of the classic TRIAD, in Eq. (12) a rotation matrix can be obtained in the 3-2-1 direction cosine sequence.

It is observed that attitude matrix is weighted twice by the first vector, indicating that the first vector is the major vector and plays a critical role in determining the attitude accurately. If the noise associated with the major vector is higher than that of the secondary vector, the classic TRIAD algorithm may not provide the optimized solution.

This issue is addressed by the proposed algorithm in this study. The error covariance for TRIAD is presented in Ref. [5] as;

$$P_{\text{TRIAD}} = \sigma_1^2 I + \frac{1}{|\hat{w}_1 + \hat{w}_2|^2}\left[\sigma_1^2(\hat{w}_1 \cdot \hat{w}_2)\left(\hat{w}_1\hat{w}_2^T + \hat{w}_2\hat{w}_1^T\right) + \left(\sigma_2^2 - \sigma_1^2\right)\hat{w}_1\hat{w}_1^T\right] \quad (22)$$

The attitude covariance matrix is denoted by P_{TRIAD}, and I represents the identity matrix. The variances of two vector sensors are denoted by σ_1^2 and σ_2^2. Equation (6) shows that if σ_1^2 is greater than σ_2^2, the error matrix is larger than the case when σ_1^2 is less than σ_2^2.

3.2 Optimized TRIAD

Different types of optimized TRIAD is proposed in this section and there is a comparison study between these algorithms in the simulation section. There are several variations of the optimized TRIAD algorithm, which aim to improve the accuracy and computational efficiency of the original algorithm. Main idea for the optimized TRIAD is mentioned below.

The optimized TRIAD algorithm is an mixture of two TRIAD algorithms, namely TRIAD-I and TRIAD-II. The former generates matrix A_1, while the latter generates matrix A_2. The approach involves creating two sets of vector triads using four vector measurements, two in the orbital reference frame (v_1 and v_2) and two in the body reference frame (w_1 and w_2). To implement the optimized TRIAD algorithm, one must compute the matrices r_i in the body coordinate and the corresponding column matrices s_i in the orbital frame. For the TRIAD-I algorithm, a specific definition is assigned

$$r_1 = \frac{v_1}{|v_1|} r_2 = \frac{(r_1 \times v_2)}{|r_1 \times v_2|} r_3 = r_1 \times r_2 \quad (23)$$

$$s_1 = \frac{w_1}{|w_1|} s_2 = \frac{(s_1 \times w_2)}{|s_1 \times w_2|} s_3 = s_1 \times s_2 \quad (24)$$

The expression for matrix A_1 is given by;

$$A_1 = r_1 \cdot s_1^T + r_2 \cdot s_2^T + r_3 \cdot s_3^T \quad (25)$$

The definition for the TRIAD-II algorithm is as follows;

$$r_5 = \frac{v_2}{|v_2|} r_2 = \frac{(r_1 \times v_2)}{|r_1 \times v_2|} r_4 = r_5 \times r_2 \quad (26)$$

$$s_5 = \frac{w_2}{|w_2|} s_2 = \frac{(s_1 \times w_2)}{|s_1 \times w_2|} s_4 = s_5 \times s_2 \quad (27)$$

The expression for matrix A_2 is given by,

$$A_2 = r_5 \cdot s_5^T + r_2 \cdot s_2^T + r_4 \cdot s_4^T \quad (28)$$

3.2.1 Optimized TRIAD – 1

The computation of the optimized attitude matrix $\hat{A}$ follows as;

$$\hat{A}' = \frac{\sigma_1^2}{\sigma_1^2 + \sigma_2^2} A_1 + \frac{\sigma_2^2}{\sigma_1^2 + \sigma_2^2} A_2 \quad (29)$$

Finally, the attitude matrix of the optimized TRIAD can be written according to study in Ref. [6]:

$$A_{opt1} = 0.5\left[\hat{A}' + \left((\hat{A}')^{-1}\right)^T\right] \quad (30)$$

Matrix, A_{opt1}, performs an optimized solution for the Classic TRIAD algorithm. The error covariance of the optimized TRIAD is given in the Eq. (31) as stated in Ref. [5],

$$P_{Opt\text{-}TRIAD} = \sigma_{opt}^2 I + \frac{1}{|\hat{w}_1 + \hat{w}_2|^2}\left[\sigma_{opt}^2(\hat{w}_1 \cdot \hat{w}_2)\left(\hat{w}_1\hat{w}_2^T + \hat{w}_2\hat{w}_1^T\right)\right] \quad (31)$$

where

$$\sigma_{opt}^2 = \left(\frac{1}{\sigma_1^2} + \frac{1}{\sigma_2^2}\right)^{-1} \quad (32)$$

3.2.2 Optimized TRIAD – 2

The second method is based on the sensor fusion technique presented in Ref. [8]. The second method uses covariance matrix of the TRIAD-1 and TRIAD-2 rather than using covariance of the reference direction vectors

$$\hat{x} = \frac{\sigma_{cov1}^2 x_{A_2} + \sigma_{cov2}^2 x_{A_1}}{\sigma_{cov1}^2 + \sigma_{cov2}^2} \quad (33)$$

where $\hat{x}$ is the angle that is optimized, σ_{cov1}^2 is the angle error variance when TRIAD-1 algorithm is used, σ_{cov2}^2 is the angle error variance when TRIAD-2 algorithm is used, x_{A_1} is the angle that is found by the TRIAD-1 algorithm and x_{A_2} is the angle that is found by the TRIAD-2 algorithm.

The error variance of the optimized angle $\hat{x}$ can be written as in Eq. (34),

$$D_x = \frac{\sigma_{cov1}^2 \sigma_{cov2}^2}{\sigma_{cov1}^2 + \sigma_{cov2}^2} \quad (34)$$

3.2.3 Optimized TRIAD – 3

The third method is based on the sensor fusion technique when three type TRIAD method outputs are processing as mentioned in Ref. [8]. The third method uses TRIAD-1 and

TRIAD-2 and also optimized TRIAD which is calculated in the method 1. Each angle calculated by the Eq. (35) as follows;

$$\hat{x} = \frac{\sigma_{cov1}^2 \sigma_{cov2}^2 x_{A_{opt}} + \sigma_{cov1}^2 \sigma_{cov3}^2 x_{A_2} + \sigma_{cov2}^2 \sigma_{cov3}^2 x_{A_1}}{\sigma_{cov1}^2 \sigma_{cov2}^2 + \sigma_{cov1}^2 \sigma_{cov3}^2 + \sigma_{cov2}^2 \sigma_{cov3}^2} \quad (35)$$

where $\hat{x}$ is the angle that is optimized, σ_{cov3}^2 is the angle error variance when the optimized TRIAD-1 method is used.

The error variance of the optimized angle $\hat{x}$ can be determined as in Eq. (36),

$$D_x = \frac{\sigma_{cov1}^2 \sigma_{cov2}^2 \sigma_{cov3}^2}{\sigma_{cov1}^2 \sigma_{cov2}^2 + \sigma_{cov1}^2 \sigma_{cov3}^2 + \sigma_{cov2}^2 \sigma_{cov3}^2} \quad (36)$$

where D_x is the angle error variance.

4 Simulation Results of TRIAD Algorithm

Simulations are performed in order to estimate attitude of a nanosatellite using the classic TRIAD algorithm and optimization of TRIAD with different methods.

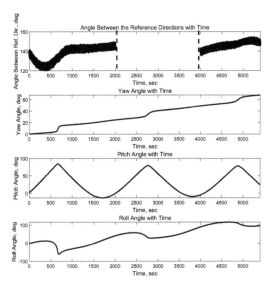

Fig. 1. Angle between the reference directions; and Euler Angles

Algorithm is run for 54000 iterations with time step of 0.1 s. Direction cosines of the standard deviations for magnetometer and sun sensor are taken as 0.08 and 0.02 respectively.

In the Fig. 1, as the angle difference for the direction sensor close to parallel which means angle difference is close to 0° or 180° TRIAD error increase. Also, as pitch angle

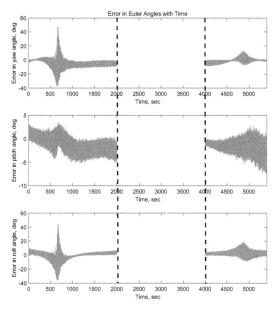

Fig. 2. Classic TRIAD error in Euler angles

is close to ±90° TRIAD error increases as in simulation results of classic TRIAD in the Fig. 2.

Classic TRIAD uses sun vector as anchor since it has high accuracy than magnetometer. As shown in Fig. 2, estimation errors are generally low but pitch angle affects the results badly when ±90° at 660th and 5000th seconds. Also at 5200th second, direction vector angle difference getting close to parallel which is another reason to get higher estimation errors.

2000 to 4000th second interval is the eclipse period so TRIAD algorithm is not capable of giving any result since the sun vector cannot measure the sun direction.

All the TRIAD algorithms are simulated and their error in Euler angles and the covariances are shown in the Figs. 3, and 4. TRIAD-1 which is also named as Classic TRIAD, uses sun vector as its anchor, TRIAD-2 uses magnetometer as its anchor and all three optimization method uses both TRIAD-1 and TRIAD-2 to get an optimized result. According to simulation result, Optimized3 which is optimized TRIAD algorithm 3 gives generally the best result both for error in Euler angles and covariance of the error model of the TRIAD algorithms.

A closer look for 1–60 s is shown in the figure below. Time to time as illustrated in the Fig. 5, lowest error of Euler angles cannot be obtained by just one algorithm, this is because the changing performance or accuracy of the direction sensors which is also simulated. Since optimization has lots of dependences such as pitch angle's degree, parallelism of the two sensors, changing performance of sensor; it is not suitable to define a definite optimized algorithm but as shown in this project it is possible to get a generally working optimized algorithm which is method 3 to have less errors and covariances (Fig. 6).

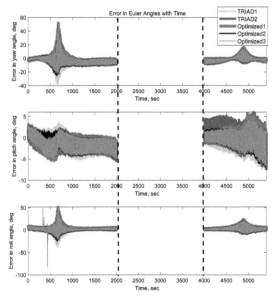

Fig. 3. Comparison of TRIAD algorithms with the error in Euler angles

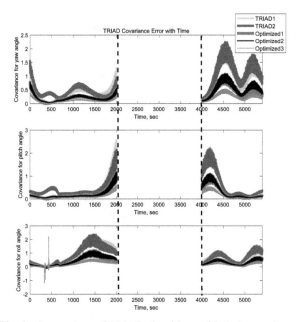

Fig. 4. Comparison of TRIAD algorithms with their covariances

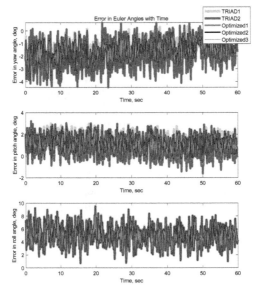

Fig. 5. Error in Euler angles for the first 60 s

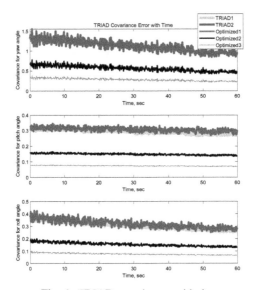

Fig. 6. TRIAD covariances with time

5 Conclusion

In this study, three distinct methods are performed to optimize TRIAD by using classic TRIAD algorithm in order to minimize the error associated with a nanosatellite's attitude. TRIAD algorithm can estimate the attitude of a nanosatellite using two vector measurements in two reference frame. In this study, orbital and body frames used as reference frames and sun and magnetic field sensors for the reference vectors. TRIAD algorithm may fail when the nanosatellite is in the eclipse position if the sun sensor is used for the one of the two direction vector and the $\pm 90°$ pitch angle and parallelism of the two reference direction vector causes to obtain higher errors.

In the classic TRIAD algorithm (TRIAD-1) which sun vector is used as anchor, less error can be obtained compared to TRIAD-2 which magnetic field vector is used as anchor. In the optimization section, it is desired to obtain a better result which has lowest error compared to both TRIAD-1 and TRIAD-2 algorithms. The results of the simulation indicated that a single algorithm is unable to yield the lowest error of Euler angles due to the simulated changing accuracy or performance of the direction sensors. It is not appropriate to define a specific optimized algorithm because optimization involves many dependencies, such as the degree of pitch angle, the parallelism of the two sensors, and changing sensor performance. However, as this study has demonstrated, it is possible to obtain a generally working optimized algorithm, which is method 3 to have less errors and covariances.

References

1. Guler, D.C., Hajiyev, C.: Gyroless Attitude Determination of Nanosatellites, pp. 45–47. LAMBET Academic Publishing (2016)
2. Black, H.D.: A passive system for determining the attitude of a satellite. AIAA J. **2**(7), 1350–1351 (1964)
3. Shuster, M.D., Oh, S.D.: Three-axis attitude determination from vector observations. J. Guid. Control **4**(1), 70–77 (1981)
4. Bar-Itzhack, I.Y., Harman, R.R.: Optimized TRIAD algorithm for attitude determination. J. Guid. Control **20**(1), 208–211 (1997)
5. Zhu, X., Ma, M., Zhou, Z.: An optimized triad algorithm for attitude determination. Artif. Satell. **52**(3) (2017)
6. Shuster, M.D.: The optimization of TRIAD. J. Astronaut. Sci. **55**(2), 245–257 (2007)
7. Sekhavat, P., Gong, Q.G.Q., Ross, I.M.: NPSAT1 parameter estimation using unscented Kalman filtering. In: 2007 American Control Conference, New York, USA, pp. 4445–4451 (2007)
8. Hajiyev, C., Soken, H.E.: Fault Tolerant Attitude Estimation for Small Satellites. CRC Press, Boca Raton (2021)

Computer System of Evaluation of the Mass Exam Results Based on Recognition of Handprinted Azerbaijani Characters

Elshan Mustafayev and Rustam Azimov(✉)

Laboratory of Recognition, Identification and Methods of Optimal Solutions,
Institute of Control Systems, Baku, Azerbaijan
`elshan.mustafayev@gmail.com, rustemazimov1999@gmail.com`

Abstract. The utilization of pattern recognition is on the rise extensively in information systems. The convergence of progress in image processing and the accessibility of open-source libraries enables the implementation of innovative solutions for diverse practical problems. One notable challenge pertains to automatically processing responses in mass large-scale exams. This paper introduces a developed system tailored for recognizing such exam results, showcasing its capacity to deliver dependable, effective, and impartial assessments.

This system can be configured on almost any type of form. Its use also allows you to abandon the expensive and difficult to use OMR scanners. To increase productivity of system we propose to use the multicore/multithreading property of modern processors to parallelize processes within a single workstation. As a result of experiments, it was found that the transition to multi-threaded recognition can increase productivity up to 3.5 times in comparison with single-threaded.

To reduce the physical size of exam cards, it is proposed to fill in the answers with handwritten symbols instead of filling in the circles. Multilayer and convolutional neural networks were used as a recognition module. A comparative evaluation of the dependence of recognition results on the architecture of neural networks and the feature extraction algorithm was carried out.

Keywords: Evaluation of Exam Results · Intelligent System · Information Processing · Multithreading Recognition · Image Processing · OCR

1 Introduction

Optical Mark Recognition (OMR) technologies, which utilize optical scanners, are frequently utilized for the automated processing of results in large-scale test examinations (see Fig. 1). OMR scanners are preferred due to their high speed and reasonably accurate performance. This methodology enables an objective evaluation of a substantial quantity of examination forms within a brief period, achieving nearly flawless recognition accuracy.

Fig. 1. Samples of recognized forms

OMR technology presents several significant limitations:

- sensitivity to paper quality and paper size: OMR scanners requires high sensitivity about both the size and quality of the paper, a specific color of the form as well;
- limitation to specific form sizes: with OMR technology custom form sizes can't be processed;
- dependence on pristine form condition: OMR processing necessitates flawless form condition, as even slight damage, crushing, or folding can result in rejection;
- inability to process handwritten text forms: OMR technology cannot cope with forms filled out by hand;
- maintenance and high cost requirements: Implementation, as well as, maintenance of OMR technology may be expensive.

To tackle these challenges, a system has been developed to facilitate independent and anonymous evaluation of responses. High-speed scanning technology is employed to convert all forms, irrespective of their dimensions, into graphical format. Using predefined templates, the system identifies forms with questionably filled marker answers. With the help of OCR technology [1, 2], handwritten fields are recognized and compared with the values of the corresponding fields filled in using circles (Fig. 2). The reason for using such a comparison module is because of inadequately filled marker answers. As a result, this helps to assure that form recognition computer system will always try to make decisions for the benefit of examinees based on this comparison. Additionally having handwritten character recognition technology in a form recognition system would help to make forms more concise since handwritten character fields hold less place.

Recognized form samples are concurrently displayed on screens to independent experts, ensuring the examinee's identity remains undisclosed. The system then generates an objective assessment of the responses based on the evaluations of these experts.

2 Description of the Processing System

At first the operational stages of the system which was developed is provided. Exam forms acquired from the exam places are brought to the scanning area. Each of the exam forms has a unique identifier and is safe against fraud. The scan operators scan the forms using high-performance scanners. After completion of form scanning, the scan operators verify scanned form count against the known number, and take some samples

Computer System of Evaluation of the Mass Exam 173

for scanning quality checking. Then the project execution group starts the process of recognition of the digital materials. Recognition process does not require presence of operators. Recognition process of the exam forms follows some steps which are common to other form recognition system in the industry [3–5]:

1. template set generation,
2. determination of reference points,
3. definition and alignment of the template,
4. recognition,
5. verification.

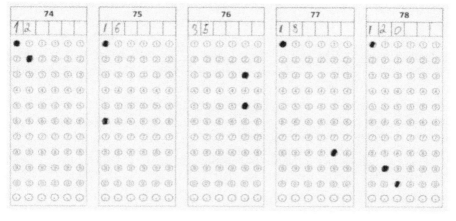

Fig. 2. Examples of confusing situations for OCR

2.1 Generation of a Set of Templates

In the stage of template set generation the developed system does scanning and then imports the blank form image. Then the form recognition system starts identifying automatically the form elements, i.e., control fields. These fields can be static repeated texts (which are on all the exam forms), reference points (black squares), barcodes, etc. The black squares are needed to ensure that the given images are aligned according the template properly. When specifying the fields, type of the field, its format and control rules have to be indicated. Control rules are used for encompassing some checks, for example, information correctness verification using data in the database in order to ensure a consistent representation of some data, like, dates or financial data; establishment of field patterns which uses regular expressions; validation using a user dictionary. If it is needed to export the results of recognition to the database, each of the fields are assigned to a corresponding column in the tables of the database.

2.2 Definition of Reference Points in the Graphic Image of the Form

Inevitable linear distortions happen almost always along width/height of the form image. In the developed form recognition system for an accurate template alignment and recognition system for an accurate template alignment and for compensating the distortions reference points are used. Usually 4 black squares are at the corners of the form image are used as points of references (Fig. 1).

Black square identification consists of the following steps:

- Colored form image is converted to a binary image (with only completely black and white pixels).
- Locations of all the connected objects are determined in the presumed areas.
- Among the determined connected objects, objects that correspond the shape and size of the black squares as indicated in the template are identified as black squares. Since these processes are carried out using the binary image of the exam form the contours of black squares may not be as perfect straight lines, and thereby have slight irregularities. Because of that, some deviations from the linear dimensions which specified during the process of template generation are allowed in identifying black squares.

2.3 Definition and Alignment of the Template

Template matching is used to match the template with the form image using reference point coordinates found in the previous stage. For a given image's alignment different transformation algorithms can be used depending on the number of detected black squares.

At least 3 points of references are required to be found for template matching. If less than 3 points of references are found subsystem of manual template matching which requires intervention of operators is used. This kind of interventions are typically needed because of either incorrectly scanned image or damage of the paper. System lets a user for creation and editing of various templates.

Different geometric transformations are used to ensure precise template matching. Recognized fields' exact position coordinates are identified on the form image. This is the vital stage of form recognition which has direct impacts on the results of recognition. Certainly more precise form image alignment depends on the (detected) reference point count.

If the number of reference points is 4, bilinear transformation is used, if it is 3 affine transformation is used for alignment [6].

2.4 Recognition

In the previous step the template was established. Using this template the system recognizes each of the fields. The field types which the system recognize are below:

- checkbox – □,☒,☑,○,●;
- radio group – just one of the radio items can be filled, e.g. ○●○○;
- handwritten text character or text – text or character that is written in character blocks.

Some statistical recognition methods can be employed by the system for radio boxes and checkboxes recognition [7, 8]. In the form recognition system the template and the field's statistics are compared. Based on a certain threshold the decision whether the item is marked or not is made by the system.

Some of the common mistakes the applicants make while filling out the form noted below (Fig. 3):

- Filling multiple radio item options in a radio button.
- Leaving a circle incompletely filled.
- Usage of a special mark rather than filling the circle.

Fig. 3. Incorrectly filled answers

The system employs some threshold parameters as below to compe with the aforementioned issues for operator intervention minimization:

- Minimum filling %: This parameter is used to determine the completion level of a circle (a circle's filling percentage below the value of these parameter says the circle is empty).
- Minimum filling % for rejecting the recognition results: if a field's filling percentage is below than the value of this parameter system defines the results as uncertainly recognized filed.
- Threshold for multiple choices: if maximum of the differences between highest filling percentage and the other percentages is more than this threshold percentage the system rejects it for manual reviewing of the operator, otherwise selects the radio item with the highest filling percentage as the answer.

2.5 Verification

The results of recognition passes to the verification zone, where the procedures which can take place are the followings:

- Form templates can be deferred manually.
- Fields with doubtful results can be examined by the operators who compare the recognition results with the graphic image of the form. If necessary operators verify the result of recognition or corrects the existing result (Fig. 4).

To speed up the verification process and increase the reliability of the results, an intelligent model is used that compares the results of recognizing handwritten fields with similar results obtained based on determining the occupancy of circles. In Fig. 5 shows

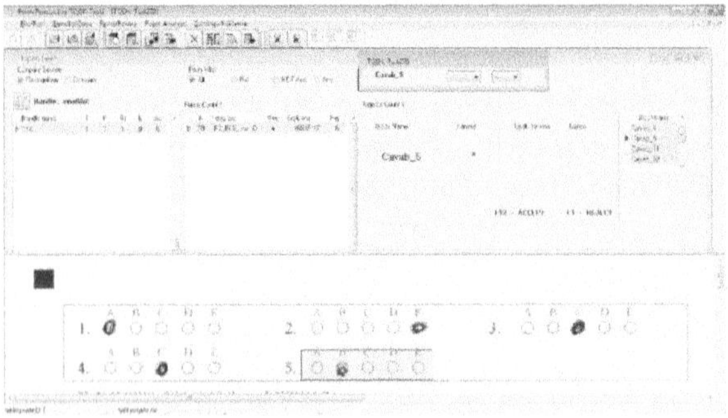

Fig. 4. Verification Screen

examples of such fields. The results of question 74 are correct, because the handwritten field and the field filled with circles match. But questions 75–78 are examples of incorrect completion by students. However, as a result of intelligent comparison of the results of the OCR (recognition of handwritten characters) and the filled circles, the answers to questions 75–77 are assigned the correct values without operator intervention. But question 78 is marked as undefined and sent to the operator for verification, because here comparison with OCR results also can't help the system to make a final decision without operator's verification.

After the recognition-verification stages for exam forms are completed, for written answers of examinees data proceeds to the expert assessment area. A written answer's graphical representation are presented to some independent experts without revealing the examinee to ensure objectivity. Once all the experts evaluated the written answers, the system make an objective decision automatically. Archive and comprehensive log of all the experts' actions are recorded.

After the recognition results of the examination form have been determined, the system processes the data to assess the accuracies. The answers which were given by the examinees are compared to the correct answers. Some specialized algorithms are used to make the final decision of the exam form.

In order to ensure transparency and objectivity, all data, including the graphical representation of the examination form, exam results, and correct answers, are made available online in the examinee's personal account.

All the data including the graphical representation of the exam form, results of the exam, correct answers are made online in the personal account of the examinees for ensuring transparency, objectivity.

3 Performance Improvement

One of the most important criterion in systems which is used for processing of forms is speed of the processing. In the developed form processing computer system for higher speed of processing some approaches can be used:

- Using parallel processes of recognition on multiple number of workstations. In order to accomplish that the workload has to be distributed across workstations which lets the system to process a large number of forms at the same time.
- Using the multicore functionality in nowadays processors. This is to be used in a workstation for process parallelizing assuming that the workstation has a capability of using multiple cores.

The second option above (using multithreading in a single workstation) is used in the developed form recognition system since this solution don't directly need to add additional hardware equipment.

Each of the different threads has template matching and form recognition stages meaning that separate threads do not work on the same form and hereby same form data. This guaranties improvement of efficiency with data consistency reliability.

The results of our conducted computer experiments shows computational benefit of using multi-threaded processing over single-threaded processing. Note that increasing thread count beyond the number of cores which hardware has capable of having efficiently does not add a significant productivity in terms of computational power [9].

4 Recognition of Handwritten Characters

Recognition of handwritten characters is extensively used for information entrance into a computer system. There are many approaches used for handwritten character recognition, and these approaches differ from each other either in terms of image representation type i.e. image features or classification method used for the pattern recognition part [10–13]. There are a lot of studies about handwritten recognition where different recognition methods were used [14–17].

Some of the classification methods are artificial multilayer feedforward neural networks and convolutional neural networks. A brief introduction to these methods were given below.

4.1 Artificial Multilayer Feedforward Neural Networks

A neural network comprises a finite number of elements, analogous to neurons, with diverse connections between them. The fundamental components of each neural network consist of relatively simple elements, typically of the same type, mimicking the functionality of neurons in the brain.

Each neuron looks like to nerve cells in the brain, possesses a current state that can be either excited or inhibited. It includes a group of synapses, unidirectional input connections linked to the outputs of other neurons, and an axon, the output connection through which signals (excitation or inhibition) are transmitted to the synapses of subsequent neurons in different layers. The activation level of a neuron is determined by multiplying each input by the corresponding weight, i.e., synaptic strength, and summing up all the products [18–20].

In the context of neural networks, each weight signifies the "strength" of a biological synaptic connection. In multilayer feedforward artificial neural networks, neurons are organized into layers, where each layer constitutes a collection of neurons with a single set of input signals (Fig. 5). The number of neurons in each layer can vary independently of other layers. Typically, a network consists of k layers, numbered from left to right. External input signals are directed to the neurons of the first layer (often designated as the input layer, numbered as zero), and the total outputs of the network are derived from the neurons in the last layer. The number of input and output elements is defined by the nature of the problem. In addition to the input and output layers, artificial neural networks may feature one or more intermediate (hidden) layers, forming cascading layers where the output of one layer serves as the input for the next layer. The neural network's function involves transforming the input vector X into the output vector Y, accomplished through the application of network weights.

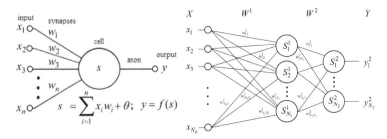

Fig. 5. Multilayer neural network

4.2 Convolutional Neural Networks

Typically, a Convolutional Neural Network (CNN) is structured as a sequence of convolution layers, subsampling layers, and fully-connected layers near the output. In this study, we will adopt the classical LeNet-5 architecture, as originally introduced by Yann Lecun [21], to serve as our CNN architecture (refer to Fig. 6).

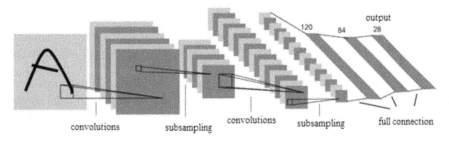

Fig. 6. Convolutional neural network LeNet-5

A convolutional layer comprises a set of feature maps, where each feature map is a two-dimensional matrix representing the outcome of convolution with an individual filter. Each neuron in a feature map is linked to a subset of neurons from the preceding layer. All feature maps within the convolutional layer share the same dimensions and are computed using the formula:

$$(w, h) = (W - k + 1, H - l + 1),$$

where (w, h) – feature map size; W and H – the width and height of the original image; k and l – width and height of convolution kernel.

A convolutional layer operates similarly to the convolution operation, utilizing a small-sized weight matrix known as the convolution kernel. This kernel is systematically moved across the entire processed layer (see Fig. 4). Notably, the kernel constitutes a set of shared weights, where all neurons within a single feature map employ the same set of weights. This shared weight system in the convolutional layer minimizes the number of connections, enabling the identification of consistent features across the entire image.

The kernel, with dimensions $k \times l$, traverses the entire image with a specified step (usually 1). At each step, it performs element-wise multiplication of the window contents with the kernel matrix, summing up the results and recording them in the result matrix (see fig. 7). Subsequently, the result matrix undergoes activation through a function, commonly ReLU, to generate the output of the convolutional layer.

The primary objective of a subsampling layer is to reduce the dimensions of the maps from the preceding layer. Two main types of subsampling layers exist: maximum pooling (max pooling) or average pooling (average pooling) (refer to Fig. 8).

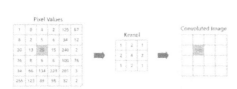

Fig. 7. Convolutional map values

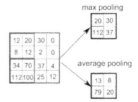

Fig. 8. Pooling layer

To achieve this, the feature map from the previous layer is partitioned into cells typically of a size like 2×2. For each cell, based on the chosen downsampling algorithm, either the maximum or average cell value is determined. This process occurs through the sequential alternation of convolutional and subsampling layers. The resulting output from the last pooling layer is then directed to the input of a fully connected 3-layer feedforward neural network, directly responsible for implementing classification. The number of neurons in the output layer is contingent on the specific task and is commonly set to match the number of classes intended for recognition.

4.3 Conditions for Carrying out Numerical Experiments.

In the computer experiments that were conducted for using its results in handwritten character recognition module of the developed form recognition system have been carried out on the alphabet of the Azerbaijani language which has 32 characters {ABCÇDEƏFGĞHIİJKLMNOÖPQRSŞTUÜVXYZ}. The recognition of handwritten characters in the developed system is accomplished in two stages. In the first stage a given character image's class are searched among the remaining 28 letter classes with excluding {ĞİÖÜ} letters because the characters of these letters can we recognized with a simple lightweight algorithm with a comparison of these characters with {GIOU}. And thereby the second stage of handwritten character recognition is recognition of the upper part of the character image in which dots or strikes are searched. So the number of letters used in the study is 28.

Original character image sizes vary, but we scaled them to 20 × 20 while preserving the original proportions. For using with CNNs white pixels were added to the around of these 20 × 20 images, and we made them 32 × 32. In the training and test sets totally 28 × 500 = 14000, 28 × 100 = 2800 character image instances were used. We did data augmentation and enlarge the number of instances in training set to 28000, 42000, 70000.

As image features raw pixel values and extracted PDC features are used. In the first case with flattened 20 × 20 grayscale pixel values to a vector of 400 elements for each image. For the case where features were with ANNs were extracted using the Peripheral Directional Contributivity feature extraction method (Fig. 9) [22]. These types of features demonstrate relative positioning, complexity and orientation of strokes in a character image. For calculating PDC features in each of the given directions (directions quantity) we find the 1st black pixel (a point) (2nd, 3rd or other points can be found, the number of these points is called depth) then we can calculate Directional Contributivity (DC) vector for each of these points. The elements of a DC vector of a point are found by moving from this point to each of the given directions (DC dimension) until when we find a black pixel. In order to reduce the number of features we divide images into segments. The parameters used in the study in PDC feature extraction are the following:

- Segment count is 8,
- Number of directions is 4,
- Deeps is 2,
- DC dimension is 4,

The resultant PDC feature set contains 8 × 4 × 2 × 4 = 256 elements.

In the study a comparative analysis of different types of NNs were carried out. These types are the followings:

- Multilayer Feedforward Artificial Neural Networks (ANN). Two types of feature sets (raw pixel values, PDC features) are used with ANNs. For each of the feature sets there were used two different ANN structures with different number of neurons in layers.
- Convolutional Neural Networks (CNN). All of the CNNs used have 2 convolutional layers. These CNNs have different structures in terms of the number of filters in the 2nd layer.

In implementation of neural networks for handwritten character recognition Keras library (it works based on the TensorFlow framework) in Python programming language was used [23]. For optimization of the objective function in parametric identification of neural networks Adam method with default arguments is used [24, 25].

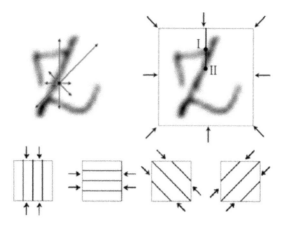

Fig. 9. Extraction of PDC feature

For each neural network architecture 5 different neural networks have been prepared. These models differ in terms of training. Each of 5 different neural networks training has been started from different initial points.

The following parameters are used in the computer experiments:

- sizes of training sets (14000, 28000, 42000, and 70000 character images);
- pooling layer method selection (maximum or average);
- feature extraction influence;
- feature map count influence in the "second" layer.

Total number of neural networks is 40:

- ANN: {quantity of models}x{quantity of DB} = 4 × 4 = 16;
- CNN: {quantity of models}x{quantity of DB} x{quantity of pooling methods} = 3 × 4 × 2 = 24.

4.4 Results

As expected, the classical convolutional network with 6–16 feature maps showed better results, but the multilayer network, which trained on the symbols images, had the worst performance. Note that a multilayer network trained on the results of feature extraction showed rather high results, comparable with convolutional networks. In general, convolutional networks performed better than multilayer networks [21, 26]. In the study maximum accuracies on the test set obtained with ANNs and CNNs are 88.21% and 89.32% respectively (on the training set the accuracy values are 99.5% and 99.64%).

As expected, neural networks, which were trained on the results of feature extraction, showed higher results than networks that were trained on the image itself [27–29]. In the study maximum accuracies on the test set obtained with raw pixel values and PDC features (extracted features) are 84.04% and 88.21% respectively (on the training set the accuracy values are 98.3% and 99.5%).

There is no definite advantage in the choice of the method in the subsampling layer. The choice of the subsampling method for a particular model can be selected experimentally.

Increase the training database did not give a tangible improvement in recognition results for convolutional networks and networks with preliminary feature extraction. However, for networks learning without feature extraction, an increase in the size of the database led to a noticeable improvement in performance.

5 Conclusion

The developed system for handling and objectively assessing test exams allows the efficient conduct of extensive mass examination results recognition, ensuring independent and unbiased evaluation. This system can be easily customized to accommodate different forms, removing the requirement for costly and cumbersome Optical Marker Recognition (OMR) scanners.

The experiments conducted revealed that adopting multi-threaded recognition can significantly enhance productivity compared to single-threaded processing. It was observed that the transition to multi-threading led to a productivity increase of up to 3.5 times. However, further increasing the number of threads beyond the physical core count did not yield significant improvements in performance.

The use of Convolutional Neural Networks makes it possible to achieve acceptable quality of recognition of hand-printed characters of the Azerbaijani alphabet.

The use of an intelligent comparison module based on handwritten character recognition results can increase the reliability of the form recognition results and reduce the number of manual operator verifications. Using hand-printed fields instead of mark fields with circles can significantly reduce the size of the forms.

References

1. Chen, J., Iii, W.W.: Multi-threading performance on commodity multi-core processors. In: Proceedings of HPCAsia, vol. 12, p. 12 (2007)
2. Koyuncu, H.: A comparative study of handwritten character recognition by using image processing and neural network techniques. Hittite J. Sci. Eng. **8**(2), 133–140 (2021)
3. Neetu Bhatia, E.: Optical character recognition techniques: a review. Int. J. Adv. Res. Comput. Sci. Softw. Eng. **4**(5), 1219–1223 (2014)
4. Hamad, K.A., Kaya, M.: A detailed analysis of optical character recognition technology. Int. J. Appl. Math. Electron. Comput. **4**(1), 244–249 (2016)
5. ABBYY FormReader Enterprise Edition. User's Guide (2016)
6. Eldan, R., Shamir, O.: The power of depth for feedforward neural networks. Conf. Learn. Theory **49**, 907–940 (2016)
7. Aida-zade, K.R., Mustafayev, E.E.: About one hierarchical handwritten recognition system on the bases neural networks. Trans. NAS Azerbaijan **2–3**, 94–98 (2002). (in Russian)
8. Aida-zade K.R., Talybov S.H., Mustafayev E.E.: Multilevel recognition system of handwritten forms. In: Proceedings of the 11th All-Russian Conference "Mathematical Methods of Pattern Recognition", Moscow, pp. 230–233 (2003)

9. https://towardsdatascience.com/metrics-to-evaluate-your-machine-learning-algorithm-f10 ba6e38234
10. LeCun Y., Doser, B., Denker, J., et al.: Handwritten digit recognition with a back-propagation network. In: Touretzky, D.S. (ed.) Advances in Neural Information Processing Systems, Denver, vol. 2, pp. 396–404 (1990)
11. LeCun, Y., Bottou, L., Orr, G., Muller, K.: Efficient BackProp. In: Orr, G.B., Müller, K.-R. (eds.) Neural Networks: Tricks of the trade, pp. 9–50. Springer Berlin Heidelberg, Berlin, Heidelberg (1998). https://doi.org/10.1007/3-540-49430-8_2
12. Mori, M., Wakahara, T., Ogura, K.: Measures for structural and global shape description in handwritten kanji character recognition. In: Document Recognition V, vol. 3305, pp. 81–89. SPIE (1998)
13. Hwang, Y.S., Bang, S.Y.: Recognition of unconstrained handwritten numerals by radial basis function network classifier. Pattern Recognit. Lett. **18**, 657–664 (1997)
14. Aida-zade, K.R., Mustafaev, E.E.: Associative multi-level systems of pattern recognition. Trans. NAS Azerbaijan **3**, 15–18 (2001). (in Russian)
15. Aida-zade, K.R., Mustafayev, E.E.: Intelligent recognition system of Azerbaijani handwritten forms. In: Proceedings of the Scientific Conference "Modern Problems of Cybernetics and Information Technologies", vol. III, Baku, pp.85–88 (2006) (in Russian)
16. Aida-zade, K.R., Mustafayev, E.E., Hasanov, J.Z.: About knowledge base usage for increasing intellectuality of recognition systems. In: Proceedings of the 11th Russian Conference "Mathematical Methods of Pattern Recognition", Moscow, pp. 6–8 (2003) (in Russian)
17. Mustafayev, E.E.: Hierarchical multilevel form recognition system. In: Proceedings of Scientific Conference "Modern Problems of Applied Mathematics", Baku, pp. 154–157 (2002)
18. Aida-zade, K.R., Mustafayev, E.E.: About neural networks' parameters optimization during learning. In: Proceedings of the Scientific Conference "Modern problems of Cybernetics and Information Technologies", vol. 1, pp. 118–121, Baku (2003)
19. Golovko, V.A., Krasnoproshin, V.V.: Neural network data processing technologies. BSU, Minsk, 263 p. (2017) (in Russian)
20. Osovsky, S.: Neural networks for information processing. Finance and Statistics, 344 p. (2002) (in Russian)
21. Forsyth, D.A., Ponce, J.: Computer Vision: Algorithms and Applications. Springer (2012)
22. Monga, P.H., Kaur, M.: A novel optical mark recognition technique based on biogeography based optimization. Int. J. Inf. Technol. Knowl. Manage. **5**(2), 331–333 (2012)
23. Rakesh, S., Atal, K., Arora, A.: Cost effective optical mark reader. Int. J. Comput. Sci. Artif. Intell. **3**(2), 44–49 (2013)
24. Lecun, Y., Bottou, L., Bengio, Y., Haffner, P.: Gradient-based learning applied to document recognition. Proc. IEEE **86**(11), 2278–2324 (1998). https://doi.org/10.1109/5.726791
25. Arif, A.F., Takahashi, H., Iwata, A., Tsutsumida, T.: Handwritten postal code recognition by neural network – a comparative study. IEICE Trans. Inf. Syst. E79-D(5), 443–449 (1996)
26. Francois, C.: Deep Learning with Python, p. 362. Manning Publications, Shelter Island, NY (2018)
27. Krizhevsky, A., Sutskever, I., Hinton, G.E.: ImageNet classification with deep convolutional neural networks. In: Advances in Neural Information Processing Systems, pp. 1097–1105 (2012)
28. Mustafayev, E.E.: Handwritten text recognition methods. Fuyuzat, 189 p. (2020) (in Russian)
29. Aida-zade, K.R., Mustafayev, E.E.: Intelligent handwritten form recognition system based on artificial neural networks. In: Proceedings of the International Conference on Modeling and Simulation, Konya, Turkey, pp. 609–613 (2006)

Empirical Analysis of Conformer Impact on the CTC-CRF Model in Kazakh Speech Recognition

Dina Oralbekova[1,2(✉)], Orken Mamyrbayev[1], Keylan Alimhan[3], and Nurdaulet Zhumazhan[4]

[1] Institute of Information and Computational Technologies, Almaty, Kazakhstan
dinaoral@mail.ru
[2] Academy of Logistics and Transport, Almaty, Kazakhstan
[3] L.N. Gumilyov, Eurasian National University, Nur-Sultan, Kazakhstan
[4] U. Joldasbekov Institute of Mechanics and Engineering, Almaty, Kazakhstan

Abstract. The enhancement of quality and efficiency in contemporary speech technology systems is increasingly reliant on the application of machine learning techniques. Recent advancements have highlighted the superior performance of end-to-end speech recognition methods. Integrating multiple end-to-end model architectures within a single network has yielded competitive outcomes. This study focuses on the development and evaluation of an integrated Connectionist Temporal Classification-Conditional Random Field system, augmented with the Conformer, to enhance the recognition accuracy of a low-resource language such as Kazakh. Due to the limited electronic and digital resources in the Kazakh language, extensive efforts are required for the collection and compilation of a comprehensive speech and text corpus. Additionally, this necessitates the development of novel mathematical models and algorithms to address the challenges in automatic speech recognition for agglutinative (Turkic) languages. The research findings demonstrate that the incorporation of the Conformer model alongside CTC-CRF frameworks significantly improves speech recognition outcomes. The system, utilizing the ResNet network and RNN-based language models, further enriched with the Conformer, attained optimal performance, enhancing the recognition accuracy by 1.6%.

Keywords: Automatic Speech Recognition · CTC-CRF Model · Conformer · Kazakh Language · Low-resource Language

1 Introduction

Speech is a powerful tool not only when communicating with people, but also with machines and automated systems. Therefore, it can be considered as a mechanism used by a person of acoustic signals, written signs and symbols for the processing, presentation, storage and dissemination of information. The development of an automatic speech recognition system is associated with an increasing need on the part of users of mobile devices and stationary computers to ensure the best quality and performance of modern speech technologies Traditional speech recognition systems consist of different components, if properly configured, they can give good results, i.e. clear readable text. These components include an acoustic model that maps a speech segment into a likelihood at the audio level, a language model determines the most probable sequence of words, and a decoder, based on the data obtained from the acoustic and language models, displays the final result as a sequence of words/phrases. The modularity design is based on many independent assumptions, which leads to computational complexity and model tuning, and the traditional acoustic model is trained on frames that depend on the Markov model. Traditional systems demonstrate a fairly high recognition accuracy, but at the same time they consist of several independent complex modules, which can cause problems when building models. In addition, building system models requires a lot of labeled data, which entails not only manual work, but also funds. With the advent of machine learning, the situation has changed a lot for the better. Now there is no need to build separate elements to get the desired result. The use of various architectures of artificial neural networks has greatly facilitated the work in designing a speech recognition system. Thus, the researchers [1] implemented a language model using RNN, in [2] others obtained a dictionary based on LSTM networks, and other scientists [3] used deep neural networks to extract acoustic features and create an acoustic model. In addition, new architectures of neural networks are emerging to improve the quality of recognition, which makes this research topic relevant.

Ensuring the best quality and performance of modern speech technologies is possible based on the widespread use of machine learning methods. Over the past few years, with the increase in computing power of computers, the use of machine learning methods in speech recognition systems has become very popular. The development of a platform for parallel computing using GPUs has reduced the training time for deep neural networks on large-scale data. When training a neural network, only one model can produce the desired result without additional components, which is called the end-to-end model.

The connection temporal classification (CTC) model works without initial alignment of the input and output sequences. CTC was designed for language decoding [4, 5]. An alternative method is the Sequence to Sequence (Seq2Seq) models with the Attention mechanism [6]. Such models consist of one encoder and one decoder. The encoder compresses the audio frame information into a more compact vector representation by reducing the number of neurons from layer to layer, and the decoder takes this compressed representation and decodes a sequence of characters and words based on a recurrent neural network. Another end-to-end approach is the Conditional Random Fields (CRF) model. The results show that the CRF-based function in the speech-to-text process improves recognition performance by reducing the number of phoneme deletions.

The basic principle of work is that modern end-to-end models are trained on the basis of big data. From this, one can discover the main problem, this concerns the recognition of languages with limited training data, which includes the Kazakh language. There is very little electronic and digital information in the Kazakh language, which requires a lot of time to collect and create a corpus of speech and texts.

A significant difference of this work from previous studies [7] is a complex and generalizing nature, aimed at creating an improved hybrid end-to-end speech recognition technology for the Kazakh language. In comparison with the work in this study, the following improvements were made: 1) an increase in training data, 2) an improvement in the jointly trained CTC + CRF models with the Conformer to improve the quality of Kazakh speech recognition.

The work is framed as follows: in Sect. 2, an analysis of existing works on the topic under study is made. Section 3 describes the applied models, their improvements and architecture. Section 4 then presents the data used, the main settings of the model, experiments, and analysis of the results obtained.

2 Literature Review

J. Lafferty et al. [8] proposed an algorithm for estimating parameters for conditional random fields and showed that CRF has a greater advantage over HMM and MEMM (maximum entropy Markov models) for natural language data. The CRF model is used to evaluate the measurement accuracy in the phonetic recognition task, as well as the accuracy of detecting the boundaries between them. The results show that when using transition functions in a CRF-based recognition framework, recognition performance is significantly improved by reducing the number of phoneme deletions. Edge detection performance is also improved, mainly for transitions between silence, stop, and click phonetic classes. In addition, the CRF model gives a lower error rate than the HMM and Maxent models in the task of detecting sentence boundaries in speech.

Segmental and linear CRFs have become widespread in speech recognition [9]. The segment approach is based on the use of acoustic sequences as initial data and the automatic construction of a universal set of functions at the segment level [10].

In [11] (2016), the authors implemented a different training scheme for the CRF model to evaluate ASR validity with only limited annotated data and large partially annotated data. The training data was increased and the experimental evaluations showed that the implemented method can improve the CMS compared to using only a small amount of annotated data and completely using automatically generated invalid full annotations.

Chen et al. [12] (2017) presented the main challenge of Dialogue Act Recognition in terms of extending richer CRF structural dependencies while taking advantage of end-to-end learning. The authors combined hierarchical semantic inference with a memory mechanism when modeling an utterance. Then we expanded the network of structured attention to a layer of a conditional random field with a linear chain, which captures contextual statements and corresponding dialogue actions. The experiments were carried out on the basis of the Switchboard Dialogue Act and Meeting Recorder Dialogue Act cases. As a result, it was shown that the proposed method provides better performance than other modern solutions to the problem.

An et al. (2019) [13] developed a CAT toolkit that combines the two models CTC and CRF. To train the model, 2 cases were used - Switchboard (Chinese) and Aishell (English). This approach benefits from a reduction in the number of configurable parameters and high performance compared to other speech technology models. Then the same authors [14] (2020) improved their model, which includes contextualized soft forgetting, which makes it possible to recognize speech without loss with limited training data. For the experiment, Chinese and English tests like Switchboard and Aishell were applied, thus obtaining the most up-to-date results among existing end-to-end models with fewer parameters and competitive compared to DNN-HMM hybrid models.

Researchers [15] (2019) have developed a Conditional Random Field (CRF) acoustic model with a one-stage architecture that is based on CTC (Connectionist Temporal Classification). They conducted scoring experiments using the WSJ, Switchboard, and Librispeech datasets. When compared directly with the CTC-CRF model using simple bidirectional LSTMs, their model consistently performed best on all three benchmark datasets for both mono and mono symbols. This was especially noticeable in comparison with the lattice-free maximum-mutual-information method.

In [16] (2021), methods have been explored that can successfully combine newly developed text word modeling modules and Conformer-type neural networks into a CTC-CRF architecture. The experiments were carried out on the English Switchboard and Librispeech datasets, as well as on the German CommonVoice dataset. The obtained experimental results showed that the use of models of the Conformer type significantly improves the quality of speech recognition.

In [17] (2019), a text processing model after Chinese speech recognition is presented, which combines a bidirectional network with long-term short-term memory (LSTM) and a conditional random field (CRF) model. This model was developed to address the issue of dialects and accents in speech recognition and requires text to be corrected before it is displayed. The task is divided into two stages: detection of text errors and correction of text errors. In this paper, a Bi-LSTM and a Conditional Random Field are applied in two steps, respectively. Through validation and system testing on the SIGHAN 2013 Chinese Spelling Check dataset, experimental results show that this model can effectively improve text accuracy after speech recognition.

3 Methodology

This section describes the CTC and CRF models, as well as the Conformer and their joint model.

3.1 Extracting Features

Feature extraction algorithms and recognition methods constitute crucial components of an ASR system. Feature extraction entails deriving a condensed set of data crucial for problem resolution [18]. Mel-frequency cepstral coefficients (MFCC) and perceptual linear prediction (PLP) algorithms are predominantly employed for feature extraction from audio signals, with MFCC being the more prevalent [19, 20].

In speech recognition tasks, the incoming audio signal is transformed into a series of feature vectors, which then serve as the basis for subsequent classification.

The mel-frequency cepstral coefficient (MFCC) algorithm stands as the most efficacious method for feature extraction. The MFCC algorithm mimics the human auditory system and is designed to replicate the ear's processing mechanism. The relationship between the mel scale and the frequency of audio signal oscillations is expressed as follows (1):

$$M = 1125 ln(1 + f/700) \quad (1)$$

For the analysis of audio signals, spectral analysis methods are employed, utilizing the fast Fourier transform to ascertain the speech signal's spectrum [21]. However, post-Fourier transform, the output signal often exhibits non-linear distortions at frame junctions. To address this issue, the results of the fast Fourier transform should be multiplied by a weighting function. In digital signal processing, these functions, known as windows, are used to mitigate such distortions. An exemplar of such a function is the Hanning window (2):

$$X_f = \sum_{k=0}^{K-1} x_k e^{-\frac{2\pi i}{K} fk}, f = 0, \ldots, K-1 \quad (2)$$

here K is the dimension of the discrete segment of the signal, x_k is the amplitude of the k-th signal, X_f is the K amplitudes of sinusoidal signals, which represent the main signal.

In signal analysis, the data derived from a single signal spectrum is often insufficient. To address this, a cepstrum, defined as the logarithm of the input signal's spectrum, is utilized. This approach transforms the spectrum into a quasi-independent signal, thereby presenting the significant spectral information in a more compact form and simplifying its analysis. The MFCC algorithm is instrumental in extracting acoustic features. This involves converting the signal frequency to the mel scale and subsequently computing feature vectors for neural network processing.

The initial step in the MFCC computation involves segmenting the input signal into sections known as frames or windows. The length of each frame has a direct impact on the algorithm's efficacy, with 25 ms being the most optimal duration for speech recognition tasks.

For each frame, its spectrum is computed using the fast Fourier transform. The spectrum coefficients of these frames are then mapped onto the mel-frequency scale. Given that the human auditory system has heightened sensitivity to lower frequencies, these mel-frequency scales are more densely concentrated at lower frequencies.

Subsequently, a discrete cosine transform is applied. The outcome of this transformation is a multi-dimensional vector comprising the mel-frequency cepstral coefficients. These feature vectors enable the identification of specific phonemes present in the incoming signal.

3.2 Connectionist Temporal Classification

Connectionist Temporal Classification (CTC) is a loss function used to train a neural network that converts speech into a sequence of words. Suppose we have an output sequence $y = S_w(x)$, where each element of the output is a probability distribution vector for each letter V' at time t. We define y_k^t as the probability of saying character k from the alphabet V' at time t. If μ is a sequence of characters, including the space character, for input x, then the probability $P(\mu|x)$ can be calculated as follows (3):

$$P(\mu|x) = \prod_t y^t_{\mu_t} \quad (3)$$

From the above equation, you can see that the elements of the output sequence are independent of each other. To align the data, you need to add an auxiliary character that will remove duplicate letters and spaces. Denote it as B. Thus, the total probability of the output sequence can be represented by the following expression (4):

$$P(y|x) = \sum_{\mu \in B^{-1}(y)} P(\mu|x) \qquad (4)$$

The above equation determines the sum over all alignments using dynamic programming, and helps to train the ANN on unlabeled data (5):

$$CTC(x) = -logP(y|x) \qquad (5)$$

It follows from the above that the ANN can be trained on any gradient optimized algorithm. In the CTC architecture, any kind of ANN can be used as an encoder, like LSTM and BLSTM.

To decode the CTC model, the assumption (6) was presented in [22]:

$$argmaxP(y|x) \approx B(\mu^\circ) \qquad (6)$$

where $\mu^\circ = argmaxP(y|x)$.

CTC eliminates the need for data alignment and allows you to apply quite a number of layers, a simple network structure to implement a model that maps audio to text.

3.3 Conditional Random Field

The Conditional Random Field (CRF) is a statistical method for classification that uniquely incorporates the context of the object being classified. Distinguished as a non-directional discriminative probabilistic graphical model, CRF is notable for its ability to bypass the need for modeling probabilistic interdependencies among observed variables [23]. This sets it apart from the Markov maximum entropy model, as CRF is not prone to the label bias issue, a notable advantage in statistical modeling [8, 24]. CRF and its diverse adaptations find applications across multiple domains, including but not limited to natural language processing, computer vision, and speech recognition (Fig. 1).

In the realm of machine learning, linear and segmental conditional random fields (linear chain CRF and segmental CRF) are prevalently employed, particularly in scenarios demanding the labeling or segmentation of variable sequences. These models excel in handling data characterized as variable sequences. The application of linear and segmental CRFs is extensive in fields associated with natural language processing and other areas involving sequential data processing.

Focusing on Segmental Conditional Random Fields, the Segment CRF, or the semi-Markov CRF [25], leverages segment recurrent neural networks. This model expands the scope of traditional conditional random fields to operate at the segment level instead of the individual frame level. In this framework, each segment is annotated with a corresponding word. Subsequently, features are derived that quantify the congruence between the acoustic signal and the word hypothesis for that segment. These features are integrated into a log-linear model to calculate the posterior probability of the word sequence given the sound.

In the context of speech recognition, given a sequence of input vectors X and a corresponding sequence of output labels Y (Fig. 1), zero-order linear CRF determines the conditional probability at the sequence level (7) using auxiliary segments as follows

$$P(y, s|x) = \frac{1}{Z(x)} \prod_{i=1}^{n} expf(y_i, s_i, x_i) \quad (7)$$

In the formula, Z(X) is a normalization term, and the full calculation of this value is given in [26]. In this work, features were investigated at the segmental level using recurrent neural networks (RNN) and segmental RNN (SRNN) model. The SRNN model was applied as an implementation of the SCRF acoustic model for multitasking learning.

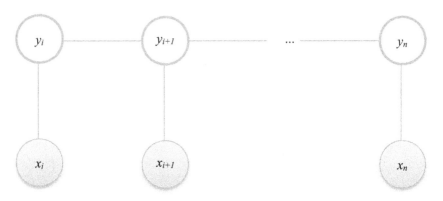

Fig. 1. Architecture of the CRF model.

3.4 Conformer

The Conformer model is a modern deep neural network architecture used for speech recognition problems [27]. It combines the advantages of Convolutional Neural Networks (CNN) and Transformer to achieve high performance in processing sequences of audio frames (Fig. 2).

The Conformer model consists of several elements: an input block, a convolutional and Transformer blocks, and a final block. Input audio frames are represented as a sequence of feature vectors, denoted as $X = \{x_1, x_2,..., x_T\}$, where T is the length of the frame sequence. Convolutional blocks are used to capture local contextual dependencies in the time domain. Convolution blocks have convolution, normalization, activation and attention mechanism. The convolution operation in the convolution block performs a linear transformation of the input features using convolution (8):

$$Conv(h) = W \cdot h + b \quad (8)$$

where conv(h) is the result of the convolution operation, W is the convolution matrix, h are the input features, b is the bias. Next, normalization of the output data and improvements in training stability are applied (9):

$$norm(h) = LayerNorm(conv(h)) \quad (9)$$

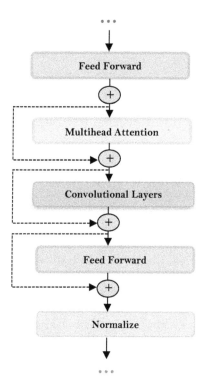

Fig. 2. Model Conformer.

After that, activation is applied after normalization to introduce non-linearity into the model. The Conformer model typically uses the ReLU (Rectified Linear Unit) activation function. Next comes the attention mechanism, which in a convolutional block allows you to model the interaction between different temporal contexts. The Conformer model uses the Multihead Attention attention mechanism, which allows you to take into account the importance of different positions in the input sequence (10):

$$att(h) = MultiheadAttention(act(h)) \qquad (10)$$

here att(h) is the output features after applying the attention mechanism.

Convolutional blocks in the Conformer model are used to capture local context dependencies in the time domain of audio frames. After the application of convolutional blocks, the sequence of features passes through transformer blocks, which provide modeling of long-range dependencies and global context. This combination of convolutional and transformer blocks allows the Conformer model to achieve high performance in speech recognition tasks.

One way to improve the CTC-CRF model and add a Conformer to further improve the quality of speech recognition is to include an acoustic Conformer model as a CTC-CRF component.

3.5 Proposed Model

The proposed co-learning model of end-to-end models can be extended using the Conformer architecture to further improve the quality of speech recognition. Instead of using only LSTM and bidirectional LSTM, the model can include convolutional Conformer blocks, which allow you to model dependencies at the feature sequence level and handle long sequences efficiently.

Thus, the proposed model can be modified as follows (11):

$$L_{CTC/SCRF/Conformer} = \tau L_{CTC}(x) + (1 - \tau)L_{SCRF}(x) + \lambda L_{Conformer}(x) \qquad (11)$$

where τ is a tunable parameter and satisfies the condition $- 0 \leq \tau \leq 1$, λ is a tunable parameter that satisfies the condition $\lambda \geq 0$. In this formula, the term $\lambda L_{Conformer}(x)$ is added, which represents the Conformer loss function. This allows the model to take into account dependencies at the level of feature sequences using the Conformer architecture to improve the quality of speech recognition.

Co-training CTC, SCRF, and Conformer models allows you to take advantage of each model and improve speech recognition results. Adjustable parameters τ and λ allow you to control the contribution of each model to the overall loss function, which allows you to more flexibly adjust the importance of each component in the learning process.

Conformer uses the Transformer attention mechanism, which allows you to model dependencies over long time intervals. This is especially important for low-resource languages, where words and phonetic features can be difficult to recognize due to the small amount of training data. The Conformer is better able to capture long-term contextual dependencies, resulting in more accurate speech recognition. In addition, the use of the Conformer model provides better handling of noisy audio data and speech variability. This is important for languages where the quality of recordings may be poor or speech may vary depending on the speaker or context.

4 Experiments and Results

This section contains case descriptions, preliminary model settings, experimental data, as well as the results obtained and a comparative analysis of the obtained data.

4.1 Data Preparation

The performance and quality of an APP system depends almost entirely on the variety and size of the speech database. There are many factors that influence the quality of an RDA system. These include recording conditions, environment, audio duration, recording devices, gender, age and area of residence of the speaker, etc. Depending on the recording condition of speech data, you can get different results of the automatic speech recognition system.

To expand the corpus, students and undergraduates from Kazakhstani universities, as well as researchers from the Institute of Information and Computing Technologies were involved [28]. A link to a fragment of the corpus is given at the link.

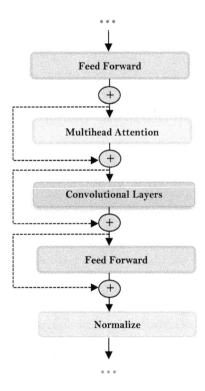

Fig. 2. Model Conformer.

After that, activation is applied after normalization to introduce non-linearity into the model. The Conformer model typically uses the ReLU (Rectified Linear Unit) activation function. Next comes the attention mechanism, which in a convolutional block allows you to model the interaction between different temporal contexts. The Conformer model uses the Multihead Attention attention mechanism, which allows you to take into account the importance of different positions in the input sequence (10):

$$att(h) = MultiheadAttention(act(h)) \qquad (10)$$

here att(h) is the output features after applying the attention mechanism.

Convolutional blocks in the Conformer model are used to capture local context dependencies in the time domain of audio frames. After the application of convolutional blocks, the sequence of features passes through transformer blocks, which provide modeling of long-range dependencies and global context. This combination of convolutional and transformer blocks allows the Conformer model to achieve high performance in speech recognition tasks.

One way to improve the CTC-CRF model and add a Conformer to further improve the quality of speech recognition is to include an acoustic Conformer model as a CTC-CRF component.

In the experimental phase, the end-to-end model, when operated using the CTC function devoid of any external language model, achieved a CER of 17.3% and a Word Error Rate (WER) of 29%. Subsequently, the incorporation of an external language model into the CTC framework led to an enhancement in both CER and WER metrics, reducing them by 14% and 19%, respectively.

Further advancements were noted in our conjoined CTC and CRF model, which employed the ResNet architecture without normalization modulation (NM). This configuration yielded a CER of 11.4% and a WER of 18.2%. The subsequent inclusion of NM in the model architecture resulted in a modest improvement in performance, exhibiting an approximate 1.5% enhancement in accuracy (Fig. 3).

Moreover, the integration of the Conformer model during the training phase significantly bolstered the system's performance, leading to a 1.6% improvement in the recognition rate. These comprehensive results and comparative analyses are systematically presented in Table 1.

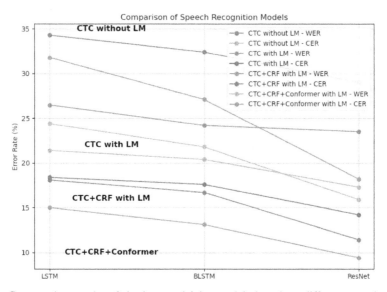

Fig. 3. Comparative graphs of the base and joint models based on different neural network architectures with LM in terms of CER.

Table 1 shows that the models that applied the ResNet network showed the best result in terms of the coefficients of correctly recognized words and symbols.

Table 1. Results of experiments on the application of end-to-end models.

Model	WER, %	CER, %
CTC without LM		
LSTM	34,3	21,4
BLSTM	32,4	20,4
ResNet	29	17,3
CTC with LM		
LSTM	26,5	18,4
BLSTM	24,2	17,6
ResNet	23,5	14,2
CTC + CRF without LM		
CTC (LSTM) + CRF	31,8	18,1
CTC (BLSTM) + CRF	27,1	16,7
CTC (ResNet) + CRF	18,2	11,4
CTC + CRF with LM		
CTC (LSTM) + CRF	26,2	15,1
CTC (BLSTM) + CRF	23,5	14,3
CTC (ResNet) + CRF	16,7	9,8
CTC (LSTM) + CRF + Conformer	24,4	15
CTC (BLSTM) + CRF + Conformer	21,8	13.1
CTC (ResNet) + CRF + Conformer	**15,9**	**9,4**

5 Conclusion

This research explores the amalgamation of end-to-end models comprising CTC, CRF, and the Conformer model for the purpose of Kazakh speech recognition. The Conformer model, distinguished by its convolutional blocks, is adept at effectively modeling dependencies in feature sequence levels and managing extended sequences. The study employed various RNN architectures, including LSTM and BiLSTM, in conjunction with the ResNet network. MFCC were utilized for feature extraction.

Experimental evaluations were conducted on a comprehensive Kazakh speech corpus, amassing a total duration of 315 h. The findings indicated that the system, which integrates the ResNet network with RNN-based language models and the additional implementation of the Conformer, demonstrated high efficiency. Decoding processes utilizing these models did not result in substantial increments in computational demands, thereby enabling rapid speech decoding capabilities. An impressive CER of 9.4% was attained, marking a competitive milestone.

Conclusively, the deployment of the end-to-end CTC-CRF-Conformer model, in conjunction with the ResNet network and RNN-based language models, has proven to be highly effective in achieving advanced speech recognition accuracy in languages with constrained data availability, such as Kazakh. This approach holds considerable potential for application in other languages possessing similar linguistic traits, particularly in agglutinative languages.

Acknowledgement. This research has been funded by the Science Committee of the Ministry of Science and Higher Education of the Republic Kazakhstan (Grant No. AP19174298).

References

1. Graves, A., Mohamed, A., Hinton, G.: Speech recognition with deep recurrent neural networks. In: ICASSP, IEEE International Conference on Acoustics, Speech and Signal Processing – Proceedings, vol. 38 (2013). https://doi.org/10.1109/ICASSP.2013.6638947
2. Khandelwal, S., Lecouteux, B., Besacier, L.: Comparing GRU and LSTM for automatic speech recognition. Research Report LIG. 2016. ffhal-01633254f (2016)
3. Hinton G., et al.: Deep neural networks for acoustic modeling in speech recognition: the shared views of four research groups. Sign. Process. Mag. **29**, 82–97. IEEE (2012). https://doi.org/10.1109/MSP.2012.2205597
4. Watanabe, S., Hori, T., Kim, S., Hershey, J.R., Hayashi, T.: Hybrid CTC/attention architecture for end-to-end speech recognition. IEEE J. Sel. Top. Sign. Process. **11**(8), 1240–1253 (2017). https://doi.org/10.1109/JSTSP.2017.2763455.(2017)
5. Mamyrbayev, O., Alimhan, K., Oralbekova, D., Bekarystankyzy, A., Zhumazhanov, B.: Identifying the influence of transfer learning method in developing an end-to-end automatic speech recognition system with a low data level. Eastern-Eur. J. Enterp. Technol. **1**, 84–92 (2022). Available at SSRN: https://ssrn.com/abstract=4070075
6. Bahdanau, D., Chorowski, J., Serdyuk, D., Brakel, P., Bengio, Y.: End-to-end attention-based large vocabulary speech recognition. Acoust. Speech Sign. Process. (ICASSP), 4945–4949. IEEE (2016)
7. Oralbekova, D., Mamyrbayev, O., Othman, M., Alimhan, K., Zhumazhanov, B., Nuranbayeva, B.: Development of CRF and CTC based end-to-end kazakh speech recognition system. In: Nguyen, N.T., Tran, T.K., Tukeyev, U., Hong, T.P., Trawiński, B., Szczerbicki, E., (eds.) Intelligent Information and Database Systems. ACIIDS 2022. Lecture Notes in Computer Science, vol. 13757. Springer, Cham (2022). https://doi.org/10.1007/978-3-031-21743-2_41
8. Lafferty, J., McCallum, A., Pereira, F.: Conditional random fields: probabilistic models for segmenting and labeling sequence data. In: Proceedings of the International Conference on Machine Learning (ICML 2001), pp. 282–289, Williamstown, MA, USA (2001)
9. Fosler-Lussier, E., He, Y., Jyothi, P., Prabhavalkar, R.: Conditional random fields in speech, audio, and language processing. In: Proceedings of the IEEE, vol. 101, no. 5, pp. 1054–1075 (2013)
10. Markovnikov, N.M., Kipyatkova, I.S.: Analytical review of integrated speech recognition systems. In: Proceedings of SPIIRAN, vol. 58, pp. 77–110 (2018)
11. Li, S., Lu, X., Mori, S., Akita, Y., Kawahara, T.: Confidence estimation for speech recognition systems using conditional random fields trained with partially annotated data. In: 2016 10th International Symposium on Chinese Spoken Language Processing (ISCSLP), pp. 1–5, Tianjin, China (2016). https://doi.org/10.1109/ISCSLP.2016.7918419.
12. Chen, Z., Yang, R., Zhao, Z., Cai, D., He, X.: Dialogue act recognition via CRF-attentive structured network. In: The 41st International acm sigir Conference on Research\& Development in Information Retrieval, pp. 225–234 (2018). https://doi.org/10.1145/3209978.3209997

13. An, K., Xiang, H., Ou, Z.: CAT: CRF-based ASR Toolkit. ArXiv, abs/1911.08747 (2019)
14. An, K., Xiang, H., Ou, Z.: CAT: A CTC-CRF based ASR toolkit bridging the hybrid and the end-to-end approaches towards data efficiency and low latency. INTERSPEECH (2020)
15. Xiang, H., Ou, Z.: CRF-based single-stage acoustic modeling with CTC topology. In: 2019 IEEE International Conference on Acoustics, Speech and Signal Processing (ICASSP), pp. 5676–5680 (2019)
16. Zheng, H., Peng, W., Ou, Z., Zhang, J.: Advancing CTC-CRF Based End-to-End Speech Recognition with Wordpieces and Conformers. ArXiv, abs/2107.03007 (2021)
17. Li, Y., Li, Y., Wang, J., Tang, Z.: Post text processing of Chinese speech recognition based on bidirectional LSTM networks and CRF. Electronics **8**(11), 1248 (2019). https://doi.org/10.3390/electronics8111248
18. Mamyrbayev, O., Kydyrbekova, A., Alimhan, K., Oralbekova, D., Zhumazhanov, B., Nuranbayeva, B.: Development of security systems using DNN and i & x-vector classifiers. Eastern-Eur. J. Enterp. Technol. **4**, 32–45 (2021)
19. Mohan, J., Babu, R.: Speech recognition using MFCC and DTW. In: 2014 International Conference on Advances in Electrical Engineering, pp. 1–4 (2014)
20. Dave, N.: Feature extraction methods LPC, PLP and MFCC in speech recognition. Int. J. Adv. Res. Eng. Technol. (ISSN 2320–6802), **1**(6), 1–5 (2013)
21. Ernawan, F., Abu, N., Suryana, N.: Spectrum analysis of speech recognition via discrete Tchebichef transform. In: Proceedings of SPIE - The International Society for Optical Engineering, vol. 8285, pp. 1-8. (2011)
22. Graves, A., Fernández, S., Gomez, F., Schmidhuber, J.: Connectionist temporal classification: labelling unsegmented sequence data with recurrent neural 'networks. In: ICML 2006 - Proceedings of the 23rd International Conference on Machine Learning, pp. 369–376 (2006). https://doi.org/10.1145/1143844.1143891
23. Sutton, C., McCallum, A.: An Introduction to Conditional Random Fields for Relational Learning. MIT Press (2006)
24. Bottou, L.: Une approche theorique de l'apprentissage connexionniste: Applications a la reconnaissance de la parole. Doctoral dissertation, Universite de Paris XI, 1991 31 In: Culotta, A., Wick, M., Hall, R., McCallum A. First-Order Probabilistic Models for Coreference Resolution Proc. of HLT-NAACL (2007)
25. Kong, L., Dyer, C., Smith, N.A.: Segmental recurrent neural networks. Preprint: arXiv: 1511.06018. https://arxiv.org/abs/1511.06018 (2021)
26. Lu, L., Kong, L., Dyer, C., Smith, N., Renals, S.: Segmental recurrent neural networks for end-to-end speech recognition. In: Proc. INTERSPEECH (2016)
27. Gulati A., et al.: Conformer: Convolution-augmented Transformer for Speech Recognition, pp. 5036–5040 (2020). https://doi.org/10.21437/Interspeech.2020-3015
28. Laboratory of computer engineering of intelligent systems – https://iict.kz/laboratory-of-computer-engineering-of-intelligent-systems/ (data of request: 12.09.2023)
29. Zhao, G., Zhang, Z., Guan, H., Tang, P., Wang, J.: Rethinking ReLU to Train Better CNNs. 603–608 (2018). https://doi.org/10.1109/ICPR.2018.8545612
30. Ioffe, S., Szegedy, C.: Batch normalization: accelerating deep network training by reducing internal covariate shift. In: Proceedings of the 32nd International Conference on Machine Learning, PMLR, vol. 37, pp. 448–456 (2015)
31. Levenshtein, V.I.: Binary codes capable of correcting deletions, insertions, and reversals. Soviet Phys. Doklady **10**, 707–710 (1996)

Ontological Approach to Describing the Interaction of Information Processes in Intelligent Systems

Valerian Ivashenko(✉) and Daniil Shunkevich

Belarusian State University of Informatics and Radielectronics, Minsk, Belarus
`{ivashenko,shunkevich}@bsuir.by`

Abstract. The paper considers an ontological approach to describing the interaction of information processes in intelligent systems. A hierarchical family of ontologies is proposed that investigate both fundamental aspects of representing temporal entities in knowledge bases of intelligent systems and higher-level aspects related to describing the interaction of information processes. The hierarchy of such processes from the processes performed within the hardware platform to the processes of problem solving within a distributed system of agents is considered. Approaches to synchronization of such processes are considered. Special attention is paid to the ontology of spatio-temporal entities, which specifies the semantics of concepts describing changes in various entities in time, such as events, situations, cause-and-effect relations, and so on. Of the three levels of information processes, the paper discusses in the most detail the lower level related to the interpretation of information processes within the hardware platform. Two approaches to organizing of the such interpretation are considered.

Keywords: Ontology · Information Process · Intelligent System · Dynamic Subject Domain · Multi-agent Systems · OSTIS Technology · SC-code

1 Introduction

1.1 Problem Statement and State of Art

When developing modern computer systems, in particular, intelligent systems, developers are increasingly faced with the need to represent dynamic or temporal knowledge, i.e., knowledge that changes over time. Examples of such types of knowledge are various phenomena: situations, events, temporal relations between entities, as well as processes describing various kinds of changes. Such processes include both informational processes describing changes occurring in the memory of an information system (first of all, processes of solving various problems) and processes occurring in the external environment, such as, for example, production, social, physical, chemical and others. The necessity to describe various processes is connected with the need to solve problems in dynamic subject domains, which include most areas of human activity in general.

Examples of tasks that require description of various external processes within the information system are tasks of automation of human activity (in the field of production, education, science, health care, law and others), analysis, evaluation and optimization of various external entities (in particular, construction of so-called digital twins), forecasting tasks and many others.

In its turn, the description of processes occurring in the memory of an information system is necessary both for solving basic problems to ensure the vitality of this system itself, such as analyzing the correctness of process execution, providing the possibility of parallel problem solving (first of all, synchronization of parallel process execution), etc., and for analyzing and correcting the behavior of the system itself, including the possibility of self-analysis (reflection) of an intelligent system, which is especially important in the case of intelligent computer systems. The main attention in this paper will be paid to the information processes occurring in the memory of an intelligent computer system or a collective of such systems.

At present, a number of approaches have been developed that allow to take into account the time factor (dynamics) in knowledge representation [1–3], as well as means of describing information processes and their synchronization [4–7]. Special mention should be made of the PDDL family of languages [8–10] intended to describe the plans in the field of AI planning, which closely overlaps with the description of the processes of realization of these plans.

Ontological models have been proposed for a unified and universal description of dynamic knowledge in general including information processes at different levels [3, 11]. In models similar to those developed, it is important to strive for their optimization, within the framework of knowledge management tasks [12] associated with design tasks, the formation of structures [12] in these models and the main tasks of cognition [12] and their subtasks, in accordance with the information requests being formed. However, in addition to means of describing processes, models are needed that describe possible conflict situations associated with the interaction of problem-solving processes, ensuring synchronization and consistency of the interaction of processes, allowing to resolve emerging conflict situations in order to maximize the number of successfully solved problems. Also, to date, there is no developed ontology for describing processes integrated into any complex technology, which does not allow the widespread use of existing results for the development of applied intelligent systems.

The proposed model of process interaction must be consistent with other models that provide solutions to the following problems: problems of ensuring and managing access to the memory of an intelligent system, tasks of managing memory (processor-memory, machine) and data and knowledge processing processes, tasks of ensuring memory consistency and knowledge bases of the intelligent system [12].

1.2 Proposed Approach

General Points of the Proposed Approach. This work proposes an ontology approach and models using a unified knowledge representation, involving the development of a consistent family of ontologies that allows:

- describe in the memory of an intelligent computer system the operational semantics of knowledge (including dynamic knowledge) related to the concept of causal change (becoming), allowing to form a concept of space-time, events, phenomena, to distinguish permanent entities from temporary entities, to identify their spatio-temporal properties and temporal relations between them;
- develop a model of synchronization and interaction of information processes at various levels listed above, including both description of conflict situations and description of mechanisms for their resolution.

Application of the ontological approach will allow to unify the used means of description of information processes at different levels, which in turn will allow:

- ensure the compatibility of various intelligent computer systems, as well as their components, in particular, tools for analyzing and managing information processes, which will ensure the possibility of reuse of such components;
- simplify the development of means of analysis and control of information processes at different levels, in particular, means of synchronization of parallel processes execution due to unification of principles of description and interaction of processes at these levels;
- provide the possibility to analyze and optimize information processes at different levels by unifying the description of these processes.

Principles of OSTIS Technology. As a basis for the development of the specified family of ontologies within the framework of this work it is proposed to use OSTIS Technology [13], corresponding to the model of unified semantic representation of knowledge. Intelligent systems developed on the basis of this technology are called ostis-systems. The base of OSTIS Technology is a universal way of semantic representation (coding) of information in the memory of intelligent computer systems, called SC-code. SC-code texts (sc-texts, sc-constructions) represent unified semantic networks with basic theoretical-multiple interpretation. The elements of such semantic networks are called sc-elements (sc-nodes and sc-connectors, which, in turn, depending on their orientation can be sc-arc or sc-edges). The universality and unified nature of the SC-code allows describing on its basis any types of knowledge and any methods of problem solving, which, in turn, greatly simplifies their integration both within one system and within a team of such systems. The basis of the knowledge base developed by OSTIS Technology is a hierarchical system of semantic models of subject domains and ontologies, among which there is a universal Core of semantic models of knowledge bases and a methodology of development of semantic models of knowledge bases, providing semantic compatibility of developed knowledge bases. The basis of the ostis-system problem solver is a set of agents interacting exclusively by means of specification of information processes they perform in semantic memory (sc-agents).

All of the above principles together allow to ensure semantic compatibility and simplify integration of both different components of computer systems and such systems themselves.

Within the framework of *OSTIS Technology* several universal variants of visualization of *SC-code* constructs have been proposed, within this work examples in *SCg-code*, a graphical variant of visualization of SC-code constructs, will be used.

This paper will present fragments of the developed ontologies as well as some examples of the application of the obtained ontologies to describe specific information processes.

It is important to note that the development of the proposed family of ontologies will allow to record the specification of all information processes performed in the memory of ostis-systems (including the constructions providing synchronization of parallel processes) and corresponding tasks by means of SC-code in the same knowledge base as the actual subject knowledge of the system. The description of all the above information in a single memory will allow, firstly, to ensure the independence of the developed components of ostis-systems from ostis-systems implementation platforms, and secondly, to ensure the ability of the system to analyze the processes occurring in it, to optimize and synchronize their execution, i.e. to ensure the reflexivity of the designed intelligent systems.

Architecture of Ostis-Systems and OSTIS Ecosystem. One of the key principles of OSTIS Technology is to ensure platform independence of ostis-systems, i.e. strict separation of the logical-semantic model of the system (sc-model of a computer system) and the platform of interpretation of such sc-models (ostis-platforms) [13].

In any variant of ostis-platform realization there is always both software and hardware part. Thus, any software variant of the ostis-platform [14] assumes its subsequent interpretation on some hardware basis, for example, on a personal computer with a traditional architecture. At the same time, it is relevant to develop an ostis-platform in the form of a specialized *associative semantic computer* [13, 15] focused on hardware implementation of sc-memory.

Thus, speaking about information processes occurring within the ostis-system, it is necessary to speak about at least two fundamentally different levels – information processes occurring at the level of *ostis-platform* and information processes occurring at the level of *sc-model of computer system*. Within each of these levels there can be its own internal hierarchy of information processes, as well as its own means of synchronization of parallel processes execution and means of describing the current state of these processes. At the same time ontological unification of these means will allow to realize the advantages of the ontological approach considered above.

The next stage in the development of OSTIS Technology is the transition from individual ostis-systems to a community of interoperable ostis-systems called the global *OSTIS Ecosystem*.

OSTIS ecosystem is a socio-technical ecosystem, which is a collective of interacting semantic computer systems and provides permanent support for the evolution and semantic compatibility of all its constituent systems throughout their life cycle [13, 16].

Within the OSTIS Ecosystem, both *individual ostis-systems* and *collectives of ostis-systems* are distinguished, which, in turn, may include other *collectives of ostis-systems*. Thus, a hierarchy of ostis-systems is formed within the OSTIS Ecosystem. In this case, ostis-systems, which are part of the OSTIS Ecosystem, interact in collective solution of various tasks, combining the *knowledge* and *skills* available to them. Thus, in addition to the previously discussed two levels of information processes, another level appears – information processes within the distributed team of interacting *ostis-systems*. The key feature of this level is the fact that different fragments of such information process in general case can be performed within different ostis-systems, while the process will remain logically integral. At the same time at this level it is also necessary to speak about means of describing the state of processes and means of synchronization of execution of parallel processes.

2 Ontology of Spatio-Temporal Entities

The basis of the model of a unified semantic representation of knowledge [3, 17] is a family of compatible languages that can be interpreted as generalized formal languages [17]. All these languages have a common alphabet [17] and are classified as sc-languages.

The basic concepts are:

- *elementary events*, which are sets and membership (non-membership) in accordance with basic semantics, closely related to their designations (sc-elements) [12, 17];
- *cause-and-effect formation* [3, 18, 19] (binary relation), the connectives of which connect the designations of these events.

Elementary events of belonging include events of actual and irrelevant belonging and non-belonging. Belonging can be a consequence of non-belonging, and non-belonging can be a consequence of belonging.

The connectives of the relationship of becoming are divided into two kinds: being (the first and second components of the connective are the same) and change (the first and second components of the connective are different) [12].

The next concept is the concept of *phenomenon*. A phenomenon is understood as a set of designations for specific events and connections between their formation.

Among the phenomena, types of phenomena are distinguished, such as: elementary phenomenon, complex phenomenon, situation, movement, action, coaction, interaction, system and others [11].

Relative concepts between phenomena are: state, process, impact, outcome, source, consequence, aftereffect, predecessor and others [3, 11].

The changes set the dynamics of events and phenomena in the language texts of the model of a unified semantic representation of knowledge, in accordance with which the operational semantics of designations in these texts is determined. Based on the dynamic properties of events and phenomena on their set, spatio-temporal concepts are introduced, including the concept of time [3], which in turn is associated with processes (including wave and information) and with the concept of entropy (non-information), and spatial-temporary relationships [20]. Since the concept of plurality and belonging is related to the concept of becoming [3], possible dynamic phenomena can be reconstructed from the static structure of ontology [20]. Static and dynamic ontological structures are considered as substructures of semantic space within the framework of the metamodel of semantic space [17].

Space-time relations include: (immediately) "previously" (meet), (immediately) "next" (met-by), "before", "after", (starts) "since", (finishes) "to", "from", (upon) "until", "when", "then", "withing", "during", "over", (inside) "in", "outside", "undefined" [3]. All of them are defined in accordance with the relation of becoming. Together with actual and irrelevant belonging, they are used to indicate the spatiotemporal position of events in processes [3]. All of the above concepts form an ontology, which is the basis for describing the dynamics in general and in particular for information processes.

Some private concepts associated with the concept of time, temporary and dynamic entities, can be considered as components of private, specialized ontologies that describe time and dynamics of an applied nature in particular subject domains.

Figures 1, 2 and 3 show examples of the application of the developed ontologies in SCg-code.

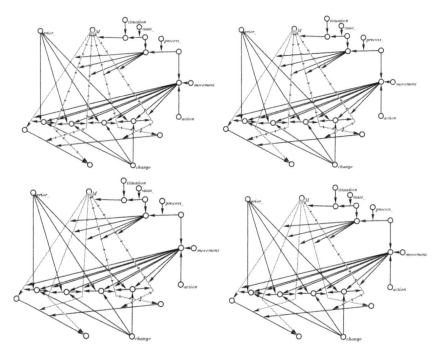

Fig. 1. Representation of the action process that forms the situational (distensible) set (M) of five sc-elements in dynamics and its state.

3 Subject Domain and Ontology of Information Processes in Ostis-Systems

3.1 The Ontology of the Semantic Logging Model

In addition to the ontology of spatiotemporal entities, including events and phenomena, ontological means are proposed for the model of semantic logging (events) of processes in knowledge processing systems. The ontology of the semantic logging model is focused

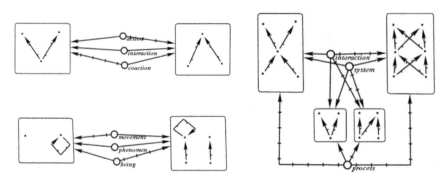

Fig. 2. Examples of actions, coactions, interactions, systems, processes and other phenomena as beings and movements.

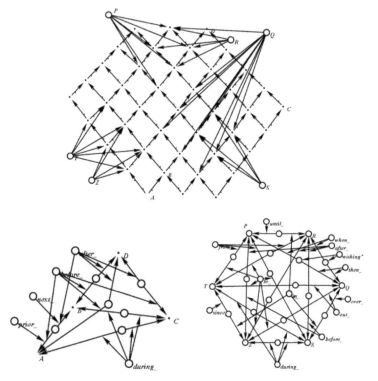

Fig. 3. Example of using space-time relationships for events A, B, C, D and phenomenon P, Q, R, S, T, Z.

on supporting optimal, fast mechanisms for recording and establishing spatiotemporal relationships between events, including the relationship of becoming [12, 21].

In this way, processes that have already taken place can be recorded in the form of protocols. Processes that have yet to go through can be specified by plans in multi-agent systems and programs.

The process interaction model defines mechanisms for managing processes and their interaction in order to prevent or resolve incorrect, conflict situations. As previously noted, in addition to describing processes, it is important to prevent and/or resolve incorrect interactions and incorrect states (situations) of the knowledge base.

The main cases of incorrect (unacceptable) interaction are: starvation and cyclic (mutual) waiting or blocking [12].

To overcome the corresponding problems, a means of describing the binding of a process to the plan (program) that it executes is required, including the binding and inheritance of modes and access rights by processes within the framework of the access control model. Also required are means (flags) to describe the state (or relationship) of waiting or blocking a process, as well as the cause-and-effect relationships of processes. In addition, to overcome the problem of starvation, means of local and/or global synchronization of processes are required, for example, such as mutexes and/or barriers [12].

The main incorrect (or unacceptable) states are associated with the creation of contradictions: the creation of semantically contradictory constructions, including generation of several (multiple) elements, in quantities greater than allowed, creation of an element during its deletion, etc.

Overcoming the corresponding problems also requires means of describing the connection of processes with plans and tasks, the state of processes (flags) and/or mechanisms for their synchronization [12].

All these tasks can arise for processes at various levels of their hierarchy associ-ated with vertical integration or interpretation of (programs for) corresponding in-formation processing models [12, 17, 22].

3.2 Principles of Interpreting Information Processes Within a Hardware Platform

Different variants of realization of associative *semantic computer are* possible, which are considered in more detail in [13]. As shown in the mentioned work, the most promising implementation variant looks like the variant with fine-grained architecture, in which the processor-memory of the associative semantic computer consists of single-type modules, which we will call processor *elements*. Each processor element corresponds to one sc-element (stores one sc-element or stores nothing at the current moment of time).

Processor elements are interconnected by two types of communication channels – *physical communication channels, the* number of which is limited, and *logical communication channels*, which correspond to the incident links between sc-elements and the number of which is potentially unlimited. The configuration of physical communication channels is fixed and in general does not depend on the configuration of logical communication channels.

Each processor element can send messages (microprograms) to other processor elements and receive messages from other processor elements via logical communication channels and has corresponding receptor-effector submodules. At the physical level, message transmission is carried out, in turn, via physical communication channels.

Thus, processor elements form a homogeneous processor-memory in which there are no separately allocated modules intended only for information storage and separately allocated modules intended only for its processing. The key feature of information processing stored within such processor-memory is that in such processor-memory architecture there is no common memory available for all modules processing information. Due to this, parallel processing of information is considerably simplified, but realization of a set of microprograms for interpreting commands of information processing in such memory becomes more complicated, as each processor element becomes very "myopic" and "sees" only those processor elements which are connected with it by logical communication channels.

Thus, a language for describing microprograms for interpreting commands of an associative semantic computer cannot be constructed as a traditional programming language, for example, of the procedural type, since all such languages assume the possibility of direct addressable or associative access to arbitrary memory elements.

As a basis for the development of such a language and means of describing the corresponding information processes, it is proposed to take the ideas of wave programming languages, developed for many years by the school of P. S. Sapaty [23, 24]. Later variants of the wave language theory development were called Spatial Grasp Technology within the framework of which Spatial Grasp Language is developed. The basic principles of the proposed approach to the development of a firmware language for the *associative semantic computer are* described in [13].

Let us consider two main approaches to the organization of information processing within such a processor-memory.

Processing in a Situation where the Sc-construct is Stacked Arbitrarily in the Processor-memory. In this case, the sc-elements incident to each other are connected by a logical communication channel, but their corresponding processor elements may be at an arbitrary distance from each other in terms of physical location in the processor-memory. In other words, the path through physical communication channels from one such processor element to another may have an arbitrary length, and it is not possible to construct it in advance using, for example, traditional algorithms on graphs, because the initial processor element does not have access to the whole memory, as it was shown above. The following principles are suggested for information processing at such variant of sc-constructions storage:

- Messages containing information processing programs are transmitted in the form of global waves through all physical communication channels starting from the processor element initiating the processing. Further speaking about the fact that the sender or addressee of some message is an sc-element, we will imply that the sender or addressee is a processor element storing this sc-element;
- each message contains information about which sc-elements are to be processed, so that each processor element decides on the necessity to execute the corresponding commands independently. In general, the addressee of the message can be either one particular sc-element (having a corresponding identifier) or a set of sc-elements defined on the basis of specified criteria (for example, sc-elements having a certain syntactic type);
- Regardless of whether the corresponding instructions have been executed within a particular processor element, each processor element transmits the message further

along the physical communication channels. Thus, a "wave" is formed that propagates through the processor-memory. The exception to this is if the message only described the transformation of a given specific sc-element stored in a given processor element, and thus there is no point in transmitting the message further;
- Each message has a unique wave identifier, which is duplicated when the same message is transmitted further. This allows, on the one hand, to link messages with each other (for example, by forming response messages containing information about the success or failure of program execution from the original message), and on the other hand – allows processor elements to understand which waves have already been processed by them and not to perform the same actions repeatedly;
- In this case, each processor element stores information about which message was received over which physical communication channel in order to avoid sending the same message again to its source. In this way, each wave is unidirectional;
- As a result of execution of the program contained in the message, a response message is generated in the processor element, the content of which is generally determined by the semantics of the specified program. At least the response message contains information that the program from the specified wave was executed for the specified sc-element and the fact of success/failure of execution with possible indication of reasons, if known. If a search command is executed, the response message may contain the identifier of the sc-element (or set of sc-elements) that were found;
- A processor element that has received some message and executed the corresponding program may initiate a new wave of messages and thus continue processing information independently of the processor element that sent the initial wave;
- To reduce the load on physical communication channels, transmitted messages may be complex, i.e. they may contain commands addressed to different sc-elements or different sets of sc-elements. To simplify the process of interpreting such messages, parallelism within the messages itself is not assumed at the moment;
- Each processor element has an incoming message queue and an outgoing message queue that holds messages waiting to be processed and sent, respectively.

The advantages of this approach to organizing information processing are:

- Described approach allows to organize information processing in conditions when the principles of sc-elements placement in memory are not specified in any way. Thus, this approach can be considered as basic and universal, although it is far from being the most efficient;
- Relative simplicity of implementation – no additional algorithms of stacking sc-constructions into processor-memory are required, while the algorithms of work of processor elements themselves are relatively simple. Besides, no additional mechanisms of synchronization of information processes executed in parallel are required, except for the situation when synchronization is assumed by the algorithm of solving a particular task.

At the same time, this approach has a number of significant disadvantages:

- Significant load on physical communication channels due to mass sending of waves of messages not addressed to specific addressees. In this case, in general, most of the sent messages may be unclaimed, because the required addressee has already received the message and performed the necessary transformations, but other processor elements do not know about it;
- High dependence of the system speed on the bandwidth of physical communication channels and the speed of message transmission through these channels.

Processing in the Situation when the Configuration of Logical Communication Channels Coincides with the Configuration of Physical Communication Channels. The second proposed processing variant is designed to eliminate some of the drawbacks of the first one by organizing the stacking of sc-constructs in processor-memory in such a way that the configuration of logical communication channels coincides with the configuration of physical communication channels. The key features of this approach are as follows:

- Since logical communication channels correspond to the incident links between sc-elements, the number of logical communication channels corresponding to one processor element is potentially unlimited. At the same time, the number of physical communication channels corresponding to one processor element is limited by the possibilities of hardware realization. In order to bring the configuration of physical communication channels to correspond to the configuration of logical ones it is suggested to use a part of processor elements as *switches* (switching element) if necessary. The switch does not store any sc-element, but is a "virtual copy" of the corresponding "parent" processor element, which is connected with one of the physical communication channels (the fact of such connection is explicitly fixed). The other physical channels of this switch are considered to be physical channels of the specified processor element. If necessary, a chain of switches corresponding to one and the same processor element can be formed, thus actually eliminating the limitation on the number of physical communication channels corresponding to one processor element. In this approach, one sc-element is generally represented not by a single processor element, but by a set of processor elements;
- When stacking the sc-structure into the processor-memory sc-elements are placed into the processor elements so that the incident sc-elements are in the processor elements connected by a physical communication channel. At the same time, if the number of sc-elements incident to a given one exceeds the possible number of physical communication channels incident to a single processor element, a new switch is created, and so on until all incident links are put in line with physical communication channels. The addition of a new fragment to the original sc-construction already stacked in processor-memory is accomplished in a similar manner. Obviously, the algorithm for such stacking is generally non-trivial and its development is a separate task;
- message transmission between processor elements and interpretation of programs written in messages is performed similarly to the variant with arbitrary stacking of

sc-constructions in processor-memory, but messages are transmitted not in global waves through physical communication channels, but through purposefully logical communication channels. Each processor element "knows" which processor elements are connected to it and independently decides which communication channels to use to send this or that message. Thus, the load on physical communication channels and dependence on their bandwidth is greatly reduced;

Thus, the key advantage of this option is to reduce the load on communication channels and dependence on their bandwidth. At the same time, this approach also has a number of disadvantages:

- stacking an sc-construction into processor-memory in accordance with the above principles may take a long time. In addition, there is a limitation related to the principal possibility of such stacking (the incident graph corresponding to the sc-construction must be planar). If stacking is fundamentally impossible, it is necessary to specify the algorithms of stacking and message distribution between processor elements taking this fact into account;
- When adding a new fragment to an already laid sc-structure, it may be necessary to re-lay a significant portion of that sc-structure in order to comply with the physical link configuration requirements discussed above, which in turn can generally make the process of adding new fragments time-consuming.

Taking into account advantages and disadvantages of the considered approaches to organization of information processing in associative semantic computer with fine-grained architecture, we can say that in practice it will be expedient to combine both considered approaches. In particular, it is possible to take into account logical communication channels when sending messages, and when stacking the sc-construction into memory try to maximize the configuration of logical communication channels to the configuration of physical ones, but do not require their full correspondence.

In the context of this paper, the task of developing a family of ontologies describing the principles of organizing the execution of information processes within the associative semantic computer is relevant, taking into account the principles outlined in this section.

3.3 Principles of Interpreting Information Processes Within Distributed Teams

It is urgent to move from the development of problem solvers of single intelligent systems to problem solvers of interacting interoperable intelligent systems, including the development of principles of problem solving in such distributed collectives and means of describing information processes performed in such a distributed environment. Basic principles of problem solving in distributed collectives of ostis-systems were considered in [13], in this paper we will briefly review the main ideas and unsolved problems of the proposed approach.

The key difference between a distributed ostis-system and an internal sc-agent system within an individual ostis-system is the absence of a common memory storing a common knowledge base for all sc-agents and acting as a environment for sc-agent communication. In general, as a means of communication between the participants of a distributed collective of ostis-systems can be used:

- Shared unallocated (monolithic) memory, as in the case of sc-agents over sc-memory;
- Shared distributed memory. In this case, from a logical point of view, agents may think that they are still working on a shared memory that stores the entire available knowledge base, but in reality the knowledge base will be distributed among several ostis-systems and the performed transformations will have to be synchronized among these ostis-systems;
- Specialized communication channels. Obviously, when solving a problem in a distributed collective of ostis-systems, there should be language and technical means allowing to transfer messages from one ostis-system to another.

All of the above means of communication can be combined depending on the class of the problem to be solved, the knowledge and skills required for its solution, and the currently existing (available) set of ostis-systems.

For realization of interaction between ostis-systems it is proposed to use the ideas of wave programming and Spatial Grasp Technology [23, 24] as well as insertion programming [25] as a basis, as well as at the hardware level. The realization of such interaction requires the development of languages of at least two levels:

- transport layer defining the principles of recording SC-code constructs in some format convenient for network transmission. SCs-code [13] is proposed to be used as a variant of such language;
- semantic level defining the content of messages transmitted through the network. The SCP language, which is the basic programming language for ostis-systems [13], is proposed as a basis for such a language.

When solving a particular problem by a distributed team of ostis-systems, in general, before directly starting to solve the problem within the subject domain, the following "organizational" subproblems related to the communication of the ostis-systems themselves must be solved:

- which ostis-system will provide the environment for other ostis-systems to interact;
- what set of ostis-systems will be involved in solving this problem (knowledge and skills of which ostis-systems will be required);
- where the general plan of problem solving will be stored and how it will be interpreted, on the basis of what strategy of problem solving it will be formed, how this strategy will be chosen;
- where the result of the task solution will be formed and intermediate results of the solution will be stored;
- how the sub-tasks will be distributed among the participants in the solution process;
- how the intermediate results of individual subtasks will be synchronized with each other.

More detailed principles of problem solving within distributed teams of ostis-systems and approaches to solving the formulated problems will be considered in future works of the authors.

4 Conclusion

The paper considers a family of ontologies that allow describing knowledge in dynamic subject domains, in particular, spatio-temporal relationships, phenomena, processes, situations and events. Special attention is paid to the description of information processes. Examples of application of some of the considered ontologies are given.

An important feature of the work is its comprehensiveness – comprehensive unified approach to the description of all types of dynamic knowledge, including not only information processes, but also lower-level concepts and orientation to all levels of description of information processes, from the level of hardware platform to the level of individual intelligent systems and distributed collectives of such systems.

The disadvantages of the current state of the presented results can be attributed to the low level of formalization and detailing in describing the principles of information processes within the hardware platform and at the level of the distributed collective of interacting intelligent systems. It is caused first of all by the limitation of the volume of the article and the corresponding results will be presented in the following works of the authors. At the same time in the work the actual tasks on development of the presented results are formulated.

The authors express their gratitude to the staff of the Department of Intelligent Information Technologies of BSUIR for their help in the work and valuable assistance in the development of this paper.

References

1. Allen, J.F.: Time and time again: the many ways to represent time. Intern. J. of Intell. Syst. **6**, 341–355 (1991)
2. Narinyani, A.S.: Non-factors: inaccuracy and underdetermination - difference and interrelation. Izv. RAN (RAS). Ser. Teoriya i sistemy upravleniya, **5**, 44–56 (2000) (in Russian)
3. Ivashenko, V.P.: Ontological model of space-time relations for events and phenomena in the processing of knowledge. Vestnik BrGTU **5**(107), 13–17 (2017). (in Russian)
4. Pavlin, G., Kamermans, M., Scafeş, M.: Dynamic process integration framework: toward efficient information processing in complex distributed systems. In: Papadopoulos, G.A., Badica, C. (eds.) Intelligent Distributed Computing III. Studies in Computational Intelligence, vol. 237, pp. 161–174. Springer, Berlin, Heidelberg (2009). https://doi.org/10.1007/978-3-642-03214-1_16
5. Brookes, S.D., Hoare, C.A.R., Roscoe, W.: A theory of communicating sequential processes. JACM (1984)
6. Baeten, J.C.M., Basten, T., Reniers, M.A.: Algebra of Communicating Processes. Cambridge University Press (2005)
7. Dijkstra, E.W.: Co-operating sequential processes. In: Genuys, F. (eds.) Programming languages: NATO Advanced Study Institute: lectures given at a three weeks Summer School held in Villard-le-Lans, 1966, pp. 43–112. Academic Press Inc. (1968)
8. Brenner, M.: A Multiagent planning language. In: Proceedings of the Workshop on PDDL. 13th International Conference on Automated Planning and Scheduling (ICAPS-200,3), Trento, Italy (2003)
9. McDermott, D., et al.: PDDL-The Planning Domain Definition Language. Technical Report CVC TR98003/DCS TR1165. New Haven, CT, Yale Center for Computational Vision and Control (1998)

10. Kovacs, D.L.: A multi-agent extension of PDDL3.1. In: Proceedings of the 3rd Workshop on the International Planning Competition (IPC) 22nd International Conference on Automated Planning and Scheduling (2012)
11. Ivashenko, V.P.: Ontological modeling of event-based causal relationships. In: Shilin, L. Yu. [and others] (eds.) Information Technologies and Systems (ITS 2017): Proceedings of the International Scientific Conference (Republic of Belarus, Minsk, October 25, 2017), pp. 138–139. BSUIR, Minsk (2017). (in Russian)
12. Ivashenko, V.P.: Models of problem solving in intellectual systems. In: 2 parts. P. 1: Formal Models of Information Processing and Parallel Models of Problem Solving: Textbook. BSUIR, Minsk (2020). (in Russian)
13. Golenkov, V., (eds.): Technology of complex life cycle support of semantically compatible intelligent computer systems of new generation. Bestprint, Minsk (2023)
14. Zotov, N.: Design principles, structure, and development prospects of the software platform of ostis-systems. Open Seman. Technol. Intell. Syst. **7**, 67–76. Minsk, Bestprint (2023)
15. Golenkov, V., Shunkevich, D., Gulyakina, N., Ivashenko, V., Zahariev, V.: Associative semantic computers for intelligent computer systems of a new generation. Open Seman. Technol. Intell. Syst. **7**, 39–60. Minsk, Bestprint (2023)
16. Zagorskiy, A.: Principles for implementing the ecosystem of next-generation intelligent computer systems Open Seman. Technol. Intell. Syst. **6**, 347–356. Minsk, Bestprint (2022)
17. Ivashenko, V.: General-purpose semantic representation language and semantic space. Open Seman. Technol. Intell. Syst. **6**, 41–64. Minsk, Bestprint (2022)
18. Hegel, G.W.F.: The Science of Logic. Cambridge University Press (2010). https://doi.org/10.1017/9780511780240
19. Pearson, K.: The Grammar of Science. Cambridge University Press (2014). https://doi.org/10.1017/cbo9781139878548
20. Ivashenko, V.: Structures and measures in knowledge processing models. In: Pattern Recognition and Information Processing (PRIP'2023): Proceedings of the 16th International Conference, October 17–19, 2023, Minsk, pp. 16–21. United Institute of Informatics Problems of the National Academy of Sciences of Belarus, Minsk (2023)
21. Ivashenko, V., Zotov, N., Orlov, M.: Semantic logging of repeating events in a forward branching time model. In: Pattern Recognition and Information Processing (PRIP'2021): Proceedings of the 15th International Conference, 21–24 Sept. 2021, Minsk, Belarus, pp. 149–152. United Institute of Informatics Problems of the National Academy of Sciences of Belarus, Minsk (2021)
22. Ivashenko, V.: Semantic space integration of logical knowledge representation and knowledge processing. Open Seman. Technol. Intel. Syst. **7**, 95–114. Minsk, Bestprint (2023)
23. Sapaty, P.S.: Spatial Grasp as a Model for Space-based Control and Management Systems (1st ed.). CRC Press, https://doi.org/10.1201/9781003230090 (2022)
24. Sapaty, P.S.: The Spatial Grasp Model. Emerald Publishing Limited (2023). https://doi.org/10.1108/9781804555743
25. Letichevsky, A.A., Letychevskyi, O.A., Peschanenko, V.S.: Insertion modeling system. In: Clarke, E., Virbitskaite, I., Voronkov, A. (eds.) Perspectives of Systems Informatics. PSI 2011. Lecture Notes in Computer Science, vol. 7162. Springer, Berlin, Heidelberg (2012). https://doi.org/10.1007/978-3-642-29709-0_23

LANet: Lightweight Attention Network for Medical Image Segmentation

Yi Tang(✉), Dmitry Pertsau, Di Zhao, Dziana Kupryianava, and Mikhail Tatur

Belarusian State University of Informatics and Radioelectronics, Minsk, Belarus
tangyijcb@163.com

Abstract. Medical image segmentation plays a crucial role in quantitative analysis, clinical diagnosis and medical interventions. However, extracting valuable information from medical images is challenging due to the presence of different object types and scales, complex backgrounds and tissue similarities. To address these issues, LANet, a Lightweight Attention Network which incorporates an Efficient Fusion Attention (EFA) block and an Adaptive Feature Fusion (AFF) decoding block was presented in the paper. The EFA block enhances the model's feature extraction capability by capturing task-relevant information while reducing redundancy in channel and spatial locations. The AFF decoding block fuses the purified low-level features from the encoder with the sampled features from the decoder, enhancing the network's understanding and expression of input features. Additionally, the model adopts MobileViT as a lightweight backbone network with a small number of parameters, facilitating easy training and faster predictive inference. The efficiency of LANet was evaluated using four public datasets: kvasir-SEG, CVC-clinicDB, CVC-colonDB, and the Data Science Bowl 2018. Experimental results demonstrate that LANet outperforms state-of-the-art methods in terms of intersection over union (mIoU) and Dice scores.

Keywords: LANet · Lightweight Attention Network · Efficient Fusion Attention Block · Adaptive Feature Fusion Decoding Block · Attention Mechanism

1 Introduction

The field of medical image segmentation plays a crucial role in various clinical applications by enabling accurate and automated analysis of anatomical structures and pathological regions. The goal of medical image segmentation is to divide it into specific subregions and extract regions of interest (ROI) based on attributes like color, texture, and shape. This process forms the foundation for subsequent tasks such as lesion area extraction, specific tissue measurement, and 3D reconstruction [1]. In the past few decades, there have been significant advancements in object segmentation in medical images. Initially, manual segmentation by trained experts like radiologists or pathologists was considered the standard for delineating anatomical structures and pathological abnormalities. However, manual segmentation is prone to inconsistencies and lacks reproducibility, making it unsuitable for handling large volumes of medical data. Traditional technology-based methods showed improvement compared to manual segmentation, but they require setting appropriate thresholds for grayscale threshold-based method [2] and manual selection of seed points for region growing-based method [3]. However, with the advent and development of specialized systems for training deep neural networks, the focus of research has shifted towards the application of deep learning methods, which have demonstrated excellent results in automatic segmentation of objects of interest.

In the field of medical image segmentation, several challenges persist. First, existing CNN-based segmentation models often exhibit high computational complexity, which impedes their deployment on resource-constrained devices or real-time clinical applications. This limitation hampers the widespread adoption of automated medical image segmentation system [4]. Second, effectively capturing task-relevant information while minimizing redundancy in features is crucial for achieving accurate segmentation results. Traditional CNN architectures may struggle to strike a balance between capturing relevant information and eliminating redundant features, potentially leading to poor segmentation performance and inefficient feature representation [5]. Additionally, the fusion of low-level with high-level features poses significant challenges to medical image segmentation. The fusion process needs to be adaptive and context-aware to ensure the seamless integration of complementary information, thereby enhancing the network's ability to accurately comprehend and represent input features.

In this study, we aim to address these issues by introducing LANet, a lightweight attention network for medical image segmentation. LANet utilizes incorporates both an Efficient Fusion Attention (EFA) block and an Adaptive Feature Fusion (AFF) decoding block. EFA blocks enhance feature extraction capabilities by capturing task-specific information while minimizing redundancy in both channel and spatial locations. The AFF decoding block effectively combines low-level with sampled features from the decoder, enhancing the network's comprehension and representation of input features. By developing LANet, our aim is to achieve state-of-the-art segmentation performance while retaining a lightweight architecture that is well-suited for resource-constrained environments. This research is significant as it addresses the need for more efficient and accurate medical image segmentation models, thereby enhancing clinical decision-making, patient care, and medical research in various healthcare applications.

2 LANet Model

In this section, an overview of the proposed segmentation framework is given in Sect. 2.1. Then we introduce two key components, the Efficient Fusion Attention block (denoted as EFA) in Sect. 2.2 and Adaptive Feature Fusion decoding block (denoted as AFF) in Sect. 2.3. Finally, loss function presented in Sect. 2.4.

2.1 Overview

LANet model is presented in Fig. 1, which is mainly based on an encoder-decoder network. MobileVit [4] is used as the backbone of the encoder network. The encoder consists of five encoding blocks (E-Blocks). Each encoding block corresponds to a specific feature map resolution, denoted as F_i ($i = 1,2,\ldots,5$). Specifically, the feature

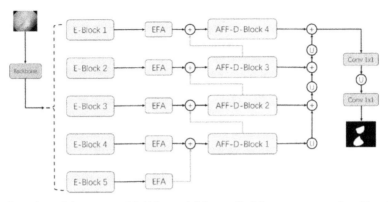

Fig. 1. Overview of the proposed LANet model for medical image segmentation. Here '*Conv*' denotes the convolutional layer followed by batch normalization and ReLU activation function. The blue arrow denotes the combination of 1×1 convolutional layer and upsampling layer. '*U*' and '*+*' represent the upsampling and addition layers, respectively.

resolution of $\frac{W}{2} \times \frac{H}{2}$ for the first level, and general resolution is $\frac{W}{2^i} \times \frac{H}{2^i}$ for each i value. While the backbone has demonstrated its efficacy in extracting feature information for image classification, it may also yield redundant information [6]. Consequently, each E-Block processes the features through an EFA block to extract significant features. Subsequently, the purified low-level feature maps are concatenated with the upsampled feature maps through skip connections, forming a combined input that passes through the AFF decoding block. The purpose of AFF is to preserve the distinctive characteristics of the advanced features in the E-Block and augment its representation capabilities. Many existing methods employ the output of the highest decoder layer as the prediction result directly, disregarding the loss of spatial information during the upsampling process of the decoding block. To address this issue, we conduct a scale transformation and combine the results of AFF obtained at different resolutions. Specifically, to accommodate the variations in shape and size across diverse decoding layers, the output of each AFF is subjected to global average pooling, enabling the aggregation of global context information. Furthermore, two distinct nonlinear activation functions, i.e., ReLU and Sigmoid, are applied using 1×1 convolutional layers to estimate inter-layer correlations and generate weights along the channel dimension. The resulting weights are then multiplied by the output to amplify the significance of features that contribute significantly to result prediction. Lastly, the outcomes achieved by concatenating the four decoding blocks undergo a continuous 1×1 convolutional layer, an upsampling layer, and another 1×1 convolutional layer to yield the ultimate prediction result.

2.2 Efficient Fusion Attention

Efficient Fusion Attention (EFA) is an attention block specifically designed to enhance the model's capacity for learning channel and spatial attentions, thereby significantly improving performance in visual tasks. Figure 2 shows the EFA structure.

The EFA block achieves efficient utilization of channel and spatial information through two branches. Firstly, the spatial attention branch aims to enhance the model's

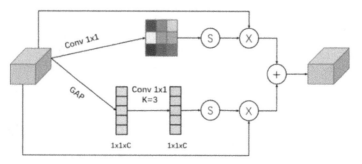

Fig. 2. Structure of Efficient Fusion Attention block

ability to focus on the spatial location within the input. It captures the significance of different spatial locations by applying a 1 × 1 convolution. Subsequently, the convolved results are transformed into attention weights ranging from 0 to 1 using a sigmoid function. Ultimately, these weights are multiplied by the input to emphasize task-relevant spatial features while attenuating task-irrelevant spatial locations. Secondly, the channel attention branch extracts channel features from the input. It captures the significance of different channels through a series of operations. Typically, this involves performing a global average pooling on the input to obtain overall feature information, followed by a convolution operation to derive attention weights for each channel location. Similar to the spatial attention branch, these weights are also mapped by a sigmoid function to ensure their range is between 0 and 1. These weights are utilized to emphasize task-relevant channel information and suppress task-irrelevant channels in the input.

Finally, the EFA block fuses the outputs of the channel attention and spatial attention branches by adding them element-wisely.

This fusion enables the model to comprehensively utilize both channel and spatial information, thereby enhancing the model's understanding and utilization of input features. By adaptively learning the importance of different channels and spatial locations, the EFA block significantly improves the model's perception and representation capabilities, thereby enhancing performance in various image analysis and understanding tasks. The computation of the proposed block can be denoted as below in (1):

$$
\begin{cases} O_c = Input \cdot (\partial(Conv_{1\times1}(GAP(Input)))) \\ O_s = Input \cdot (\partial(Conv_{1\times1}(Input))) \\ O = O_c + O_s \end{cases}, \quad (1)
$$

where ∂ denotes sigmoid activation function, GAP and $Conv_{1\times1}$ denotes global average pooling layer and 1 × 1 convolutional layer respectively.

2.3 Adaptive Feature Fusion Decoding Block

In a deep neural network, the encoder plays a crucial role in extracting features from the original data, while the decoder is responsible for reconstructing these features to represent the original data. However, the traditional encoder-decoder structure may have certain limitations in terms of feature fusion. Many variants of the U-Net model [7–10] have focused on designing improved encoders to extract better features. Nevertheless, these approaches often overlook the significance of the decoder, which directly contributes to the prediction process and requires efficient extraction of vital information from the feature map. To address this issue, we propose an Adaptive Feature Fusion (AFF) decoding block that aims to enhance the feature representation and expression capabilities of deep neural networks when processing features. The structure of the decoding block is shown in Fig. 3.

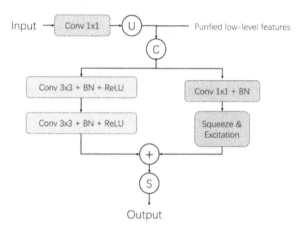

Fig. 3. Structure of proposed Adaptive Feature Fusion Decoding Block

The AFF decoding block receives input comprising the encoder's purified low-level and the decoder's features. These two features are concatenated via skip connections to form a fused feature representation. Subsequently, the fused feature representation enters two parallel branches. The first branch comprises two 3 × 3 convolutional layers, each followed by batch normalization and ReLU activation function. This branch is designed to extract higher-level feature representations through convolution operations, enabling the model to capture more complex and abstract features. The second branch initially employs a 1 × 1 convolutional layer and batch normalization layer to reduce feature dimensionality and computational complexity. Then, the attention mechanism is implemented by introducing the Squeeze & Excitation layer. This layer enables adaptive feature weighting, allowing the model to emphasize important features and enhance their expressive capability. Finally, the outputs of the two branches are added and mapped through a sigmoid function to obtain the final output. This approach to attention and feature fusion exhibits enhanced flexibility and robustness.

2.4 Loss Function

The widely adopted binary cross-entropy loss is commonly used in various segmentation tasks; however, it exhibits several limitations. Firstly, it fails to consider the global structure of the image when computing the loss for individual pixels independently. Secondly, in medical image datasets, there is typically an imbalance between positive and negative samples, where background pixels outnumber boundary pixels. This imbalance often leads to a bias towards predicting background categories when using binary cross-entropy. To address this issue, the Lovasz-Softmax loss function offers a better solution by effectively handling imbalanced categories through sorting error processing. Lastly, by leveraging the gradient of the ranking error, the Lovasz-Softmax loss function can more accurately reflect the significance of boundary pixels. Consequently, it enhances the accuracy of boundary localization and segmentation tasks.

Therefore, we adopt Lovasz-Softmax as our loss function. It can be formulated in (2):

$$Loss = \frac{1}{T} \sum_{t \in T} \overline{\Delta J_t}(m(t)) \text{ and } m_i(t) = \begin{cases} 1 - p_i(t), t = y_i(t) \\ p_i(t), otherwise \end{cases}, \quad (2)$$

where T and $\overline{\Delta J_t}$ denotes the number of classes and Lovasz extension of the Jaccard index respectively, $y_i(t) \in \{-1,1\}$ and $p_i(t) \in [0,1]$ are the ground truth label and predicted probability of pixel i for class t [11].

3 Experiments

In this section, we present a comparative analysis of our proposed model with state-of-the-art models for medical image segmentation, utilizing publicly available datasets. We begin by introducing the datasets and discussing the evaluation metrics (see Sect. 3.1) employed to assess the effectiveness of our model. Subsequently, we provide comprehensive details on the training and optimization procedures in Sect. 3.2. Then in Sect. 3.3, we conduct a thorough comparison of our proposed model with existing state-of-the-art segmentation methods.

3.1 Datasets and Evaluation Metrics

Datasets. To validate proposed LANet model, we conducted experiments on four publicly available medical image datasets. These datasets are described as follows:

Kvasir-SEG [12]: it's an open-access dataset consisting of gastrointestinal polyp images and corresponding segmentation masks. The annotations were carefully done by experienced gastroenterologists. The dataset comprises 1000 polyp images, with resolutions ranging from 332 × 487 to 1920 × 1072 pixels.

CVC-clinicDB [13]: this dataset contains 612 RGB images with a resolution of 384 × 288 pixels. It includes various structures and lesions such as normal colon tissue, polyps, and ulcers. The dataset accounts for different patients and lens settings to ensure diversity and realism.

CVC-ColonDB [14]: specifically curated for colonoscopy image analysis and polyp detection. It consists of 380 colonoscopy images from different patients. The images in this dataset have a size of 500 × 570 pixels.

2018 Data Science Bowl [15]: this dataset comprises 670 images with resolution 256 × 256 pixels. It provides a diverse collection of cell images, including cancer cells, normal cells, and more.

To preprocess the data, we standardized the dataset by resizing all images to a uniform scale. Subsequently, we partitioned the preprocessed dataset into separate sets for training, validation, and testing, maintaining an 8:1:1 ratio. Table 1 provides a comprehensive overview of the image partitioning and the resized resolutions for different training scenarios. During the testing stage, the test images were resized to 256 × 256 pixels and fed into the LANet model to generate segmentation maps. These maps were then rescaled back to the original size for final evaluation.

Table 1. Details of medical dataset used in experiments. Columns "Images", "Size", "Train", "Validation", "Test" denote the number of total images, size of input images, number of training samples, number of validation samples and number of test samples.

Dataset	Images	Size	Train	Validation	Test
Kvasir-SEG [12]	1000	256 × 256	800	100	100
CVC-ClinicDB [13]	612		490	61	61
CVC-ColonDB [14]	380		300	40	40
2018 Data Science Bowl [15]	670		530	67	67

Evaluation Metrics. In this research, we employed a range of standard evaluation metrics [16] to validate the effectiveness of LANet. These metrics encompass mean Precision (mPrec), mean Dice Coefficient (mDC), mean Recall (mRec), and mean intersection over union (mIoU).

To calculate these metrics, we compared the model's predictions against the ground truths across the entire dataset of M images. All evaluation metrics can be calculated using numbers of true positives (TP), false positives (FP) and false negative (FN). These metrics can be denoted as shown in (3):

$$\begin{cases} mPrec = \frac{1}{M} \sum_{i=1}^{M} \frac{TP_i}{TP_i + FP_i} \\ mDC = \frac{1}{M} \sum_{i=1}^{M} \frac{2 \cdot TP_i}{2 \cdot TP_i + FP_i + FN_i} \\ mRec = \frac{1}{M} \sum_{i=1}^{M} \frac{TP_i}{TP_i + FN_i} \\ mIoU = \frac{1}{M} \sum_{i=1}^{M} \frac{TP_i}{TP_i + FP_i + FN_i} \end{cases} \quad (3)$$

3.2 Implementation Details

The proposed framework was implemented with PyTorch and trained on a GeForce RTX 4060 Ti graphic card with 16 GB memory. We utilized MobileVit as the backbone network, which was pre-trained on ImageNet [17]. The Adam algorithm was employed for model optimization, with an initial learning rate set to $1e-4$. The batch size was set to 16, and the model underwent 300 epochs of training.

In deep learning research today, researchers often rely on large datasets to mitigate issues related to overfitting and data imbalance. However, biomedical image datasets typically suffer from small sample sizes. To address this, we employed data augmentation techniques during training. Specifically, we applied random cropping and rotations of [90°, 180°, 270°] to augment the training dataset. This approach helped generate new datasets, enhancing the robustness of the trained model. Notably, data augmentation was not applied to the test set.

3.3 Comparison with State-of-the-Art Methods

Kvasir-SEG. Based on the data presented in Table 2, we compared our LANet model with other state-of-the-art (SOTA) models using the Kvasir-SEG dataset and assessed their performance based on four evaluation metrics. The best results in the table are highlighted in bold font. Our LANet model outperforms other models on all evaluation metrics except mean Recall. Specifically, our model achieved the highest score of 0.911 on the mDC metric, while other models attained a maximum score of 0.880. Regarding the mIoU metric, the LANet model obtained a score of 0.851, surpassing the highest score of other models, which was 0.815. Moreover, introduced model achieved high scores on the mRec and mPrec indicators, with values of 0.903 and 0.949, respectively.

Table 2. Quantitative results of evaluation metrics for LANet in comparison to SOTA models on the Kvasir-SEG. The best results are shown in bold fonts.

Method	mDC	mIoU	mRec	mPrec
U-Net [18]	0.596	800	100	100
ResUNet ++ [8]	0.690	0.572	0.724	0.745
FCN [19]	0.831	0.736	0.834	0.881
DoubleU-Net [10]	0.812	0.733	0.840	0.861
U-Net ++ [9]	0.800	0.700	0.871	0.799
Attention U-Net [7]	0.794	0.695	0.838	0.828
FANet [20]	0.880	0.815	**0.905**	0.900
LANet (ours)	**0.911**	**0.851**	0.903	**0.949**

CVC-ClinicDB. We trained proposed network on the CVC-ClinicDB dataset and conducted a comprehensive comparison with other well-known segmentation models. The results, presented in Table 3, highlight the effectiveness and efficiency of our method in accurately segmenting polyps within the CVC-ClinicDB dataset. Proposed model demonstrates competitive performance and surpasses several state-of-the-art models in terms of segmentation accuracy and other evaluation metrics. Notably, LANet model exhibits a significant improvement over the DoubleU-Net [5], with mRec and mPrec improving by 8.0% and 2.6%, respectively.

Table 3. Quantitative results of evaluation metrics for LANet in comparison to SOTA models on the CVC-ClinicDB. The best results are shown in bold fonts

Method	mDC	mIoU	mRec	mPrec
U-Net [18]	0.823	0.755	0.647	0.788
ResUNet ++ [8]	0.795	0.455	0.668	0.887
SFA [21]	0.700	0.607	–	–
PraNet [22]	0.899	0.849	–	–
Double U-Net [10]	0.924	0.861	0.846	0.940
LANet (ours)	**0.944**	**0.896**	**0.926**	**0.966**

CVC-ColonDB. Based on the quantitative evaluation metric results presented below in Table 4. LANet achieved impressive scores of 0.771 for the average Dice coefficient (mDC), 0.712 for the mIoU, 0.758 for the average recall (mRec), and 0.894 for the average precision (mPrec). LANet showcases significant improvements across all metrics compared to other models. Notably, as shown in next subsection when compared to the most recent state-of-the-art methods such as C2FNet, MSNet, and PraNet, LANet exhibits superior performance in feature extraction and segmentation tasks, thereby validating the effectiveness of proposed method.

Table 4. Quantitative results of evaluation metrics for LANet in comparison to SOTA models on the CVC-colonDB

Method	mDC	mIoU	mRec	mPrec
U-Net [18]	0.504	0.436	0.533	0.626
U-Net ++ [9]	0.482	0.408	0.525	0.607
C2FNet [23]	0.724	0.650	–	–
MSNet [24]	0.751	0.671	–	–
PraNet [22]	0.712	0.640	0.749	0.852
LANet (ours)	**0.771**	**0.712**	**0.758**	**0.894**

Data Science Bowl 2018. Data Science Bowl 2018 is a prestigious competition focusing on the challenging task of nuclei segmentation. In this study, we developed a network specifically designed to participate in the competition, and it was trained using the provided dataset. To evaluate the performance of our proposed model, we compared results with those of other leading models in the challenge. As depicted in Table 5, proposed method showcases outstanding performance on the dataset, achieving exceptional segmentation accuracy and demonstrating its proficiency in accurately detecting and segmenting nuclei. Among them, mDC and mIoU reached 93.0% and 87.1% respectively, scoring higher than other methods.

Table 5. Quantitative results of evaluation metrics for LANet in comparison to SOTA models on the 2018 Data Science Bowl

Method	mDC	mIoU	mRec	mPrec
U-Net [18]	0.757	0.910	0.606	0.901
DoubleU-Net [10]	0.913	0.841	0.641	**0.950**
U-Net ++ [9]	0.897	0.926	–	–
Attention U-Net [7]	0.908	0.910	0.918	0.924
FANet [20]	0.918	0.857	**0.922**	0.919
LANet (ours)	**0.930**	**0.871**	0.918	0.946

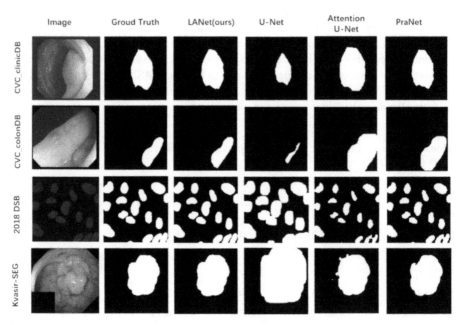

Fig. 4. Visualization of predictions of our LANet and other three SOTA methods on four medical image segmentation datasets

Figure 4 depicts the segmentation results by comparing LANet model with three other medical image segmentation methods. Based on the visual results, our model achieves the closest results to the ground truth, outperforming the other compared methods in handling different challenging factors. Particularly in the second row, despite the small predicted size, proposed method is still able to segment it accurately, whereas U-Net, Attention U-Net, and PraNet are unable to segment it. In the case of the Data Science Bowl 2018 dataset, the image comprises multiple small-sized targets, with insignificant color changes between the target to be predicted and the background. In such case, as observed in the third row of the figure, LANet model yields more accurate predictions compared to other methods. In the first and fourth rows, the polyp exhibits a large size and complex boundaries. Meanwhile, the images reveal unclear boundaries between the polyp and the background due to lighting and shadows. With such challenging instances, proposed model can segment the polyps more accurately compared to U-Net, Attention U-Net and PraNet.

Overall, the visual results provide further evidence of our model's strong performance in addressing various challenging factors across diverse datasets. Although there may be minor deficiencies in capturing fine details, proposed LANet, incorporating EFA and AFF, effectively captures both subject parts and edge information, thereby enhancing segmentation performance.

Model Complexity and Inference Time Comparison. To investigate the model complexity and inference time, we present the model sizes and inference times of LANet model and other compared methods in Table 6. The model size is measured in million (M) parameters, floating-point operations (FLOPs) are measured in Giga (G), and the inference time is reported in frames per second (FPS).

Upon observation, proposed model stands out with the second lowest number of parameters compared to the other methods. Furthermore, it ranks second in terms of inference speed. By combining these findings with previous results obtained from different datasets, it is evident that LANet model achieves fast inference and prediction capabilities while maintaining a high level of segmentation accuracy.

Table 6. Comparison of model size and inference time

Models	Unet	Unet ++	PraNet	ResU-Net	MSNet	Attention U-Net	LANet (ours)
Speed (FPS)	123.11	82.51	25.05	23.50	31.08	55.93	89.29
FLOPs (G)	123.87	262.16	13.15	55.36	17.00	266.54	46.53
Params (M)	34.52	36.63	30.50	18.22	27.69	34.88	20.92

3.4 Ablation Study

To investigate the significance of the key components in proposed LANet, we perform ablation experiments on two main datasets Kvasir-SEG and CVC-ClinicDB. The ablative results are shown in Table 7. The visual comparisons are shown below in Fig. 5.

Effectiveness of EFA. To validate the importance of EFA, we replace it with the original skip connection, denoted as "w/o EFA". As reported in Table 7, we can observe that although the performance of "w/o EFA" does not degrade much on both datasets. However, combining the comparison plots in Fig. 5, it can be seen that the model missing EFA has missing edge details of segmentation in prediction. The results validate the effectiveness of EFA, which can improve the performance of image analysis and comprehension tasks by enhancing the model's ability to learn channel and spatial attentions, thus making enhancements over medical image segmentation details.

Table 7. Detailed ablation study of the LANet architecture, "mDC", "mIoU", "mRec", "mPrec", "mAcc" stands for mean dice coefficient, mean intersection over union, mean recall, mean prediction and mean accuracy, respectively

Dataset	Method	mDC	mIoU	mRec	mPrec	mAcc
Kvasir-SEG	w/o EFA	0.889	0.830	0.897	0.915	0.959
	w/o AFF	0.901	0.841	0.901	0.923	0.937
	LANet(ours)	0.911	0.851	0.903	0.949	0.977
CVC-clinicDB	w/o EFA	0.911	0.865	0.903	0.958	0.936
	w/o AFF	0.907	0.859	0.893	0.926	0.978
	LANet(ours)	0.944	0.896	0.926	0.966	0.991

Effectiveness of AFF. To investigate the contribution of AFF, we use two 3 × 3 convolutional layers with batch normalization and ReLU activation function as decoding block to replace AFF, namely "w/o AFF". As can be seen in Table 7, the absence of the proposed AFF results in a noticeable performance decline, particularly on the CVC-ClinicDB dataset. Specifically, the metrics of mDC, mIoU, mRec, mPrec, and mAcc show significant decreases of 3.7%, 3.7%, 3.3%, 4.0%, and 1.3%, respectively. Furthermore, Fig. 5 clearly demonstrates a substantial discrepancy between the predictions of the model without the AFF block and the ground truth. This observation indicates that AFF plays a crucial role in fusing cross-level features to leverage local and global contextual information, thereby enhancing the feature representation and expressive capabilities of deep neural networks for input features.

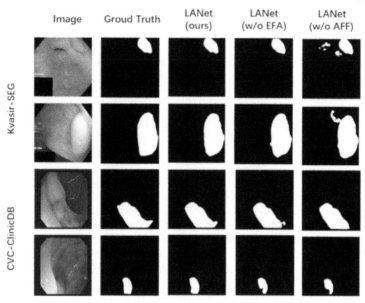

Fig. 5. Visual comparison on the Kvasir-SEG, and CVC-ClinicDB datasets. For each dataset, we have included two diverse images. The series of images, sequenced from left to right, encompass the following elements: (1) Original Image; (2) Ground Truth; (3) Our network with EFA and AFF; (4) Variant network without EFA; and (5) Variant network without AFF

4 Conclusion

In the paper, we propose a Lightweight Attention Network (LANet) for medical image segmentation. Specifically, we use MobileVit as a lightweight backbone to efficiently extract feature information from input images. Additionally, we proposed two key components, namely the Efficient Fusion Attention (EFA) block and the Adaptive Feature Fusion (AFF) decoding block. The EFA block enhances the model's feature extraction capability by capturing task-relevant information while minimizing redundancy in both channel and spatial locations. On the other hand, the AFF decoding block effectively combines the purified low-level features with the sampled features from the decoder, thereby improving the network's ability to comprehend and express the input features.

To assess the segmentation performance of LANet, extensive evaluations were conducted on four public datasets: kvasir-SEG, CVC-ClinicDB, CVC-ColonDB, and Data Science Bowl 2018. The experimental results on four public datasets show that proposed LANet outperforms other state-of-the-art methods. Furthermore, ablation experiments were conducted to investigate the significance of the key components in LANet. The results of these experiments validated the effectiveness of the proposed EFA and AFF blocks.

In conclusion, LANet, with its lightweight architecture and integrated EFA and AFF blocks, presents a promising solution for medical image segmentation tasks. Future research endeavors can focus on further enhancements of LANet and explore its applicability in the domain of 3D medical images.

Acknowledgments. This work was supported by the China Scholarship Council (CSC).

Data Availability. Official implementation for the article was published on GitHub: https://github.com/tyjcbzd/LANet.

Declaration of Competing Interest. The authors declare that they have no known competing financial interests or personal relationships that could have appeared to influence the work reported in this paper.

References

1. Asgari Taghanaki, S., Abhishek, K., Cohen, J.P., Cohen-Adad, J., Hamarneh, G.: Deep semantic segmentation of natural and medical images: a review. Artif. Intell. Rev. **54**(1), 137–178 (2020). https://doi.org/10.1007/s10462-020-09854-1
2. Tong, L., Luo, J., Adams, J., Osinski, K., Liu, X., Friedland, D..: A clustering-aided approach for diagnosis prediction: a case study of elderly fall. In: 2022 IEEE 46th Annual Computers, Software, and Applications Conference, pp. 337–342 (2022)
3. Yu, H., He, F., Pan, Y.: A scalable region-based level set method using adaptive bilateral filter for noisy image segmentation. Multimedia Tools Appl. **79**(9), 5743–5765 (2020)
4. Mehta, S., Rastegari, M.: Mobilevit: light-weight, general-purpose, and mobile-friendly vision transformer. arXiv preprint arXiv: 2110.02178 (2021)
5. Li, J., Wen, Y., He, L.: SCConv: spatial and channel reconstruction convolution for feature redundancy. In: Proceedings of the IEEE/CVF Conference on Computer Vision and Pattern Recognition, pp. 6153–6162 (2023)
6. Qiu, J., Chen, C., Liu, S., et al.: Slimconv: reducing channel redundancy in convolutional neural networks by features recombining. IEEE Trans. Image Process. **30**, 6434–6445 (2021)
7. Oktay, O., Schlemper, J., Folgoc, L., et al.: Attention u-net: Learning where to look for the pancreas, arXiv preprint, arXiv:1804.03999 (2018)
8. Jha, D., Smedsrud, P.H., Riegler, M.A., et al.: Resunet++: an advanced architecture for medical image segmentation. In: 2019 IEEE International Symposium on Multimedia (ISM), pp. 225–235 (2019)
9. Zhou, Z., Rahman Siddiquee, M.M., Tajbakhsh, N., et al.: Unet++: a nested u-net architecture for medical image segmentation. deep learning in medical image analysis and multi-modal learning for clinical decision support. In: 4th International Workshop, DLMIA 2018, and 8th International Workshop, ML-CDS 2018, 4, pp. 3–11 (2018)
10. Jha, D., Riegler, M.A., Johansen, D., et al.: Doubleu-net: a deep convolutional neural network for medical image segmentation. In: 2020 IEEE 33rd International Symposium on Computer-Based Medical Systems (CBMS), pp. 558–564 (2020)
11. Berman, M., Rannen Triki, A., Blaschko, M.B.: The Lovasz-Softmax loss: a tractable surrogate for the optimization of the intersection-over-union measure in neural networks. In: CVPR (2018)
12. Jha, D., et al.: Kvasir-SEG: a segmented polyp dataset. In: Ro, Y., et al. MultiMedia Modeling. MMM 2020. Lecture Notes in Computer Science(), vol. 11962 , pp. 451–462. Springer, Cham (2020). https://doi.org/10.1007/978-3-030-37734-2_37
13. Bernal, J., Sánchez, F.J., Fernández-Esparrach, G., Gil, D., Rodríguez, C., Vilariño, F.: WM-DOVA maps for accurate polyp highlighting in colonoscopy: validation vs. saliency maps from physicians. Comput. Med. Imaging Graph. **43**, 99–111 (2015)

14. Tajbakhsh, N., Gurudu, S.R., Liang, J.: Automated polyp detection in colonoscopy videos using shape and context information. IEEE Trans. Med. Imaging **35**(2), 630–644 (2015)
15. Caicedo, J.C., Goodman, A., Karhohs, K.W., et al.: Nucleus segmentation across imaging experiments: the 2018 data science bowl. Nat. Methods **16**(12), 1247–1253 (2019)
16. Yang, C., Guo, X., Zhu, M., Ibragimov, B., Yuan, Y.: Mutual-prototype adaptation for cross-domain polyp segmentation. IEEE J. Biomed. Health Inform. **25**(10), 3886–3897 (2021)
17. Deng, J., Dong, W., Socher, R., et al.: Imagenet: a large-scale hierarchical image database. In: 2009 IEEE Conference on Computer Vision and Pattern Recognition, pp. 248–255 (2009)
18. Ronneberger, O., Fischer, P., Brox, T.: U-Net: convolutional networks for biomedical image segmentation. In: Navab, N., Hornegger, J., Wells, W., Frangi, A. (eds.) Medical Image Computing and Computer-Assisted Intervention – MICCAI 2015. MICCAI 2015. Lecture Notes in Computer Science(), vol. 9351, pp. 234–241 . Springer, Cham (2015). https://doi.org/10.1007/978-3-319-24574-4_28
19. Long, J., Shelhamer, E., Darrell, T.: Fully convolutional networks for semantic segmentation. In: Proceedings of the IEEE Conference on Computer Vision and Pattern Recognition, pp. 3431–3440 (2015)
20. Tomar, N.K., Jha, D., Riegler, M.A., et al.: Fanet: a feedback attention network for im-proved biomedical image segmentation. IEEE Trans. Neural Netw. Learn. Syst. (2022)
21. Ni, J., Liu, J., Li, X., et al.: SFA-Net: scale and feature aggregate network for retinal vessel segmentation. J. Healthcare Eng. (2022)
22. Fan, D.P., et al.: PraNet: parallel reverse attention network for polyp segmentation. In: Martel, A.L., et al. Medical Image Computing and Computer Assisted Intervention – MICCAI 2020. MICCAI 2020. Lecture Notes in Computer Science(), vol. 12266, pp. 263–273. Springer, Cham (2020). https://doi.org/10.1007/978-3-030-59725-2_26
23. Sun, Y., Chen, G., Zhou, T., et al.: Context-aware cross-level fusion network for camouflaged object detection. In: Proceedings of the International Joint Conference on Artificial Intelligence, pp. 1025–1031 (2021)
24. Zhao, X., Zhang, L., Lu, H.: Automatic polyp segmentation via multi-scale subtraction network. In: de Bruijne, M., et al. Medical Image Computing and Computer Assisted Intervention – MICCAI 2021. MICCAI 2021. Lecture Notes in Computer Science(), vol. 12901, pp. 120–130 (2021). Springer, Cham. https://doi.org/10.1007/978-3-030-87193-2_12

Lightweight Hybrid CNN Model for Face Presentation Attack Detection

Uğur Turhal[1], Asuman Günay Yilmaz[2], and Vasif Nabiyev[2](✉)

[1] Bayburt University, Bayburt, Turkey
uturhal@bayburt.edu.tr
[2] Karadeniz Technical University, Trabzon, Turkey
{gunaya,vasif}@ktu.edu.tr

Abstract. Today, face recognition systems are widely used in many areas that require biometric-based verification, especially because they are contactless and require low user cooperation. Despite their ease of implementation, these systems are vulnerable to attacks. Especially the increasing use of social media, makes it easier to spoof face recognition systems. Therefore, it is very important to develop new methods for Face Presentation Attack (FPA) detection. In this study, FPA detection performances of the lightweight Convolutional Neural Network (CNN) models -MobileNetV3-Small, ShuffleNet, SqueezeNet- and lightweight hybrid CNN models designed with combinations of these networks were investigated. In the experiments, the change in FPA detection performance was examined by giving the cropped face regions and the whole images as input to the networks. Using the whole image instead of the cropped face region significantly improved the FPA detection performance of the CNN models. These results showed that non-facial areas carry important information in FPA detection. Also, the FPA detection performance of the proposed lightweight hybrid CNN models were better than the single use of the networks for both inputs. 0.17% HTER performance was achieved with the hybrid CNN model built from the combination of ShuffleNet + SqueezeNet on the Replay-Attack dataset.

Keywords: Face Presentation Attack Detection · Face Spoofing · Deep Learning · Lightweight CNN Model · Fake Image Detection

1 Introduction

Over the last two decades, technological advances in electronics and computer science have provided a significant portion of the world's population with access to high-end technology devices. Consequently, various biometric systems, such as online payment and e-commerce security, smartphone-based authentication, secure access control and biometric passports, are widely used in real-life applications [1]. Therefore, studies to improve the accuracy of biometric recognition applications have become very important. Unfortunately, these systems are vulnerable to attacks. Especially the increasing use of social media, makes it easier to spoof face recognition systems.

The problem of face presentation attack (FPA) detection can basically be evaluated through two types of attacks: impersonation attacks and obfuscation attacks. In impersonation attacks, a person who does not have access authorization tries to spoof the system by using the face image or video of the authorized person. Today, the increasing use of social media and the increasing frequency of uploading images and videos containing face images to these platforms make it easier for malicious users to access this data and use this data to spoof face authentication systems. In obfuscation attacks, the attacker uses various tricks to avoid being recognized by the system and manipulates the parts of input image to force the detection system to fail. Previous studies have generally focused on impersonation attacks.

FPA detection problem can be analyzed under two headings: hardware and software based. While features such as the sensor characteristics of the capture device come to the fore in hardware-based applications, computational complexity and time criteria are important in software-based applications. Software-based approaches started with applications such as motion and liveness detection with hand-crafted features and have been replaced by deep learning-based methods in recent years. The integration of deep learning models into the FPA detection problem has significantly advanced the field, leading to improved accuracy, robustness and the development of new techniques and databases.

Jin et al. discussed the use of pre-trained deep learning models for object detection and face recognition for benchmarking [2]. Ma et al. proposed a method that learns deep texture features from aligned face images and whole frames and combined with blink detection for cross-database FPA detection [3]. In addition, Benlamoudi et al. presented a FPA detection method using deep background subtraction and emphasized the use of convolutional neural networks for this purpose [4]. Sanghvi et al. introduced MixNet, a deep learning-based network for FPA detection in cross-database and unseen attack environments [5]. Denisova emphasized that the best FPA detection methods rely on computationally intensive deep learning-based features [6]. Singh et al. presented a real-time system for detecting and identifying individuals in surveillance broadcasts using deep learning and face recognition algorithms, demonstrating the practical implications of deep learning in security applications [7]. Tu and Fang used a Resnet-50 network with transfer learning to detect high-level semantic information and meaningful hidden features. Then, they use LSTM to learn temporal features of image sequences. Finally, they perform spoofing detection based on the learned features [8]. Li et al. designed an LBP-based end-to-end learnable network by combining learnable convolutional layers with fixed-parameter LBP layers, which can greatly reduce the number of network parameters. The network consists of sparse binary filters and derivable simulated output functions. Compared with existing deep learning approaches, it is concluded that the proposed Convolutional Neural Network (CNN) structure can reduce the number of parameters by up to 64 times [9]. Chen et al. combined texture features extracted via Rotation-Invariant Local Binary Pattern (RI-LBP) with deep-level features extracted using CNN network to discover deep-level features as well as textural features. The size of the deep-level features was reduced using the Principal Component Analysis (PCA) algorithm. Finally, these feature sets were classified using SVM algorithm with RBF kernel and spoofing detection was performed [10]. Li et al. proposed a CompactNet structure to overcome the problem of overlapping samples in color spaces. The model consists of three stages: a space generator, a feature extractor and a triple loss function. Initially, the RGB image is given to the compact space generator with fewer parameters and helps to map the available color spaces to another new space. Then, the generated face image is given as input to the feature extractor module for the calculation of deep-level features. Finally, the point-to-center integration mechanism is applied to select training samples with a triple loss function, which is used to maximize the inter-class variation and minimize the intra-class variation [11].

In this study, FPA detection was performed using lightweight CNN models. For this purpose, the FPA detection performance of lightweight CNN models, MobileNetV3-Small, ShuffleNet and SqueezeNet, were analyzed. Then, hybrid CNN models were designed to improve the FPA detection performance. The facial region was cropped from the images in the dataset and given as input to the CNN models. In addition, the change in FPA detection performance was examined when the whole images were given as input to the CNN models.

The main contributions of the study are as follows.

- FPA detection was performed with 3 lightweight CNN models (MobileNetV3-Small, ShuffleNet and SqueezeNet).
- A hybrid FPA detection model is proposed by combining lightweight CNN models.
- The networks were fed with the cropped facial images and whole images, so the effect of background information on FPA detection was analyzed.

The rest of the paper is organized as follows. Section 2 describes the proposed FPA detection model in detail. Section 3 presents and discusses the experimental results and Sect. 4 concludes the paper.

2 Proposed Method

In this section, the lightweight CNN models used for FPA detection (MobileNetV3-Small, ShuffleNet and SqueezeNet) and the proposed hybrid model are explained in detail.

2.1 ShuffleNet

ShuffleNet is a highly efficient CNN architecture specifically designed for mobile devices with limited computational resources [12]. It achieves this efficiency by using group convolution and channel mixing, which significantly reduces the number of multiply-add operations required for processing, making it suitable for deployment on devices with low computing power [13]. The architecture's memory efficiency also makes it suitable for use in robotics and other mobile applications [14].

ShuffleNet has been widely recognized for its computational efficiency, making it a popular choice for a variety of applications, including disease identification in agricultural settings, automatic detection of plant diseases, and even medical applications such as detection of premature ventricular contractions [15–17]. Furthermore, the architecture's channel mixing process, which allows channels to be subdivided into subgroups, was highlighted as an important feature contributing to its efficiency [18]. Figure 1 shows a schematic diagram of the channel shuffle operation in the ShuffleNet architecture.

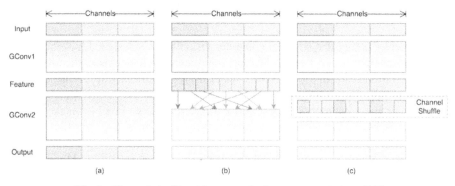

Fig. 1. Channel shuffle with two stacked group convolutions [12]

2.2 SqueezeNet

SqueezeNet is CNN architecture designed to achieve high accuracy with significantly reduced parameter number and computational cost. It is specifically designed to address the challenge of deploying deep neural networks on resource-constrained platforms such as mobile devices and embedded systems. SqueezeNet's architecture is characterized by a series of fire modules consisting of a squeeze convolution layer (1×1 convolutions) followed by expansion convolution layers (1×1 and 3×3 convolutions) [19]. Figure 2 shows the structure of the fire module in the SqueezeNet architecture. This design allows SqueezeNet to strike a balance between model size and computational efficiency, making it well suited for real-time applications such as fire detection [20].

SqueezeNet has been shown to achieve competitive accuracy on benchmark datasets such as ImageNet while using significantly fewer parameters compared to traditional CNN architectures such as AlexNet [19]. This reduction in model size and computational complexity makes SqueezeNet particularly attractive for applications with resource constraints, such as medical imaging for COVID-19 diagnosis, glaucoma severity diagnosis and document image classification [21–23].

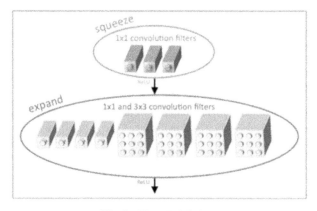

Fig. 2. Fire Module [19]

2.3 MobileNet

The MobileNet family is a series of computer vision models designed for efficient in-device vision applications. One of the most important developments in the MobileNet series is the introduction of MobileNetV2, which brings significant improvements in performance and efficiency compared to its predecessor. MobileNetV2 introduced inverted residuals and linear bottlenecks that improve the performance of mobile models across a variety of tasks and benchmarks, while addressing different model sizes [24].

Building on the success of MobileNetV2, MobileNetV3 was developed, offering further improvements in computational efficiency and feature extraction. MobileNetV3 is introduced as a lightweight network architecture that can be adapted to specific tasks by changing the model size [25]. Furthermore, MobileNetV3 includes squeeze and excitation (SE) blocks that enhance feature extraction capabilities [26]. This structure, seen in Fig. 3, allows MobileNetV3 to efficiently extract feature information, making it suitable for a wide range of applications, including image segmentation, medical image classification and object detection [27–30].

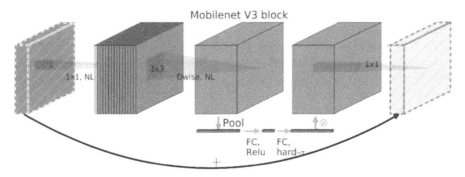

Fig. 3. MobileNetV3 block structure [13]

The MobileNetV3 family includes different versions, such as MobileNetV3-Small and MobileNetV3-Large, which meet different application requirements. These versions offer a balance between accuracy and computational speed [31]. MobileNetV3-Small model is used in the study.

2.4 Lightweight Hybrid CNN Model

In this study, the FPA detection performance of lightweight CNN models, MobileNetV3-Small, ShuffleNet and SqueezeNet are analyzed. Then, hybrid models are designed to improve the FPA detection performance. The proposed lightweight hybrid CNN model is shown in Fig. 4.

As shown in the figure, input images are given to all networks and feature extraction is performed in each network. The features extracted from the networks are passed through the ReLU activation function, then Global Average Pooling (GAP) is applied to these features and then all the features are concatenated. Dense layer and Sigmoid

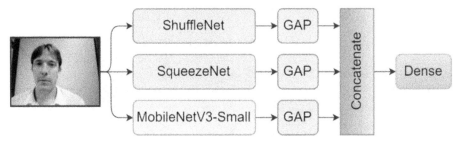

Fig. 4. Proposed lightweight hybrid FPA detection model

function are used for classification. In this study, each combination of networks is used to generate this hybrid model and the best hybrid model with the highest FPA detection performance is investigated.

3 Experimental Results

3.1 Replay-Attack Dataset

The REPLAY-ATTACK dataset [32] consists of real and fake access videos of 50 subjects taken with the MacBook Air 13" built-in camera in two distinct lighting conditions. In controlled environment fluorescent lamps were used for illumination and the videos were taken using a uniform background. In uncontrolled environment there is a non-uniform background and daylight was used for illumination. High-resolution photos and videos were obtained under the same conditions with the iPhone 3GS, and Canon PowerShot SX150 IS devices. There are three different types of attacks in the database. In Print Photo Attack the high-resolution photos printed on A4 paper were showed to the camera. In Mobile Attack the high-resolution photos and videos were showed to the camera using the iPhone 3GS screen. And in High-Resolution Attack the high-resolution photos and videos were showed to the camera using the iPad screen. Attack types are divided into two subgroups, Hand (attacker holds the device) and Fixed (device is positioned on a fixed support), according to the presentation method of the images/videos. The dataset was divided into separate training, development, and testing subgroups. In this study, the images that facial regions detected by the Dlib framework [33] were used. For each video, 30 frames were selected at equal intervals to provide data diversity. Figure 5 shows sample images from the REPLAY-ATTACK dataset.

Fig. 5. Sample images from REPLAY-ATTACK dataset

3.2 Experimental Setup

Studies in the field of FPA detection usually focus on face regions in input images due to limited resources. The face image is the key data for an FPA detection model, but details such as the background, lighting features, etc. of a fake face can provide information in FPA detection. Therefore, in this study, the FPA detection performance of deep learning models is analyzed using images provided by dataset providers and images containing only face regions. Sample input images used in the training phase are given in Fig. 6.

All experiments were performed on a standard desktop computer with 64G RAM, AMD RyzenTM 5 5600 CPU @ 3.5 GHZ processor and NVIDIA GeForce RTXTM 3060 12G GDDR6 @ 1.78 GHz graphics card on Python 3.10 software platform running on Windows 11. The deep learning parameters used in the experiments are given in Table 1. All networks were trained by learning from scratch method.

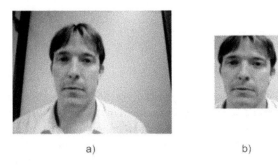

Fig. 6. Input images used in training: a) Whole image, b) Cropped image

Table 1. Deep learning parameters used in the experiments

Parameter	Value
Input Size	Face Image 128 × 128 Whole Image 224 × 224
Batch Size	32
Data Augmentation (in training set)	Vertical and Horizontal Flip: 0.2 Random Brightness: 0.2 Random Contrast: 0.2 Random Rotation: 0.2
Optimizer	Adam
Learning Rate	0.00001
Epochs	50
Loss Function	Binary Cross-Entropy

FPA detection systems are subject to two types of errors: denial of real accesses (false reject) and acceptance of attacks (false accept). Equal Error Rate (EER) and Half

The FPA detection performance obtained by using the whole images is given in Table 3. When the results in the table are analyzed, it is seen that there is a significant increase in the FPA detection performance of lightweight networks, especially MobileNetV3-Small and ShuffleNet. Even when the whole image is used, the FPA detection performance is improved by combining the features extracted from different networks. In particular, the proposed lightweight hybrid CNN model, which is a combination of ShuffleNet and SqueezeNet, has a very high FPA detection performance with 0.26% EER and 0.17% HTER.

Table 3. FPA detection performances of CNN models using whole images

Model	Parameter Count	Performance Metrics		
		EER	HTER	TIME Epoch/sec
MobileNetV3-Small	939,697	6.65	4.57	4.11
SqueezeNet	723,009	0.21	10.03	3.35
ShuffleNet	970,105	0.00	4.69	3.21
ShuffleNet + MobileNetV3-Small	1,909,801	0.79	10.13	6.81
ShuffleNet + SqueezeNet	1,693,113	0.26	**0.17**	6.04
MobileNetV3-Small + SqueezeNet	1,662,705	16.89	5.94	6.96
ShuffleNet + SqueezeNet + MobileNetV3-Small	2,632,809	0.21	2.29	9.69

The HTER values of the lightweight CNN models and the proposed hybrid models when trained using the cropped face image and the whole image are shown in Fig. 7. As can be seen in the figure, the use of the whole image in all six models except the ShuffleNet + MobileNetV3-Small model has greatly improved the FPA detection performance. Especially in the ShuffleNet + SqueezeNet hybrid model, using the whole image brought the HTER value closer to zero.

These results show that non-facial areas carry important information in FPA detection and positively affect the performance. The use of the whole image increases the amount of memory used, the training time of the network and the response time. However, with the lightweight hybrid CNN models proposed in this study, it is possible for the system to respond in an acceptable time.

The comparison of the proposed lightweight hybrid FPA detection model with other studies in the literature is given in Table 4. As can be seen from the table, the CNN model proposed in this study outperforms most of the other studies in the literature on the Replay-Attack dataset. In addition, most studies do not explain which frames were used in the experiments. This situation has a direct impact on the fair comparison of the results. It can be said that the proposed model is successful in FPA detection due to its low number of parameters and ease of implementation.

Total Error Rate (HTER) metrics are generally used to evaluate the performance of FPA detection models. EER is the value at the point where False Acceptance Rate (FAR) and False Rejection Rate (FRR) are equal to each other. The threshold τ corresponding to this value is obtained from the development set and HTER is calculated from the test set using this threshold value. HTER is calculated by equation below.

$$HTER(\tau) = \frac{FAR(\tau) + FRR(\tau)}{2} \quad (1)$$

3.3 Results and Discussion

In this study, firstly, FPA detection is performed with lightweight CNN models MobileNetV3-Small, ShuffleNet and SqueezeNet. Then, hybrid models are designed with to improve the performance of FPA detection. In the FPA detection process, both the cropped face regions and the whole images were used. Accordingly, the FPA detection results obtained by using cropped face regions are given in Table 2. When the results in the table are analyzed, it is seen that lightweight networks do not show sufficient FPA detection performance. On the other hand, combining the features extracted from different networks improved the FPA detection performance. The best EER value with 5.56% was obtained from the hybrid ShuffleNet + SqueezeNet network model. The best HTER value 5.70% was obtained with ShuffleNet + MobileNetV3-Small hybrid CNN network. The results shows that the combination of different deep features produced by different networks increases the FPA detection performance. However, this performance is higher in binary combinations. Although the hybrid model obtained with the combination of three networks exceeds the performance of the networks alone, it remained lower than the ShuffleNet + MobileNetV3-Small model.

Table 2. FPA detection performances of CNN models using cropped face images

Model	Parameter Count	Performance Metrics		
		EER	HTER	TIME Epoch/sec
MobileNetV3-Small	939,697	28.74	22.07	1.42
SqueezeNet	723,009	10.92	20.61	1.16
ShuffleNet	970,105	6.86	14.20	1.23
ShuffleNet + MobileNetV3-Small	1,909,801	6.65	**5.70**	2.41
ShuffleNet + SqueezeNet	1,693,113	5.56	8.38	2.14
MobileNetV3-Small + SqueezeNet	1,662,705	6.45	9.68	2.38
ShuffleNet + SqueezeNet + MobileNetV3-Small	2,632,809	6.03	6.68	3.48

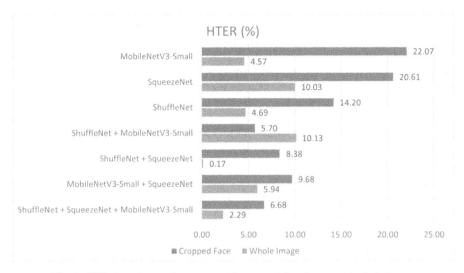

Fig. 7. FPA detection performance with cropped face image and whole image

Table 4. Comparison of the proposed model with the literature

Method	REPLAY-ATTACK	
	EER	HTER
Tu X., et al. (2017) [8]	1.03	1.18
Li L., et al. (2018) [34]	0.60	1.30
Li H., et al. (2018) [9]	0.30	1.20
Chen F-M., et al. (2019) [10]	2.30	2.60
Chen H., et al. (2020) [35]	0.21	0.39
Li L., et al. (2020) [11]	0.80	0.70
Shi L., et al. (2021) [36]	0.00	0.50
Shu X., et al. (2023) [37]	4.70	0.39
Pei M., et al. (2023) [38]	0.00	0.00
Proposed Method	**0.26**	**0.17**

4 Conclusion

In this study, the FPA detection performance of lightweight CNN networks and lightweight hybrid CNN models designed with combinations of these networks are investigated. In this study, three lightweight CNN models (MobileNetV3-Small, ShuffleNet and SqueezeNet) are used for FPA detection on Replay-Attack dataset. Then, hybrid CNN models (ShuffleNet + MobileNetV3-Small, ShuffleNet + SqueezeNet,

MobileNetV3-Small + SqueezeNet and ShuffleNet + SqueezeNet + MobileNetV3-Small) were designed by combining the features extracted from these networks and FPA detection was performed. Results were generated for the "All" test scenario, which combined all attack types in the dataset (printed photo, digital photo, video replay, in controlled and adverse environment, with device positionings fixed and hand), since it is not possible to know which attack type the attacker will use in real-world applications. On the other hand, the FPA detection performance of the deep learning models was analyzed using whole images provided by the dataset providers and only face regions.

The results show that the FPA detection performance of the proposed lightweight hybrid CNN models is much higher than the networks' individual performances, when both the cropped face image and the whole image are used as input. In general, using the whole image instead of the cropped face region greatly improved the FPA detection performance. Especially in the ShuffleNet + SqueezeNet hybrid model, the use of the whole image converges the HTER to zero. These results show that non-facial areas carry important information in FPA detection. The proposed lightweight model performs successful FPA detection in environments with limited resources. In future studies, the FPA detection performance of other lightweight networks can be investigated. The performance of the proposed model can be evaluated on other face spoofing datasets.

References

1. Ming, Z., Visani, M., Luqman, M.M., Burie, J-C.: A survey on anti-spoofing methods for face recognition with RGB cameras of generic consumer devices (2020)
2. Jin, B., Cruz, L., Gonçalves, N.: Deep facial diagnosis: deep transfer learning from face recognition to facial diagnosis. IEEE Access **8**, 123649–123661 (2020)
3. Ma, Y., Xu, Y., Liu, F.: Multi-perspective dynamic features for cross-database face presentation attack detection. IEEE Access **8**, 26505–26516 (2020). https://doi.org/10.1109/ACCESS.2020.2971224
4. Benlamoudi, A., Bekhouche, S.E., Korichi, M., et al.: Face presentation attack detection using deep background subtraction. Sensors **22**, 3760 (2022)
5. Sanghvi, N., Singh, S.K., Agarwal, A., et al.: Mixnet for generalized face presentation attack detection. In: 2020 25th International Conference on Pattern Recognition (ICPR), pp. 5511–5518. IEEE (2021)
6. Denisova, A.: An improved simple feature set for face presentation attack detection. Comput. Sci. Res. Notes **3201**, 16–23 (2022). https://doi.org/10.24132/CSRN.3201.3
7. Singh, A., Bhatt, S., Nayak, V., Shah, M.: Automation of surveillance systems using deep learning and facial recognition. Int. J. Syst. Assur. Eng. Manag. **14**, 236–245 (2023)
8. Tu, X., Fang, Y.: Ultra-deep Neural Network for Face Anti-spoofing. Lect Notes Comput Sci (including Subser Lect Notes Artif Intell Lect Notes Bioinformatics) 10635 LNCS, pp. 686–695 (2017). https://doi.org/10.1007/978-3-319-70096-0_70/COVER
9. Li, H., He, P., Wang, S., et al.: Learning generalized deep feature representation for face anti-spoofing. IEEE Trans. Inf. Forensics Secur. **13**, 2639–2652 (2018). https://doi.org/10.1109/TIFS.2018.2825949
10. Chen, F.-M., Wen, C., Xie, K., et al.: Face liveness detection: fusing colour texture feature and deep feature. IET Biometrics **8**, 369–377 (2019). https://doi.org/10.1049/iet-bmt.2018.5235

11. Li, L., Xia, Z., Jiang, X., et al.: CompactNet: learning a compact space for face presentation attack detection. Neurocomputing **409**, 191–207 (2020). https://doi.org/10.1016/j.neucom.2020.05.017
12. Zhang, X., Zhou, X., Lin, M., Sun, J.: Shufflenet: An extremely efficient convolutional neural network for mobile devices. In: Proceedings of the IEEE Conference on Computer Vision and Pattern Recognition, pp. 6848–6856 (2018)
13. Howard, A., Sandler, M., Chen, B., et al.: Searching for MobileNetV3. In: 2019 IEEE/CVF International Conference on Computer Vision (ICCV), pp. 1314–1324 (2019)
14. Neary, P.L., Watnik, A.T., Judd, K.P., et al.: CNN classification architecture study for turbulent free-space and attenuated underwater optical OAM Commun. Appl. Sci. **10** (2020)
15. Jiang, J., Liu, H., Zhao, C., et al.: Evaluation of diverse convolutional neural networks and training strategies for wheat leaf disease identification with field-acquired photographs. Remote Sens. **14** (2022)
16. Bayram, H.Y., Bingol, H., Alatas, B.: Hybrid deep model for automated detection of tomato leaf diseases. Trait du Sign. **39**, 1781 (2022)
17. De Marco, F., Ferrucci, F., Risi, M., Tortora, G.: Classification of QRS complexes to detect premature ventricular contraction using machine learning techniques. Plos One **17**, e0268555 (2022)
18. Özyurt, F.: Automatic Detection of COVID-19 Disease by Using Transfer Learning of Light Weight Deep Learning Model. Trait du Sign. **38** (2021)
19. Iandola, F.N., Han, S., Moskewicz, M.W., et al.: SqueezeNet: AlexNet-level accuracy with 50x fewer parameters and <0.5 MB model size. arXiv Prepr arXiv160207360 (2016)
20. Zhang, J., Zhu, H., Wang, P., Ling, X.: ATT squeeze U-Net: a lightweight network for forest fire detection and recognition. IEEE Access **9**, 10858–10870 (2021)
21. Ucar, F., Korkmaz, D.: COVIDiagnosis-Net: deep bayes-SqueezeNet based diagnosis of the coronavirus disease 2019 (COVID-19) from X-ray images. Med. Hypotheses **140**, 109761 (2020)
22. Yi, S., Zhang, G., Qian, C., et al.: A multimodal classification architecture for the severity diagnosis of glaucoma based on deep learning. Front. Neurosci. **16**, 939472 (2022)
23. Hassanpour, M., Malek, H.: Learning document image features with SqueezeNet convolutional neural network. Int. J. Eng. **33**, 1201–1207 (2020)
24. Sandler, M., Howard, A., Zhu, M., et al.: Mobilenetv2: inverted residuals and linear bottlenecks. In: Proceedings of the IEEE Conference on Computer Vision and Pattern Recognition, pp. 4510–4520 (2018)
25. Li, T., Yin, Y., Yi, Z., et al.: Evaluation of a convolutional neural network to identify scaphoid fractures on radiographs. J. Hand Surg. (Eur. **48**, 445–450 (2023)
26. Wu, Q., Wang, Z., Fang, H., et al.: A lightweight electronic water pump shell defect detection method based on improved YOLOv5s. Comput. Syst. Sci. Eng. **46**, 961–979 (2023)
27. Xia, Y., Li, Y., Ye, Q., Dong, J.: Image segmentation for blind lanes based on improved SegNet model. J. Electron. Imaging **32**, 13038 (2023)
28. Susanto, P.E., Kurniawardhan, A., Fudholi, D.H., Rahmadi, R.: A mobile deep learning model on Covid-19 CT-scan classification. Int. J. Artif. Intell. Res. **6** (2022)
29. Cai, L., Wang, C., Xu, Y.: A real-time FPGA accelerator based on winograd algorithm for underwater object detection. Electronics **10**, 2889 (2021)
30. Deng, T., Wu, Y.: Simultaneous vehicle and lane detection via MobileNetV3 in car following scene. Plos One **17**, e0264551 (2022)
31. Ye, N., Zhang, S.: Lightweight object detection network for helmets on the road. In: International Conference on Image, Signal Processing, and Pattern Recognition (ISPP 2023). SPIE, pp. 852–858 (2023)

32. Chingovska, I., Anjos, A., Marcel, S.: On the effectiveness of local binary patterns in face anti-spoofing. In: 2012 BIOSIG - Proceedings of the International Conference of Biometrics Special Interest Group (BIOSIG), pp. 1–7 (2012)
33. King, D.E.: Dlib-ml: a machine learning toolkit. J. Mach. Learn. Res. **10**, 1755–1758 (2009)
34. Li, L., Feng, X., Xia, Z., et al.: Face spoofing detection with local binary pattern network. J. Vis. Commun. Image Represent. **54**, 182–192 (2018). https://doi.org/10.1016/j.jvcir.2018.05.009
35. Chen, H., Hu, G., Lei, Z., et al.: Attention-based two-stream convolutional networks for face spoofing detection. IEEE Trans. Inf. Forensics Secur. **15**, 578–593 (2020). https://doi.org/10.1109/TIFS.2019.2922241
36. Shi, L., Zhou, Z., Guo, Z.: Face anti-spoofing using spatial pyramid pooling. In: 2020 25th International Conference on Pattern Recognition (ICPR), pp. 2126–2133. IEEE (2021)
37. Shu, X., Li, X., Zuo, X., et al.: Face spoofing detection based on multi-scale color inversion dual-stream convolutional neural network. Expert Syst. Appl. **224**, 119988 (2023). https://doi.org/10.1016/J.ESWA.2023.119988
38. Pei, M., Yan, B., Hao, H., Zhao, M.: Person-specific face spoofing detection based on a siamese network. Pattern Recognit. **135**, 109148 (2023). https://doi.org/10.1016/j.patcog.2022.109148

Disaggregation Analyses of the External/Internal Parameters for Chloride Diffusion in Concrete Structures: Artificial Intelligence by k-means Algorithm

Enrico Zacchei[1,2(✉)], Emilio Bastidas-Arteaga[3], and Ameur Hamani[3]

[1] Itecons, Coimbra, Portugal
enricozacchei@gmail.com
[2] University of Coimbra, CERIS, Coimbra, Portugal
[3] Laboratory of Engineering Sciences for Environment (LaSIE) UMR CNRS 7356, La Rochelle University, Avenue Michel Crépeau, 17042 La Rochelle, France

Abstract. In this paper, the parameters involved in the chloride ions diffusion for concrete structures have been analysed. Usually, several factors for diffusivity and surface chloride concentration are introduced to consider all internal and external actions that contribute to the diffusion. However, the inter-correlations between them are neglected. This paper treats these inter-correlations. From experimental laboratory tests about 850 parameters and values have been filtered and collected. These values have been divided in 11 categories and combined with each other. More of 120 numerical analyses by using k-means algorithm have been carried out. This algorithm is considered an artificial intelligence (AI) to support the human limitations in managing and analysing an enormous quantity of parameter. Results provide disaggregated values where each parameter is correlated with other one. This allows to understand their weight and effects on chloride ions diffusion. This paper provides not only a new approach but also practical values for more accurate analyses.

Keywords: Chloride diffusion · AI · k-means · disaggregation analyses

1 Introduction

A relevant issue for reinforced concrete (RC) structures is their degradation induced by chloride ingress. In [26] a list of published studies on this issue was presented, showing the purpose and goals in a conceptual way. Structures placed in coastal zones are usually subjected to this attack. In [27] it is shown how the chloride concentration in the atmosphere reaches high values at 4.50 km from the coast due to the inter-combinations of temperature/humidity/wind during a year. Structures placed in industrial areas also suffer this phenomenon indicating that this problem regards several assets in the world.

Comprehensive lifetime assessment of RC structures subjected to chloride ingress requires adequate prediction models as well as a detailed knowledge of their main model/material parameters, e.g., concrete characteristics, water/cement (w/c) ratio, temperature, humidity, surface chloride concentrations, etc., and their inter-correlations [25].

In many works the w/c ratio was considered as a unique parameter to estimate the chloride diffusivity and thus chloride concentrations. However, it is evident that this simplification could provide non-accurate results because chloride diffusivity depends also on other factors. For example, the effects of temperature and humidity are studied in [21].

In [28] was found that the surface chloride concentration, C_s, increases in function of the concrete age, up to a defined value where it stabilizes. For this reason, C_s is usually considered constant; but this approximation could under- or over-estimate the total chloride concentration in RC structures.

Experimental assessments of chloride profiles show that the peak value of concentrations, C_s', happens inside the concrete and near the concrete surface area called "convection area" (non-saturated area), and thus C_s concentrations at surface concrete are lower than this peak [3, 4, 21].

All these parameters deserve more attentions and for this the k-mean algorithm is adopted in this paper. It can be considered as tool for the artificial intelligence (AI) [29]. K-means algorithm has been used for different applications, due to its simplicity and efficiently/versatility as mentioned in [30], where it has been used to recognize the concrete aggregates from images and for prediction and sensitivity analysis of mechanical properties of recycled aggregate concrete [32].

To the best of the authors' knowledge there are no studies on the inter-correlations of the parameters involved in the chloride ions diffusion in RC structures by using k-mean algorithm.

2 Materials and Methods

2.1 Materials: Collected Data and Filtering

From literature, about 850.0 parameters and values from experimental laboratory tests (here called "in lab") have been collected from 15 published studies [1–10, 15–19]. The main criterion to select the published studies is related to unsaturated exposure conditions where the surface chloride concentrations, C_s', reach a maximum value at x' point inside the concrete element. The collected values come from chloride profiles, i.e., chloride concentrations (C_s, C_s') versus concrete depths (x, x'), thus it was possible to estimate the values of these four key parameters.

Other parameters regard the concrete characteristics (i.e., concrete density, w/c ratio, volume) and the external/exposure conditions (i.e., NaCl concentration, temperature, T, humidity, h, exposure time). Therefore, in total 11 filtered categories have been defined.

Disaggregation Analyses of the External/Internal Parameters for Chloride Diffusion   243

In Table 1 it is possible to see, for brevity, only 8 categorise (main categories) and their mean values regarding 10 published studies. The means, μ_0, and standard deviations, σ_0, of all values are also shown, which are calculated by considering all parameters thus they are different to the sum of each mean.

Table 1. Mean values of the collected "in lab" parameters (for brevity, only [1–10]).

Reference	NaCl concentration (g/L)	Temperature, T (°C)	Humidity, h (%)	Exposure time (year)	C_s (kg/m³)	x (cm)	C_s' (kg/m³)	x' (cm)
[1]	80.0	20.0	71.33	0.08	4.59	0.10	9.72	0.30
[2]	50.0	N/A	N/A	0.58	8.64	0.50	10.56	1.50
[3]	N/A	26.30	100.0	5.0	3.10	0.30	3.29	0.70
[4]	16.50	23.0	95.0	0.25	1.03	0.06	1.15	0.31
[5]	23.0	21.70	100.0	0.27	4.81	0.77	8.14	0.43
[6]	30.0	20.0	75.0	0.08	2.99	0.15	3.95	0.49
[7]	N/A	N/A	N/A	4.17	15.54	0.25	18.90	0.60
[8]	30.0	13.30	65.0	11.25	6.71	0.40	8.92	0.93
[9]	24.0	21.80	56.50	0.16	2.57	1.06	3.69	0.38
[10]	N/A	20.0	95.0	0.49	2.12	0.10	4.25	0.35
…	…	…	…	…	…	…	…	…
$\mu_0 \pm \sigma_0$ [a]	36.71 ± 20.18	19.09 ± 3.40	77.25 ± 13.75	2.54 ± 4.13	8.61 ± 6.61	0.33 ± 0.29	10.47 ± 6.22	0.74 ± 0.47

N/A = Not available.
[a] μ_0 was calculated by considering all parameters [1–10, 15–19] thus it is different to the sum of each mean.

Figure 1 shows the non-clustered values of C_s with respect C_s'. The horizontal lines indicate the four levels of aggressiveness (from low to extreme) used as reference to estimate an input C_s for modelling purposes. They are $C_s < 1.15$ kg/m³ (low); $1.15 \leq C_s < 2.95$ kg/m³ (moderate); $2.95 \leq C_s < 7.35$ kg/m³ (high); $C_s \geq 7.35$ kg/m³ (extreme) [20, 21, 31].

It is possible to see a very good linear approximation since the R-squared value is $R^2 = 0.908$ (≈ 1.0). Therefore, the obtained liner relation (i.e., $C_s = (0.867 \times C_s') - 0.994$) could provide a good estimation for evaluating the C_s' value (usually unknown) under atmospheric exposure. In fact, it provides a peak value at the interface between convection and diffusion zone as studied in [21].

Figure 2 compares C_s (red points) and C_s' (black points) values with respect to NaCl (Fig. 2a)) and temperature (Fig. 2b)). Here all data are shown without clustering analyses. The horizontal lines, already explained for Fig. 1, indicate the four levels of aggressiveness (low to extreme).

In Fig. 2a), it is possible to see that the chloride concentration decreases by increasing the NaCl quantity. Apparently, this trend should be opposite; however, the NaCl quantity was measured externally to the studied specimen thus it represents only an input value, and it does not provide information about the concrete response in terms of its permeability. Whereas, as expected, by increasing temperature T (12.0 °C → 20.0 °C), the diffusivity thus the concentrations increase, as shown in Fig. 2b). Actually, the factor

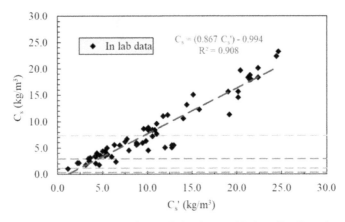

Fig. 1. C_s vs. C_s' values and their correlation by considering all collected studies.

that accounts the temperature follows this trend as shown by the common Arrhenius' law [22].

2.2 Methodology

To calculate the similarity among clusters, the most used Euclidian distance [14], dE, thus the squared Euclidian distance (i.e., square error), dE^2, is adopted [12, 23]:

$$dE = \sqrt{\sum_i^k (c_i - p_i)^2} \rightarrow dE^2 = \sum_i^k (c_i - p_i)^2 \quad (1)$$

where $c_i(x, y)$ is the centre of the cluster i (up to the total clusters number k), $p_i(x, y)$ is the point to be compared, and the subscript $i = \{1, 2, 3, \ldots, k\}$ indicates thus the dimension. Since a 2D space is considered, the points p_i and c_i are defined by two coordinates x and y. For a set, t, of cases $\Pi = \{p_1, p_2, p_3, \ldots, p_n\} \in \mathcal{R}^d$, where $\mathcal{R}^d$ is the data space of d dimensions, the k-means algorithm tries to find a set of k cluster centres $C = \{c_1, c_2, c_3, \ldots, c_k\} \in \mathcal{R}^d$ that is a solution of the minimization problem:

$$E = \sum_i^k \sum_j^{n_i} \|p_{ij} - c_i\|^2 \quad (2)$$

$$\min_n\{E\} = \min_n \left\{ \sum_i^k \sum_j^{n_i} \|p_{ij} - c_i\|^2 \right\} \quad (3)$$

where n_i is the number of cases with $j = \{1, 2, 3, \ldots, n_i\}$ included in cluster k and $\sum_i^k n_i = n$. Thus, the k-means clustering technique is considered a variance (Eq. (2)) minimization technique.

Disaggregation Analyses of the External/Internal Parameters for Chloride Diffusion 245

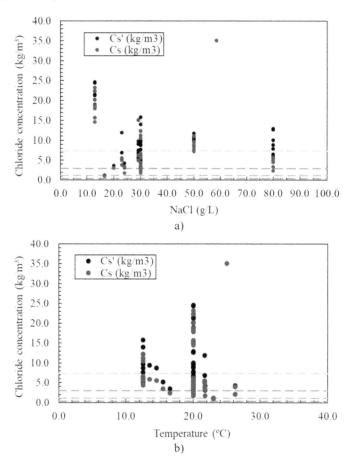

Fig. 2. Chloride concentrations (C_s, C_s') in function of a) NaCl and b) temperature values.

For each iterated c_i, the following condition must be verified [13]:

$$\frac{\partial E}{\partial c_i} = 0 \rightarrow c_i^{(t+1)} = \frac{1}{|n_i^t|} \sum_j^{n_i^t} p_{ij} \qquad (4)$$

thus, the average of the elements of each group is taken as the new centroid.

As mentioned in [11, 13], mathematically, the k-means algorithm approximates the Gaussian model with an estimation of the clusters by maximum likelihood. This model considers a cluster as a probability for each case, based on the mean, μ_i, standard deviation, $\pm \sigma_i$, and its probability density function (PDF). The k-means algorithm is a sub-case that assumes that the clusters have $\pm \sigma_i$ values and a trend PDF.

The main goal of the k-mean algorithm is to produce groups with a high degree of similarity and reduce the complexity of the data. The procedures are divided in the following resumed 5 steps: (i) collection data (see Sect. 2.1); (ii) choice of the distance function between c_i and p_i (Eq. (1)); (iii) definition of cluster numbers, k; (iv) definition of the random position of c_i and its coordinates (first iteration). Here a further determination of a k number is made, however Mathematica software [23] provides automatic solutions; (v) solution (Eqs. (2)–(3)) and verification (Eq. (4)).

Regarding step (iii), the following "intuitive approach", as stated in [24], has been adopted: the couples C_s –x and C_s' –x' have been plotted and divided in 6 clusters (from k = 1 to k = 6). Then, for each cluster, dE (Eq. (1)) has been estimated as shown in Fig. 3 (note that, for instance, dE by blue line corresponds to the evolution of the first cluster).

In Fig. 3 it is possible individuate three parts: (i) a part where the Euclidean distance leads to a constant value for k ≥ 4.0; in this area clusters are very similar to each other and they should be merged; (ii) a sharply sloping part for k < 3.0 where the high slope indicates that very dissimilar clusters should not merge together since they are no longer internally homogeneous; (iii) and a curved transition part at k = 3.0 (or curve "knee") where there is a reasonable number of optimum clusters. Therefore k = 3.0 has been adopted in this study. The dashed lines represent the exponential trend.

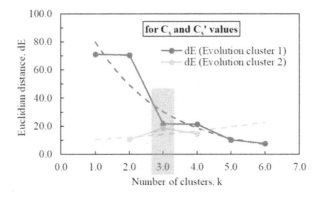

Fig. 3. Individuation of the number of clusters, k: dE trends of the clusters by C_s and C_s' values.

3 Analyses and Results

The k-means clustering technique, as already mentioned, can be described as a centroid model since one vector representing $\mu_i \pm \sigma_i$ values is used to describe each cluster [12]. From all combinations, about 120 analyses have been carried out. Table 2 list the obtained μ_i for cluster 1, μ_1 (for brevity only a matrix 8 × 8 is shown).

The values in Table 2 represent new values to be used to estimate in more accurate way the chloride concentrations in RC elements and their correlations. It is important to highlight that μ_0 values (Sect. 2.1) have been obtained independently of the other parameters, whereas μ_1 values have been estimated by coupled analyses. In this sense, results provide disaggregation values can be used to evaluate the effect and weight of a certain parameter with respect other one. This shows the great potential of the AI by k-means algorithm.

Table 2. Mean values of the Cluster 1, μ_1.

	NaCl concentration (g/L)	Temperature, T (°C)	Humidity, h (%)	Exposure time (year)	C_s (kg/m³)	x (cm)	C_s' (kg/m³)	x' (cm)
NaCl concentration	80.00	20.30	66.98	0.33	3.82	0.19	7.35	0.43
Temperature	62.80	21.57	63.80	1.10	5.44	0.26	7.82	0.54
Humidity	69.20	21.00	65.44	0.22	5.10	0.24	8.10	0.53
Exposure time	80.00	22.50	71.33	0.18	11.53	0.14	9.91	0.83
C_s	69.20	21.62	65.56	0.31	3.63	0.26	6.16	0.33
x	80.00	20.36	70.33	0.17	5.13	0.30	8.47	0.39
C_s'	69.20	20.72	66.30	0.30	4.81	0.30	6.91	0.82
x'	42.15	20.97	68.53	0.27	4.90	0.30	4.53	0.30

Figure 4a) shows an example of how Table 2 could be read by considering all 11 categories. C_s parameter is used as reference since it represents a key parameter to estimate chloride concentration in RC elements [21, 22, 25]. It is clear as cluster 3 overestimates the C_s concentrations. This indicates two main aspects: (i) the values in cluster 3 should not be used; (ii) the presence of cluster 3 allows to not include in other clusters the "outliers" avoiding contaminating them. Therefore, cluster 1 and 2 should be used; in particular, we discuss the mean value of C_s for cluster 1 (i.e., 5.32 ± 2.69 kg/m³) since it is like more common cases.

The parameter directly correlated to C_s are NaCl, T, h, and t. In Fig. 4a) there are three values smaller and one higher than the non-clustered $\mu_0 = 8.61 \pm 6.61$ kg/m³ (red horizontal line) already shown in Table 1.

Figure 4b) shows by histograms the μ_0/μ_i ratio of the cluster 1 for all combinations; some results are also shown in Table 2.

From Fig. 4b) it is possible to see the utility of the cluster algorithm. In general, when $\mu_0/\mu_i = 1.0$ the preliminary non-clustered value does not suffer alterations therefore it could be adopted during the chloride diffusion analysis. When $\mu_0/\mu_i > 1.0$ or $\mu_0/\mu_i < 1.0$ the preliminary value could be over- or under-estimated, respectively.

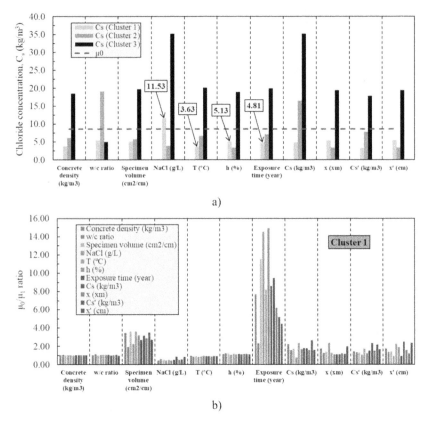

Fig. 4. Example of application: a) clusters for C_s concentrations by considering each parameter; b) μ_0/μ_1 ratio for all combinations for the cluster 1.

Finally, Fig. 5 shows the scatter diagrams with histograms. Due to the high number of diagrams (i.e., about 350 diagrams) only some combinations have been shown. Here it is also possible to count how many values are placed between a certain interval. In this sense, the a disaggregated cluster analysis is combined with a stochastic analysis. For example, C_s' - C_s' combination indicate 4 squares with histograms with 1 (blue square), 8, 11 (red square), 8 values (from left to right and bottom to top).

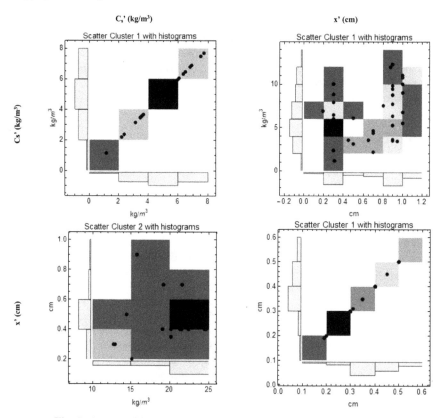

Fig. 5. Scatter with histogram for some combinations (cluster 1 and 2).

4 Conclusions

In this paper k-means algorithm has been applied for chloride ion concentrations in RC elements.

From several previous studies, −850.0 parameters have been collected and divided in 11 categories. These parameters regard internal and external conditions of concrete for chloride diffusion. Preliminary analyses from experimental laboratory tests show that there is liner relation between C_s and C_s', which could provide a good estimation for evaluating the C_s' (usually unknown), i.e., $C_s = (0.867 \times C_s') - 0.994$.

Clustered data provide new values (Table 2). These disaggregation analyses provide the weight and impact of a certain parameter with respect other one. Also, from Fig. 4b) it is possible to see the utility of the cluster algorithm. In general, when $\mu_0/\mu_i = 1.0$ the preliminary non-clustered value does not suffer alterations therefore it could be adopted during the chloride diffusion analysis. When $\mu_0/\mu_i > 1.0$ or $\mu_0/\mu_i < 1.0$ the preliminary value could be over- or under-estimated, respectively. This shows the great potential of the AI by k-means algorithm.

This paper provides not only a new approach (i.e., k-means for chloride diffusion) but also practical values for more accurate analyses.

Acknowledgements. The first author thanks the Itecons institute, Coimbra, Portugal, and the University of Coimbra (UC), Portugal, to pay the rights (when applicable) to completely download all papers in the references. The first author is also grateful for the Foundation for Science and Technology's support through funding UIDB/04625/2020 from the research unit CERIS (https://doi.org/10.54499/UIDB/04625/2020).

References

1. Win, P.P., Watanabe, M., Machida, A.: Penetration profile of chloride ion in cracked reinforced concrete. Cem. Concr. Res. **34**, 1073–1079 (2004)
2. Zhang, S.F., Lu, C.H., Liu, R.G.: Experimental determination of chloride penetration in cracked concrete beams. Procedia Eng. **24**, 380–384 (2011)
3. Farahani, A., Taghaddos, H., Shekarchi, M.: Prediction of long-term chloride diffusion in sflica fume concrete in a marine environment. Cem. Concr. Compos. **59**, 10–17 (2015)
4. Liu, J., Liu, J., Huang, Z., Zhu, J., Liu, W., Zhang, W.: Effect of fly ash as cement replacement on chloride diffusion, chloride binding capacity, and micro-properties of concrete in a water soaking environment. Appl. Sci. **10**, 6271 (2020)
5. Samson, E., Marchand, J.: Modeling the effect of temperature on ionic transport in cementitious materials. Cem. Concr. Res. **37**, 455–468 (2007)
6. Fraj, A.B., Bonnet, S., Khelidj, A.: New approach for coupled chloride/moisture transport in non-saturated concrete with and without slag. Constr. Build. Mater. **35**, 761–771 (2012)
7. Tadayon, M.H., Shekarchi, M., Tadayon, M.: Long-term field study of chloride ingress in concretes containing pozzolans exposed to severe marine tidal zone. Constr. Build. Mater. **123**, 611–616 (2016)
8. Thomas, M.D.A., Matthews, J.D.: Performance of pfa concrete in a marine environment – 10-year results. Cement Concr. Compos. **26**, 5–20 (2004)
9. Nielsen, E.P., Geiker, M.R., Chloride diffusion in partially saturated cementitious material. Cem. Concrete Res. **33**, 133–138 (2003)
10. Climent, M.A., de Vera, G., Lopez, J.F., Viqueira, E., Andrade, C.: A test method for measuring chloride diffusion coefficients through nonsaturated concrete – Part I the instantaneous plane source diffusion case. Cem. Concrete Res. **32**, 1113–1123 (2002)
11. Ramdani, F., Kettani, O., Tadili, B.: Evidence for subduction beneath gibraltar arc and andean regions from k-means earthquake centroids. J. Seismol. **19**, 41–53 (2015)
12. Morissette, L., Chartier, S.: The k-means clustering technique: general considerations and implementation in mathematica. Tutorials Quant. Meth. Psychol. **9**, 15–24 (2013)
13. Symons, M.J.: Clustering criteria and multivariate normal mixtures. Biometrics **37**, 35–43 (1981)
14. Li, Y., Min, K., Zhang, Y., Wen, L.: Prediction of the failure point settlement in rockfill dams based on spatial-temporal data and multiple-monitoring-point models. Eng. Struct. **243**, 1–12 (2021)
15. Gao, Y.H., Zhang, J.Z., Zhang, S., Zhang, Y.R.: Probability distribution of convection zone depth of chloride in concrete in a marine tidal environment. Constr. Build. Mater. **140**, 485–495 (2017)
16. Sleiman, H.: Étude du transport des chlorures dans les matériaux cimentaires non saturés: Validation expérimentale sur bétons en situation de marnage, PhD Thesis, University of La Rochelle, La Rochelle, France (2008)
17. Francy, O., François, R.: Measuring chloride diffusion coefficients from non-steady state diffusion tests. Cem. Concr. Res. **28**, 947–953 (1998)

18. Tahlaiti, M.: Etude de la pénétration des chlorures et de l'amorçage de la corrosion en zone saturée et en zone de marnage, PhD Thesis, University of La Rochelle, La Rochelle, France (2010)
19. Sergi, G., Yu, S.W., Page, C.L.: Diffusion of chloride and hydroxyl ions in cementitious materials exposed to a saline environment. Mag. Concr. Res. **44**, 63–69 (1992)
20. El Hassan, J., Bressolette, P., Chateauneuf, A., El Tawil, K.: Reliability-based assessment of the effect of climatic conditions on the corrosion of RC structures subject to chloride ingress. Eng. Struct. **32**, 3279–3287 (2010)
21. Zacchei, E., Bastidas -Arteaga, E.: Multifactorial chloride ingress model for reinforced concrete structures subjected to unsaturated conditions. Buildings **12**, 107 (2022)
22. Zacchei, E., Nogueira, C.G.: Chloride diffusion assessment in RC structures considering the stress-strain sate effects and crack width influences. Constr. Build. Mater. **201**, 100–109 (2019)
23. Wolfram Mathematica, version 12.0, Wolfram Research, Inc.: Champaign, IL, USA (2019)
24. Sugar, C.A., James, G.M.: Finding the number of clusters in a dataset: An information-theoretic approach. J. Am. Stat. Assoc. **98**, 750–763 (2003)
25. Zacchei, E., Nogueira, G.C.: Calibration of boundary conditions correlated to the diffusivity of chloride ions: an accurate study for random diffusivity. Cement Concr. Compos. **126**, 1–14 (2022)
26. Verma, S.K., Bhadauria, S.S., Akhtar, S.: Evaluating effect of chloride attack and concrete cover on the probability of corrosion. Front. Struct. Civ. Eng. **7**, 379–390 (2013)
27. Guerra, J.C., Castañeda, A., Corvo, F., Howland, J.J., Rodriguez, J.: Atmospheric corrosion of low carbon steel in a coastal zone of ecuador: anomalous behaviour of chloride deposition versus distance from the sea. Mater. Corros. **70**, 444–460 (2019)
28. Silva, C.A., Guimarães, A.T.C.: Evaluation of the diffusion model considering the variation in time of the chloride content of the concrete surface. Matéria **19**, 81–93 (2014)
29. Turco, C., Funari, M.F., Teixeira, E., Mateus, R.: Artificial neural networks to predict the mechanical properties of natural fibre-reinforced compressed earth blocks (CEBs). Fibers **9**, 1–21 (2021)
30. Chen, L., Shan, W., Liu, P.: Identification of concrete aggregates using K-means clustering and level set method. Structures **34**, 2069–2076 (2021)
31. Bastidas-Arteaga, E., Bressolette, P., Chateauneuf, A., Sánchez-Silva, M.: Probabilistic lifetime assessment of RC structures under coupled corrosion-fatigue deterioration processes. Struct. Saf. **31**, 84–96 (2009)
32. Nguyen, T.D., Cherifa, R., Mahieux, P.Y., Lux, J., Aït-Mokhtar, A., Bastidas-Arteaga, E.: Artificial intelligence algorithms for prediction and sensitivity analysis of mechanical properties of recycled aggregate concrete: a review. J. Build. Eng. **66**, 1–20 (2023)

K-means for Small Earthquakes. Alternative Disaggregation Analyses by Considering Wave Components and Soil Types

Enrico Zacchei[1,2(✉)] and Reyolando Brasil[3,4]

[1] Itecons, Coimbra, Portugal
enricozacchei@gmail.com
[2] University of Coimbra, CERIS, Coimbra, Portugal
[3] Polytechnic School of São Paulo, University of São Paulo (USP), 380 Prof. Luciano Gualberto, São Paulo, SP, Brazil
[4] Center for Engineering, Modeling and Applied Social Sciences, Federal University of ABC (UFABC), 3 Rua Arcturus, São Bernardo do Campo, SP, Brazil

Abstract. Prediction of ground motion by earthquakes is still the main issue for seismologists, geo-technicians, engineers, and researchers. In this paper k-means algorithm has been used to disaggregate seismic parameters to evaluate their inter-correlations. A goal is to quantify the weights and effects of each parameter with respect other ones. From database, about 4900.0 data, divided in 22 categories, have been collected. The main divisions regard the wave components in horizontal and vertical axis, and the soil characteristics. The studied seismic zone is the "Norpirenaica oriental", placed at Pyrenees area between Spain and France, classified as a very high seismic hazard. Numerical and analytical analyses have been carried out to implement the algorithm. These analyses could quantify the role of the sand horizontal stratigraphy, the non-linear response, and the elasticity of the soil. Results are plotted in stochastic distributions and elastic spectra accelerations. Rigorously, results are valid only for the studied seismogenic zone under predefined constrictions and range.

Keywords: AI · Machine Learning · k-means · Pyrenees Area · Seismic Analyses

1 Introduction

After the 20th century, with the gradual development of artificial intelligence (AI) several algorithms are used to simulate the input-output relationship for high-precision analyses. In [1] the k-means algorithm is considered as an "artificial neural network (ANN)" thus an AI, as shown in [2]. K-means algorithm is useful to analyse several values and dimensions since it is very difficult for humans to compare items of such complexity reliably without a support to aid the comparison (e.g., the common issue of the "big data" treatment [3]). The analyses based on human subjective judgments are affected by personal and individual experiences. In fact, k-means algorithm is considered an unsupervised machine learning that belongs to the AI in the great area of the computer science [2].

The main goals of the k-mean algorithm are: (i) to produce groups of cases/variables with a high degree of similarity within each group (called "compactness" in [4] and "cohesion" in [5]) and a low degree of similarity between groups ("separation" [4]); (ii) to reduce the complexity of the data to obtain useful outputs.

In Web of Science database, there are only 5 articles [6–10] published between 2015–2022 with the words "k-means" and "earthquake" in the title, indicating the necessity to improve this research. In general, to the best of the authors' knowledge there are not sufficient studies on the inter-correlations of the parameters involved for seismic analyses by using k-mean algorithm.

In [8], k-means algorithm has been used to classify earthquakes in Gibraltar Arc and Andean regions; it was stated that "very little research has been done on the basis of seismic events analysed from clustering point of view". In [9], it was proposed and tested, for Turkey, the ability of the k-mean clustering method to create the training dataset for earthquake vulnerability analysis. In [10], the nonlinear seismic site response for Japan has been classified by using k-means algorithm (according to [10], this study was the first to apply the machine learning clustering algorithm to address this problem). In [7], a new search space reduction algorithm using machine learning techniques for the earthquake source parameter determination was presented. Finally, in [6], a seismic prediction model based on clustering of global earthquake data was proposed.

In this paper, a unique seismogenic zone (ZS) with relative few events (in [10] only one event was considered) has been considered to try to obtain homogeneous results. The Pyrenees area has been taken as case study since it represents a seismic interesting area [11]. The considered parameters deserve more attentions, and this algorithm would allow to disaggregate each parameter with respect other one and to understand their weight/effect on a seismic analysis.

2 Materials and Methods

2.1 Materials

The studied seismogenic zone (ZS) is the ZS16 called "Norpirenaica oriental" placed at Pyrenees area between Spain and France (for more details see IGME (2015) [12]). This zone is characterized by a very high seismic hazard.

From ESM database [13] 234.0 data for 22.0 parameters have been collected ($\approx$4900.0 data in total). 8 small earthquakes have been considered. For brevity, in Table 1 are listed only some parameters (6 parameters) indicating their mean value. Other 16 parameters are: vertical peak ground acceleration (VPGA), Arias's intensity I_A, time interval T_{90}, moment magnitude M_w, soil type, wave component, sampling interval, cut-off frequency of the low- and hight-pass filter, unprocessed PGA, peak ground velocity (PGV), peak ground displacement, Housner intensity, spectral acceleration S_a at 1.0 s and 3.0 s structural period, seismic moment.

The meaning of these parameters is usually well known thus they are not explained here but they can be retrieved in literature [14, 15, 25].

Table 1. Mean value (except for "year") of some collected parameters.

Parameter	Event 1	Event 2	Event 3	Event 4	Event 5	Event 6	Event 7	Event 8
Year	2002	2003	2009	2012	2007	2014	2006	2008
Local magnitude, M_L	4.20	4.50	4.10	4.70	4.50	4.40	5.20	3.70
Depth, Δ (km)	2.0	2.0	5.0	15.10	10.0	12.70	14.80	3.40
Epicentral distance, d_e (km) [a]	29.0	67.30	60.40	49.0	79.60	82.10	93.60	71.20
HPGA (cm/s^2) [b]	−0.54	2.74	−1.14	−0.70	−0.63	−2.44	0.88	−5.37
Sa$_{0.30}$ (cm/s^2) [b]	1.53	3.02	0.68	2.42	0.92	3.15	7.12	1.94

Note: HPGA = Horizontal peak ground acceleration. Sa$_{0.30}$ = Spectral acceleration at 0.30 s structural period. All values refer to the original list without separation in terms of wave components, soil types, etc. (except for HPGA and Sa$_{0.30}$)
[a] Surface distance between the station where the event was registered and the epicentre
[b] Calculated by considering both horizontal waveform components, orthogonal to each other, i.e., N (north-south) and E (east-west)
Note that the values in Table 1 represent raw data without separating two important aspects, i.e., the wave components and the soil type

The wave components are N, E, Z, where N e E refer to the horizontal waveform components orthogonal to each other, north-south and east-west, respectively, whereas Z refers to the vertical waveform component (positive upward) [13]. The soil A and B corresponds to rock and very dense sand, respectively, in accordance with Eurocode [16].

From the collected data, some preliminary results can be obtained. Figure 1 shows (non-clustered) combinations between I_A and PGV. There are several parameters that characterize an earthquake, some of which are very significant, as for instance, I_A, which is proportional to the total energy input of an infinity set of undamped linear oscillators.

In Fig. 1, the linear trends (here plotted in logarithmic scale) with a R-squared value between R2 = 0.83–0.91 indicate a good approximation. The difference of the N axis with respect E axis for soil B (Fig. 1b)) is probably due to the sand horizontal stratigraphy that provide a non-linear response between IA and PGV, whereas the different response of Z axis could be correlated to the elasticity of the soil [17].

2.2 Methodology

To calculate the similarity among clusters, the most used Euclidian distance [3], dE, thus the squared Euclidian distance (i.e., square error), dE2, is adopted [18, 20]:

$$dE = \sqrt{\sum_i^k (c_i - p_i)^2} \rightarrow dE^2 = \sum_i^k (c_i - p_i)^2 \quad (1)$$

where $c_i(x, y)$ is the centre of the cluster i (up to the total clusters number k), $p_i(x, y)$ is the point to be compared, and the subscript i = {1, 2, 3, …, k} indicates thus the dimension.

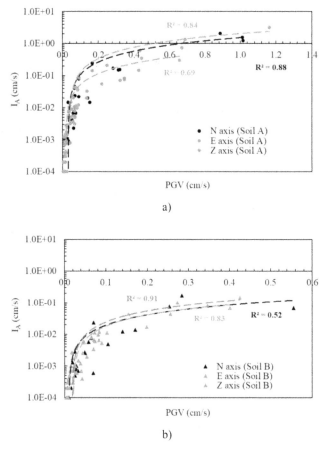

Fig. 1. Correlations between non-clustered I_A vs. PGV parameters divided in components (N, E, Z) for: a) soil A and b) soil B.

Since a 2D space is considered, the points pi and ci are defined by two coordinates x and y. For a set, t, of cases $\Pi = \{p1, p2, p3, ..., pn\} \in \mathcal{R}^d$, where $\mathcal{R}^d$ is the data space of d dimensions, the k-means algorithm tries to find a set of k cluster centres $C = \{c1, c2, c3, ..., ck\} \in \mathcal{R}^d$ that is a solution of the minimization problem:

$$\min_n\{E\} = \min_n\left\{\sum_i^k \sum_j^{n_i} \|p_{ij} - c_i\|^2\right\} \quad (2)$$

where n_i is the number of cases with $j = \{1, 2, 3, ..., n_i\}$ included in cluster k and $\sum_i^k n_i = n$. Thus, the k-means clustering technique is considered a variance (Eq. (2)) minimization technique.

For each iterated c_i, the following condition must be verified [19]:

$$\frac{\partial E}{\partial c_i} = 0; \quad c_i^{(t+1)} = \frac{1}{|n_i^t|} \sum_j^{n_i^t} p_{ij} \quad (3)$$

thus, the average of the elements of each group is taken as the new centroid.

Equation (3) represents the tolerance level between the cluster solutions. As mentioned in [8, 19], mathematically, the k-means algorithm approximates the Gaussian model with an estimation of the clusters by maximum likelihood. This model considers a cluster as a probability for each case, based on the mean, μ_i, standard deviation, $\pm \sigma_i$, and its probability density function (PDF). The k-means algorithm is a sub-case that assumes that the clusters have $\pm \sigma_i$ values and a trend PDF.

The main goal of the k-mean algorithm is to produce groups with a high degree of similarity and reduce the complexity of the data. The procedures are divided in the following resumed 5 steps: (i) collection data to be analysed (see Sect. 2.1); (ii) choice of the distance function to be used between c_i and p_i (Eq. (1)); (iii) definition of cluster numbers, k. Given that k-means algorithm is an unsupervised (i.e., a classification made by clustering) and non-hierarchical method (i.e., a construction of clusters where the objects in a cluster are more like one another than to objects in different clusters), this step is mandatory; (iv) definition of the random position of c_i and its coordinates (first iteration). Here a further determination of a k number is made, however Mathematica [20] provides automatic solutions, under multiple iterations, that are intrinsically computed; (v) estimation of E, $\min_n\{E\}$ (Eq. (2)) and verification Eq. (3). If this is not verified, it is necessary to repeat step (iv) up to convergence.

3 Analyses and Results

3.1 Clustered Stochastic Results

Figure 2 shows two combinations of clustered values for N axis component (soil A). Obviously, when the horizontal and vertical axis quantify the same parameters, all clusters are plotted in a diagonal line (see Fig. 2a)). Figure 2b) shows the scatter diagrams with histograms where it is possible to count how many values are placed between a certain interval for each axis. A green scatter area indicates 1 point up to magenta area with 11 points. The colour tone changes depending on the number of values within the coloured square. Also, the horizontal histograms, regarding de axis, correspond to the points sum, i.e., from left to right there are 9, 12, 3, 7, 5 points.

Here we adopt k = 3.0 since the used data are already sufficiently homogeneous due to the applied collecting and filtering process; and to well separate "outliers", which can contaminate other clusters. In general, the best cluster is the cluster placed at the middle with respect other ones. The general criterium was to choose a cluster that provides a mean value more like that for raw data. This is because a goal is to obtain more reliable and refined results from real registrations.

As mentioned, the k-means clustering technique can be described as a centroid model as one vector representing the mean and its standard deviation, $\mu i \pm \sigma i$, used to describe each cluster. Table 2 shows $\mu i \pm \sigma i$ outputs of the studied parameters (more significant parameters) obtained by considering all combinations, which would indicate new values for each component to be used to frame the seismic context of the ZS16 in a more correct way. This division in terms of components and soils is important since it allows to obtain outputs in a universal way.

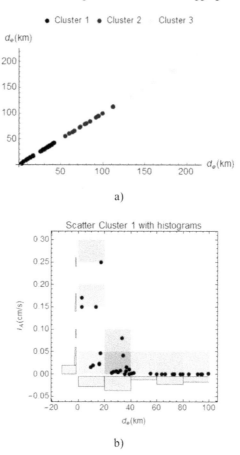

Fig. 2. Combinations of some parameters for N axis in soil A by a) 3 clusters and b) scatter with histograms (cluster 1).

Table 2. $\mu_i \pm \sigma_i$ results by considering all combinations.

	d_e (km)	PGA (cm/s^2)	I_A (cm/s)	T_{90} (s)
N axis in soil A	72.87 ± 32.72	−0.41 ± 6.30	0.07 ± 0.06	19.09 ± 6.48
E axis in soil A	68.77 ± 16.01	−3.35 ± 7.90	0.05 ± 0.14	16.96 ± 4.71
Z axis in soil A	75.94 ± 32.39	0.53 ± 2.18 [a]	0.09 ± 0.06	21.77 ± 7.09
N axis in soil B	35.73 ± 24.61	−0.40 ± 1.32	0.02 ± 0.01	18.18 ± 4.49
E axis in soil B	36.81 ± 21.26	0.17 ± 1.63	0.01 ± 0.0	15.33 ± 4.58
Z axis in soil B	47.30 ± 23.0	−0.30 ± 1.65 [a]	0.01 ± 0.01	23.93 ± 6.51

[a] It corresponds to VPGA.

In Table 2 it is possible to note some physical behaviours (some already discussed for Fig. 1). In particular, for soil A, since the bedrock schematizes an ideal soil without strong influences of the elasticity, heterogeneity, and non-linearity in accordance with classical theories [17]. In fact, the ratio between vertical and horizontal PGA for soil A is 0.16 (= 0.53/3.35), which is like the mean ratio obtained by using European code (i.e., 0.18) [16].

PGA values of the horizontal components appear a little different each other, in particular for soil A; probably due to the different horizontal frequency of the rock in two directions. The high value of 3.35 cm/s^2 (E axis in soil A) is strictly correlated to a minor T90 value indicating a relatively short and intense events. For E axis in soil B, this correlation is also verified, however a lower value of PGA was found. This difference is probably correlated to the fact that a rock provides "pure" values, whereas a non-linear sand provides damped values due to the material damping [17]. For medium-high d_e values (i.e., 75.0 km [17]), the radiation damping due to the spreading of the energy plays an important role.

Figure 3 shows the PDF curves of the clustered results (solid curves) and non-clustered results (dashed curves) for soil A and B, respectively. The formers have been plotted by using values in Table 2, whereas the latter regard raw data discussed in Sect. 2.1.

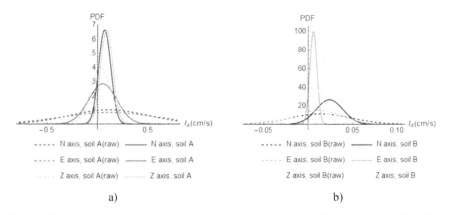

Fig. 3. PDF curves for different parameters and wave components for a) soil A and b) soil B.

3.2 New Seismic Inputs

In this section new seismic inputs in terms of (elastic and synthetic) spectral acceleration, Sa, have been shown.

The elastic spectrum, in Fig. 4a), refers to the "type 1", in accordance with Eurocode [16], where Sa trends (mean values) in function of structural period, T, have been plotted.

Figure 4b) shows synthetic spectra for a far and relative strong event (estimated as $M_w = 4.8$, $d_e = 68.77$ km). Synthetic spectra have been obtained by using four attenuation relations (i.e., SP96 [21], Am96 [22], Am05 [23], BT03 [24]). The use of these relations is justified by the fact that they have been calibrated for a seismologic and geological context like that studied in this paper. They are also valid within the range of magnitude and distance considered. In general, these equations have been largely used in the South-European area.

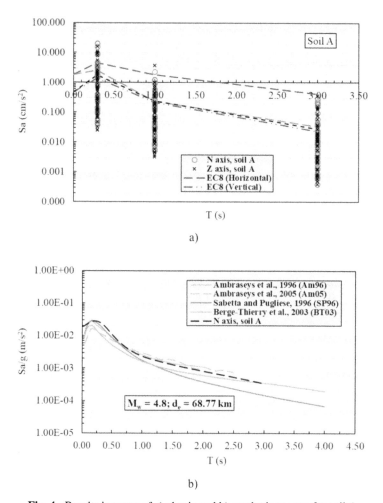

Fig. 4. Results in terms of a) elastic and b) synthetic spectra for soil A.

All curves, in Fig. 4b), provide similar results for stiff structures (i.e., T = 0–1.0 s) indicating the real goodness in using these attenuation equations for ZS16. For soil B (not shown here) these agreements are not verified, in fact 4 equations have not been calibrated for very dense sand (they are calibrated for upper and lower limit, i.e., rock, and soft or alluvium sites, respectively) and, in general, it is difficult to calibrate them for low magnitudes.

4 Conclusions

In this paper k-means algorithm has been applied to disaggregate several seismic parameters for understand possible inter-correlations. As case study, ZS16 called "Norpirenaica oriental" placed at Pyrenees area between Spain and France has been considered.

About 4900.0 data divided in 22 categories have been treated to carry out the k-means algorithm. The main divisions regard the wave components (i.e., N, E, Z) and soil characteristics (i.e., A, B). This could quantify the role of the sand horizontal stratigraphy, the non-linear response, and the elasticity of the soil.

Clustered results, considering all combinations, provide new values to be used for seismic analyses (see Table 2) in ZS16 in a more refine way. This disaggregation analysis allows to evaluate the weight and effect of a certain parameter with respect other ones. This could show the great potential of the AI by k-means algorithm.

Finally, new seismic inputs for ZS16 in terms of elastic spectra, Sa, have been plotted. Results show that horizontal Sa values by code overestimates response for soil A. Synthetic spectra by attenuation equations provide good approximation for a far and relative strong event. It is important to highlight that all results are rigorously valid only for ZS16.

Acknowledgements. The first author thanks the Itecons institute, Coimbra, Portugal, and the University of Coimbra (UC), Portugal, to pay the rights (when applicable) to completely download all papers in the references. The first author is also grateful for the Foundation for Science and Technology's support through funding UIDB/04625/2020 from the research unit CERIS (https://doi.org/10.54499/UIDB/04625/2020). The second author received support from CNPq and FAPESP, both Brazilian research funding agencies.

References

1. Afshoon, I., Miri, M., Mousavi, S.R.: Combining Kriging meta models with U-function and K-Means clustering for prediction of fracture energy of concrete. J. Build. Eng. **35**, 1–16 (2021)
2. Turco, C., Funari, M.F., Teixeira, E., Mateus, R.: Artificial neural networks to predict the mechanical properties of natural fibre-reinforced compressed earth blocks (CEBs). Fibers **9**, 1–21 (2021)
3. Li, Y., Min, K., Zhang, Y., Wen, L.: Prediction of the failure point settlement in rockfill dams based on spatial-temporal data and multiple-monitoring-point models. Eng. Struct. **243**, 1–12 (2021)
4. Di Giuseppe, M.G., Troiano, A., Troise, C., De Natale, G.: K-Means clustering as tool for multivariate geophysical data analysis. an application to shallow fault zone imaging. J. Appl. Geophys. **101**, 108–115 (2014)
5. Sheikhhosseini, Z., Mirzaei, N., Heidari, R., Monkaresi, H.: Delineation of potential seismic sources using weighted K-means cluster analysis and particle swarm optimization (PSO). Acta Geophys. **69**, 2161–2172 (2021)
6. Yuan, R.: An improved K-means clustering algorithm for global earthquake catalogs and earthquake magnitude prediction. J. Seismol. **25**, 1005–1020 (2021)
7. Lee, S., Kim, T.: Search space reduction for determination of earthquake source parameters using PCA and k-means clustering. J. Sens. **1–12** (2020)

8. Ramdani, F., Kettani, O., Tadili, B.: Evidence for subduction beneath Gibraltar arc and Andean regions from k-means earthquake centroids. J. Seismol. **19**, 41–53 (2015)
9. Shafapourtehrany, M., Yariyan, P., Ozener, H., Pradhan, B., Shabani, F.: Evaluating the application of K-mean clustering in Earthquake vulnerability mapping of Istanbul, Turkey. Int. J. Disaster Risk Reduct. **79**, 1–23 (2022)
10. Ji, K., Wen, R., Ren, Y., Dhakal, Y.P.: Nonlinear seismic site response classification using K-means clustering algorithm: Case study of the September 6, 2018 Mw6.6 Hokkaido Iburi-Tobu earthquake, Japan. Soil Dyn. Earthq. Eng. **128**, 1–14 (2020)
11. Garcia-Mayordomo, J., Insua-Arevalo, J.M.: Seismic hazard assessment for the Itoiz dam site (Western Pyrenees, Spain). Soil Dyn. Earthq. Eng. **31**, 1051–1063 (2011)
12. IGME (2015), ZESIS: Base de Datos de Zonas Sismogénicas de la Península Ibérica y territorios de influencia para el cálculo de la peligrosidad sísmica en España. http://info.igme.es/zesis. Accessed March 2023
13. Luzi, L., et al.: ORFEUS Working Group 5. Engineering Strong Motion Database (ESM) (Version 2.0). Istituto Nazionale di Geofisica e Vulcanologia (INGV) (2020). https://doi.org/10.13127/ESM.2
14. Zacchei, E., Brasil, R.: A new approach for physically based probabilistic seismic hazard analyses for Portugal. Arab. J. Geosci. **15**, 1–22 (2022)
15. Faccioli, E., Paolucci, R.: Elements of seismology applied to engineering, Pitagora Editrice, Bologna, Italy, p. 255 (2005)
16. European Committee for Standardization (CEN), Eurocode 8: Design of structures for earthquake resistance, Part 1: General rules, seismic actions and rules for buildings, BS EN 1998–1: 2004. Brussels, Belgium (2004)
17. Kramer, S.L.: Geotechnical Earthquake Engineering, first ed., Prentice-Hall, Upper Saddle River, NJ, p. 653 (1996)
18. Morissette, L., Chartier, S.: The k-means clustering technique: general considerations and implementation in Mathematica. Tutorials Quant. Meth. Psychol. **9**, 15–24 (2013)
19. Symons, M.J.: Clustering criteria and multivariate normal mixtures. Biometrics **37**, 35–43 (1981)
20. Wolfram Mathematica, version 12.0, Wolfram Research, Inc.: Champaign, IL, USA (2019)
21. Sabetta, F., Pugliese, A.: Estimation of response spectra and simulation of nonstationary earthquake ground motions. Bull. Seismol. Soc. Am. **86**(2), 337–352 (1996)
22. Ambraseys, N.N., Simpson, K.A., Bommer, J.J.: Pre- diction of horizontal response spectra in Europe. Earthq. Eng. Struct. Dyn. **25**, 371–400 (1996)
23. Ambraseys, N.N., Douglas, J., Sarma, S.K., Smit, P.M.: Equations for the estimation of strong ground motions from shallow crustal earthquakes using data from Europe and the Middle East: horizontal peak ground acceleration and spectral acceleration. Bull. Earthq. Eng. **3**(1), 1–53 (2005)
24. Berge-Thierry, C., Cotton, F., Scotti, O.: New empirical response spectral attenuation laws for moderate European earthquakes. J. Earthq. Eng. **7**(2), 193–222 (2003)
25. Zacchei, E., Lyra, P.: Recalibration of low seismic excitations in Brazil through probabilistic and deterministic analyses: application for shear buildings structures. Struct. Concr. **1–19** (2022)

Algorithm for Distributing Incoming Messages Among Handlers

Ivan Zverev and Anna Zykina(✉)

Omsk State Technical University, 11 Mira Prospect, Omsk, Russia
avzykina@mail.ru

Abstract. In modern digital world, the amount of information is increasing dramatically. All this information requires immediate processing which can be done by automated systems. To accelerate processing in automated systems, new models, methods and efficient algorithms for their implementation are needed. The article considers the problem of distributing incoming messages to handlers in an automated system of an organization. The classical assignment problem and known methods of its solution are the mathematical basis for solving the problem. A generalization of the assignment problem with invalid combinations of assignments is considered as a model. A solution algorithm is developed and a software product for solving the described problem in the Python language is implemented. Moreover, numerical experiments are carried out to verify the quality of the developed algorithm and check its feasibility. The applied problem posed in this paper is relevant as in terms of its practical application and as the study of the mathematical model of the assignment problem and methods of its solution. The developed algorithm allows us to calculate the system load and regulate the number of handlers in operation. The obtained results can serve as input parameters for the simulation model of the incoming message processing system which will allow experimenting with the load of such a system.

Keywords: Assignment Problem · Mathematical Modeling · Distributed Computing Systems · Equivalent Transformations · Simulation Modeling · Python

1 Introduction

This paper proposes an algorithm to perform the task of distributing incoming messages to handlers in an organization's automated system. This problem should be solved in a specified system due to the active interaction of organizations in the modern digital world through various channels of electronic communication.

The proposed algorithm is based on an open assignment problem with invalid combinations of assignments. The assignment problem has a large number of modifications depending on the specifics of the problem statement. Such modifications are generally solved by using equivalent transformations of mathematical models. This allows the original problem to change into a classical assignment problem in a given number of steps. The classical assignment problem is solved by well-known algorithms, which are standard algorithms. Depending on the dimension of the problem, the specifics of the constraints and the target function, one can for instance use the Hungarian algorithm or Mack's method [1, 2]. For the classical assignment problem [3], there have been developed efficient algorithms implemented in various libraries (for example, the SciPy library for solving optimization and mathematical problems in Python).

Theoretical analysis of the assignment problem and its modifications was carried out in the research process. In addition, a mathematical model of the practical problem was formulated, equivalent transformations of the constructed mathematical models were carried out, and an algorithm for solving the practical problem was developed and implemented.

Section 2 reviews existing papers on the subject, forming the basis for a theoretical study of the assignment problem. Section 3 describes the problem statement and the algorithm for solving the applied problem. Section 4 contains numerical experiments to verify the correctness of the algorithm.

2 Literature Review

The works considering the formulation and solution of the assignment problem differing from the classical model are of particular interest to researchers [3]. Alternative mathematical models of the assignment problem are based on generalizations and represent the following types of problems: problem with nonlinear target function, interval and fuzzy problems, multi-index problem, and multi-criteria problem [4–6]. The diversity of mathematical models leads to a variety of methods for solving the problem, such as graph solving algorithms, dynamic programming algorithms, genetic algorithms and algorithms based on the use of dual methods [7–10].

In [11], a solution to the assignment problem in a distributed computing system is considered. What is distinctive about the work is the statistics about node utilization. This allows the weight matrix to change over time and obtain more efficient distributions when solving the problem. In our study, the problem solution is required to model the distribution of messages in a customized system, so the weight matrix is computed just once.

In terms of applications, the most interesting paper is the one on solving the optimization problem of controlling the flow of incoming requests [12]. In this paper, a multi-criteria assignment problem and its solution algorithm based on simulated annealing are considered. An essential feature of this problem is that it is NP-complete due to the presence of handler occupancy functions.

Finding optimal solutions for the considered non-classical models can involve cumbersome algorithms that usually lead to an approximate solution. Articles [13–16] describe the use of equivalent transformations to simplify the mathematical model and the possibility of applying classical solution methods to it. However, the combination of several equivalent transformations is not considered in these publications, thus leaving part of the formulations unexplored. In this regard, the problem of combining equivalent transformations seems relevant.

Thus, the considered applied problem is relevant both in terms of its application and investigation of the mathematical model of the assignment problem and algorithms for its solution.

3 Theoretical Study of the Assignment Problem

Let us consider the classical formulation of the assignment problem.

There are n tasks, each of which can be done by any of n operators. The costs c_{ij} of the i-th operator performing the j-th task are known. It is obvious from the meaning of costs that they are non-negative finite numbers

$$c_{ij} \geq 0 (i = \overline{1,n}, j = \overline{1,n}).$$

Only one task can be assigned to each operator, and each task can be performed only by one operator. It is necessary to distribute the tasks in such a way that they are executed with minimum total costs.

The mathematical model of the classical assignment problem has the following form:

$$f(X) = \sum_{i=1}^{n} \sum_{j=1}^{n} c_{ij} x_{ij} \to min, \tag{1}$$

$$\sum_{i=1}^{n} x_{ij} = 1, j = \overline{1,n}, \tag{2}$$

$$\sum_{j=1}^{n} x_{ij} = 1, i = \overline{1,n}, \tag{3}$$

$$x_{ij} \in \{0,1\}, i,j = \overline{1,n}. \tag{4}$$

Here, the choice of variable parameters x_{ij} set by conditions (4) means the following: if the i-th operator is assigned to perform the j-th task, then $x_{ij} = 1$, otherwise $x_{ij} = 0$. Conditions (2) and (3) mean that each j-th task can be performed by only one operator and each i-th operator can be assigned only one task. Function (1) sets the amount of costs c_{ij} for which $x_{ij} = 1$.

The assignment problem described by the simplest linear model (1)–(4) belongs to the class of integer linear programming problems, so universal algorithms of linear and integer programming can be used to solve it [6]. One of the universal algorithms of linear programming is the simplex method. The algorithm for solving the linear programming problem by simplex method is based on the sequential enumeration of admissible basis solutions. Due to its features, the classical assignment problem can also be solved with algorithms for solving the transportation problem, for instance, the method of potentials which is an analogue of the dual simplex method.

Let us consider the formulation of an open assignment problem.

The open assignment problem implies that the cost matrix $C = (c_{ij})_{m \times n}$ is not square, i.e. $m \neq n$. Suppose $n > m$. This means that the number of tasks exceeds the number of operators. Then the mathematical model of the open assignment problem has the following form:

$$f(X) = \sum_{i=1}^{n} \sum_{j=1}^{n} c_{ij} x_{ij} \to min, \tag{5}$$

$$\sum_{i=1}^{m} x_{ij} \leq 1, j = \overline{1,n}, \qquad (6)$$

$$\sum_{j=1}^{n} x_{ij} = 1, i = \overline{1,m}, \qquad (7)$$

$$x_{ij} \in \{0,1\}, i = \overline{1,m}, j = \overline{1,n}. \qquad (8)$$

Suppose $m > n$. This means that the number of operators exceeds the number of tasks. Then the mathematical model of the open assignment problem has the following form:

$$f(X) = \sum_{i=1}^{n} \sum_{j=1}^{n} c_{ij} x_{ij} \to \min, \qquad (9)$$

$$\sum_{i=1}^{n} x_{ij} = 1, j = \overline{1,n}, \qquad (10)$$

$$\sum_{j=1}^{n} x_{ij} \leq 1, i = \overline{1,m} \qquad (11)$$

$$x_{ij} \in \{0,1\}, i = \overline{1,m}, j = \overline{1,n}. \qquad (12)$$

The inequality signs in the constraints (11) mean that not all operators receive task assignments.

The equivalent transformations of the rectangular cost matrix into a square matrix in the open problems (5)–(8) and (9)–(12) correspond to the rules of reducing the open model of the transportation problem, the assignment problem being its special case, to the closed model and are formulated as follows:

1) determine the order of the cost matrix Q in the closed model $p = \max\{m,n\}$,
2) proceed to the cost matrix $Q = (q_{ij})_{p \times p}$ based on the rule:

$$q_{ij} = c_{ij}, i = \overline{1,m}, j = \overline{1,n};$$

$$q_{ij} = 0, i = \overline{1,p}, j = \overline{n+1,p}, \text{ if } m > n;$$

$$q_{ij} = 0, i = \overline{m+1,p}, j = \overline{1,p}, \text{ if } m < n.$$

The result of applying the transformation is the transition to the classical model of the form:

$$\varphi(X) = \sum_{i=1}^{p} \sum_{j=1}^{p} q_{ij} x_{ij} \to \min, \qquad (13)$$

$$\sum_{i=1}^{n} x_{ij} = 1, j = \overline{1,p}, \qquad (14)$$

$$\sum_{j=1}^{n} x_{ij} = 1, i = \overline{1,p}, \qquad (15)$$

$$x_{ij} \in \{0,1\}, i, j = \overline{1, p}. \tag{16}$$

Pairs of models (5)–(8), (13)–(16) and (9)–(12), (13)–(16) are equivalent to each other.

As a result, an algorithm for solving the open assignment problem (5)–(8) (or (9)–(12)) can be formulated.

1) Apply the equivalent transformation of a rectangular cost matrix into a square cost matrix to the model (5)–(8) (or (9)–(12)). The result is the equivalent model (13)–(16).
2) Solve the problem described by the ratios (13)–(16) by the Hungarian algorithm or Mack's method.
3) In the resulting solution, select a submatrix (x_{ij}), $i = \overline{1, m}, j = \overline{1, n}$. This submatrix is the optimal solution to the open assignment problem.

Consider the formulation of the assignment problem with invalid assignments.

Let there be n tasks and n operators. Assume the condition that some assignments are invalid, e.g. an operator of one type cannot process a task of the other type.

Let us denote the set of indices for operators and tasks by $I = \{1, 2, \ldots, n\}$. Let us define a binary relation of admissible assignments on the set I

$$R_{v.a.} = \{(i, j) | assignment(i, j) valid\},$$

and impose an additional constraint on the variables x_{ij} representing the assignment of the task j to the operator i:

$$x_{ij} = 0, \text{ if } (i, j) \notin R_{v.a.}, \text{ where } i, j = \overline{1, n}.$$

Then the mathematical model of the problem with invalid assignments takes the form:

$$f(X) = \sum_{i=1}^{n} \sum_{j=1}^{n} c_{ij} x_{ij} \to \min, \tag{17}$$

$$\sum_{i=1}^{n} x_{ij} = 1, j = \overline{1, n}, \tag{18}$$

$$\sum_{j=1}^{n} x_{ij} = 1, i = \overline{1, n}, \tag{19}$$

$$x_{ij} = 0, (i, j) \notin R_{v.a.}, \tag{20}$$

$$x_{ij} \in \{0, 1\}, (i, j) \in R_{v.a.}. \tag{21}$$

The formulated model differs from the classical model in that its unknowns, corresponding to the pairs of indices $(i, j) \notin R_{v.a.}$, are predetermined to be zero. To transition to the classical model of the assignment problem, it suffices to transition to the cost matrix $Q_{n \times n}$, associated with the relation $R_{v.a.}$ as follows:

$$q_{ij} = c_{ij} \text{ if } (i, j) \in R_{v.a.} \tag{22}$$

$$q_{ij} = M \text{ if } (i,j) \notin R_{v.a.} \tag{23}$$

where M is a sufficiently large positive number in comparison with $max(c_{ij})$.

The introduction of the matrix Q has the following meaning. It is necessary to provide a system of penalties for the choice of variables x_{ij} that have $(i,j) \notin R_{v.a.}$. This is done by introducing an expression Mx_{ij}. In the optimization process, this penalty will result in a zero value of the unknown x_{ij}, $(i,j) \notin R_{v.a.}$. Theoretically, it is required that $M \to \infty$, but in terms of practical calculations, it is sufficient if the final value of the quantity M is large enough not to affect the final result of the calculations. Number $M > n \cdot max(c_{ij}), \forall (i,j) \in R_{v.a.}$ can be taken as a penalty. For the sake of certainty, it can be assumed that $M = 2n \cdot max(c_{ij}), \forall (i,j) \in R_{v.a.}$.

In this case, the equivalent transformation of the assignment problem with invalid assignments is formulated as follows:

1) find the amount of the penalty $M = 2n \cdot max(c_{ij}), \forall (i,j) \in R_{v.a.}$;
2) proceed to the cost matrix Q based on rule (22), (23).

The result of applying the transformation is the transition to the classical model of the form:

$$\varphi(X) = \sum_{i=1}^{n} \sum_{j=1}^{n} q_{ij} x_{ij} \to min, \tag{24}$$

$$\sum_{i=1}^{n} x_{ij} = 1, j = \overline{1,n}, \tag{25}$$

$$\sum_{j=1}^{n} x_{ij} = 1, i = \overline{1,n}, \tag{26}$$

$$x_{ij} \in \{0,1\}, i,j = \overline{1,n}. \tag{27}$$

The models (17)–(21) and (24)–(27) are equivalent to each other. Therefore, it is possible to apply the classical algorithm for solving the assignment problem to the model (24)–(27).

As a result, it is possible to formulate an algorithm for solving the problem of assignments with invalid assignments (17)–(21) (or (24)–(27)).

1) Apply an equivalent transformation of invalid assignments to the model (17)–(21). The result is the equivalent model (24)–(27).
2) Solve the problem described by the ratios (24)–(27) by the Hungarian algorithm or Mack's method.

4 Problem Statement and Solution Algorithm for the Applied Problem

Let us apply theoretical models to the task in the applied field. Let the organization receive electronic messages through various communication channels. Messages have different sizes and types of information. On the organization side, it is necessary to distribute the received messages to handlers having different processing cost for each message. Also, handlers can only process certain types of information. After the message has been converted in the processor, the information is passed on to the computer system, where further processing takes place and a notification of successful message processing is generated.

A minimum sufficient number of handlers must be defined in order to process all incoming messages in a timely manner. For this purpose, the problem at hand is formulated as an open assignment problem with unacceptable combinations of assignments. To do this, we need to apply equivalent transformations and construct an algorithm to solve the problem at hand.

Formal description of the applied problem.

Let the number of incoming messages be equal to m, and the number of handlers be equal to n. In general, the number of messages and handlers does not match, and the number of messages is much greater than the number of handlers ($m \gg n$).

The cost (or time) of processing the i-th message on the j-th handler is denoted as c_{ij}. Obviously, all costs (or processing time) are non–negative finite numbers.

Each of the handlers can only process certain types of messages. This condition can be denoted by the set of valid assignments

$$R_{v.a.} = \{(i,j) | assignment(i,j) valid\}$$

A natural criterion for the problem is to minimize the processing time of incoming messages.

The mathematical model of this applied problem has the following form:

$$f(X) = \sum_{i=1}^{m} \sum_{j=1}^{n} c_{ij} x_{ij} \to min, \tag{28}$$

$$\sum_{i=1}^{n} x_{ij} \leq 1, j = \overline{1, m}, \tag{29}$$

$$\sum_{j=1}^{m} x_{ij} = 1, i = \overline{1, n} \tag{30}$$

$$x_{ij} = 0, (i,j) \notin R_{v.a.}, \tag{31}$$

$$x_{ij} \in \{0,1\}, (i,j) \in R_{v.a.}. \tag{32}$$

The formulated model (28)–(32) combines the properties of two generalized assignment problem models:

- the model of an open assignment problem (5)–(8),
- the assignment problem with invalid combinations (17)–(21).

Let us apply the necessary equivalent transformations to reduce the problem to the classical assignment problem and to solve it further.

The first step is to bring the problem to a closed form.

To do this, it is necessary:

1) determine p – the order (dimension) of the cost matrix Q in the closed model $p = max\{m, n\} = m$,

2) since $m > n$, then proceed to the cost matrix (cost or processing time of messages) $Q = (q_{ij})_{p \times p}$, constructed according to the rule:

$$q_{ij} = c_{ij}, i = \overline{1, m}, j = \overline{1, n};$$

$$q_{ij} = 0, i = \overline{1, p}, j = \overline{n+1, p}.$$

The second step is to reduce the problem to the classical form using equivalent transformations of the assignment problem with invalid combinations.

It should be noted that after changing the matrix dimension at the first step (adding dummy handlers) it is necessary to change the set of valid assignments $R_{v.a.}$, forming it as follows:

$$R'_{v.a.} = R_{v.a.} \cup \{(i, j) | i = \overline{1, p}, j = \overline{n+1, p}\}.$$

Replacing the set of valid assignments $R_{v.a.}$ with the set $R'_{v.a.}$ means that dummy handlers can process any messages (practically it means the following: messages assigned to dummy handlers for processing will not be processed).

Thus, the algorithm of the second step looks like this:

1) form the set $R'_{v.a.}$;
2) find the amount of the penalty $M = 2p \cdot \max(c_{ij})$, $(i, j) \in R'_{v.a.}$;
3) proceed to the cost matrix Z based on the rule:

$z_{ij} = q_{ij}$, if $(i, j) \in R'_{v.a.}$;
$z_{ij} = M$, if $(i, j) \notin R'_{v.a.}$.

As a result, a classical assignment problem of the form (13)–(16) with the cost matrix Z will be obtained.

Thus, the algorithm for solving the assignment problem can be formulated as follows.

1) Apply an equivalent transformation of a rectangular cost matrix into a square one to the model (26)–(30).
2) Apply an equivalent transformation of invalid assignments to the resulting model. The result is the model (31)–(34) equivalent to the original one.
3) Solve the problem described by the ratios (31)–(34) by the Hungarian algorithm or Mack's method.
4) In the resulting solution, select the submatrix (x_{ij}), $i = \overline{1, m}, j = \overline{1, n}$. This submatrix is the optimal solution to the open assignment problem.

The solution of this problem will result in an assignment matrix in which not all messages will be processed, according to the formulation of the open assignment problem. But it is necessary to process all messages received by the organization. To do this, the solution algorithm is modified.

5) Delete the rows i corresponding to the processed messages on real handlers from the matrix Z according to the following rule:

$$\left\{ i | \sum_{j=1}^{n} x_{ij} = 1, i = \overline{1, m} \right\}.$$

6) Delete the columns j corresponding to dummy handlers added when converting the task to a closed form from the matrix Z, but the real handlers cannot be deleted:

$$\{j | j = \overline{p - n + 1, p}, j > n\}.$$

7) If the number of unprocessed messages happens to become less than the number of handlers, it is necessary to introduce dummy messages to bring the problem to a closed form, using appropriate equivalent transformations.

As a result of the solution, s matrices of message distribution by handlers will be obtained. Each next iteration of processing takes the message into operation when the processing of the previous iteration is finished.

8) By combining s distribution matrices into one, the matrix of the final distribution of messages among handlers is obtained, where each message is handled once, and each handler may have processed more than one message.

Thus, the processing of all messages will be completed in:

$$t = max\left(\sum_{i=1}^{m} z_{ij} x_{ij}\right), j = \overline{1, n}.$$

The message processing system does not operate around the clock, but on a specific schedule. It is necessary to process all messages in a single working day. The duration of the working day is T.

Since the number of incoming messages cannot be controlled, it is necessary to determine the minimum number of handlers sufficient to process all incoming messages per working day.

5 Numerical Experiment

The developed algorithm for solving the applied problem was implemented on Python language software using SciPy library for solving optimization and mathematical problems.

Consider the distribution of 12 messages to 5 handlers to be executed in 2 h. Table 1 presents the original cost matrix (processing time is given in minutes), and Table 2 presents the matrix of impossible assignments (cells with the value -1, indicate impossible assignments).

Table 1. Initial cost matrix.

	0	1	2	3	4
0	12	53	31	40	47
1	2	21	72	61	17
2	70	72	85	54	39

(*continued*)

Table 1. (*continued*)

	0	1	2	3	4
3	93	34	62	75	51
4	76	14	15	7	72
5	43	95	41	74	34
6	26	44	35	77	30
7	6	77	96	58	91
8	46	88	4	93	78
9	79	82	85	80	60
10	55	40	86	68	27
11	78	15	27	39	13

Table 2. Matrix of impossible assignments.

	0	1	2	3	4
0	0	0	−1	0	0
1	0	0	0	0	0
2	0	0	0	0	−1
3	−1	0	0	0	0
4	0	0	0	−1	0
5	0	0	0	0	0
6	0	−1	0	0	0
7	0	0	−1	0	0
8	0	0	0	0	−1
9	0	0	0	0	0
10	−1	0	−1	0	0
11	−1	−1	−1	−1	0

The initial transformation of the cost matrix involves reducing it to a square matrix and replacing the cost values in invalid combinations with a calculated penalty with the value of 2304. The result of performing the equivalent transformations is the matrix presented in Table 3.

After the first iteration, the following pairs of assignments were obtained: (0, 3), (1, 0), (2, 5), (3, 6), (4, 1), (5, 7), (6, 8), (7, 9), (8, 2), (9, 10), (10, 11), (11, 4). The cost matrix for the second iteration is presented in Table 4.

Table 3. Cost matrix after equivalent transformations.

	0	1	2	3	4	5	6	7	8	9	10	11
0	12	53	2304	40	47	0	0	0	0	0	0	0
1	2	21	72	61	17	0	0	0	0	0	0	0
2	70	72	85	54	2304	0	0	0	0	0	0	0
3	2304	34	62	75	51	0	0	0	0	0	0	0
4	76	14	15	2304	72	0	0	0	0	0	0	0
5	43	95	41	74	34	0	0	0	0	0	0	0
6	26	2304	35	77	30	0	0	0	0	0	0	0
7	6	77	2304	58	91	0	0	0	0	0	0	0
8	46	88	4	93	2304	0	0	0	0	0	0	0
9	79	82	85	80	60	0	0	0	0	0	0	0
10	2304	40	2304	68	27	0	0	0	0	0	0	0
11	2304	2304	2304	2304	13	0	0	0	0	0	0	0

Table 4. Cost matrix of the second iteration.

	0	1	2	3	4	5	6
0	70	72	85	54	2304	0	0
1	2304	34	62	75	51	0	0
2	43	95	41	74	34	0	0
3	26	2304	35	77	30	0	0
4	6	77	2304	58	91	0	0
5	79	82	85	80	60	0	0
6	2304	40	2304	68	27	0	0

After the second iteration, the following pairs of assignments were obtained: (0, 3), (1, 1), (2, 5), (3, 2), (4, 0), (5, 6), (6, 4). The cost matrix for the third iteration is presented in Table 5.

After the third iteration, the following pairs of assignments were received: (0, 2), (1, 4), (2, 0), (3, 1), (4, 3). After summarizing the processing time by handler, the final result of the problem solution was obtained and is presented in Table 6.

The maximum processing time is 100 min, which fits into the specified two-hour interval. The problem has been solved successfully.

Table 5. Cost matrix of the third iteration.

	0	1	2	3	4
0	43	95	41	74	34
1	79	82	85	80	60
2	0	0	0	0	0
3	0	0	0	0	0
4	0	0	0	0	0

Table 6. Final result.

0	1	2	3	4
8	48	80	94	100

6 Conclusion

In this paper, a mathematical model is constructed and an iterative algorithm is developed to solve an applied assignment problem for distributing incoming messages to handlers. The following novelty results were obtained.

1. An algorithm for solving the open assignment problem with invalid assignment combinations based on the application of equivalent model transformations and iterative solution of the equivalent model has been developed.
2. A software product has been implemented to solve the problem in Python.
3. Numerical experiments have been carried out to verify the quality of the developed algorithm and check its performance.

The developed algorithm allows the optimal system load to be calculated and the optimal number of handlers in operation to be adjusted.

The obtained results can be used as input parameters for a simulation model of the incoming message processing system, which allows conducting experiments with the effective load of such a system.

The work was carried out within the framework of the State Assignment № FSGF-2024-0006 «Development of new models and methods of decision-making in multi-level management systems».

References

1. Bunday, B.: Basic linear programming. School of Mathematical Sciences, University of Bradford Edward Arnold, London, Edward Arnold (1989)
2. Belenkiy, A.S.: Research of operations in transport systems: ideas and schemes of planning optimization methods. Mir, Moscow (1992)

3. Burkard, R.E., Dell'Amico, M., Mortello, S.: Assignment problems. SIAM, Philadelphia (2009)
4. Brun, L., Gauzere, B., Renton, G., Bougleux, S., Yger, F.: A differentiable approximation for the Linear Sum Assignment Problem with Edition. In: 26th International Conference on Pattern Recognition (ICPR), pp. 3822–3828. Montreal, QC, Canada (2022)
5. Kordyukov, R., Dopira, R.V., Ivanova, A.V.: A model and algorithmization of the assignment problem under additional constraints. Softw. Syst. **2**, 16–22 (2016)
6. Oleynikova, S.A., Menkova, E.S.: Dynamic problem on assigning a single task with time limits. Bull. Voronezh State Techn. Univ. **16**(6), 19–24 (2020)
7. Diamanti, M., Fryganiotis, N., Papavassiliou, S., Pelekis, C., Tsiropoulou, E.: On the minimum collisions assignment problem in interdependent networked systems. In: IEEE Symposium on Computers and Communications (ISCC), pp. 1–6. Rhodes, Greece (2022)
8. Horng, S., Pan, Y., Seitzer, J., Tsai, H.: Optimal algorithms for the channel-assignment problem on a reconfigurable array of processors with wider bus networks. IEEE Trans. Parallel Distrib. Syst. **13**(11), 1124–1138 (2002)
9. Malyugina, O.A., Medvedev, S.N., Chernyshova, G.D.: Using dual methods to solve one multicriteria assignment problem. In: Proceedings of Voronezh State University. Series: Systems Analysis and Information Technologies, vol. 1, 30–33 (2010)
10. Medvedeva, O.A., Poletaev, A.: Solution of the assignment problem with an additional requirement. Proc. Voronezh State Univ. **1**, 77–81 (2016)
11. Kostyukov, A.A., Makarychev, P.P.: Solving the destination problem in a distributed computing system. Models Syst. Networks Econ. Technol. Nat. Soc. **3**(19), 138–146 (2016)
12. Dyatchina, A.V., Oleinikova, S.A., Nedikova, T.N.: Developing a prototype of software-hardware system for synchrophasor measurements based on GPS-disciplined analog-to-digital converter. Bull. Voronezh State Techn. Univ. **19**(4), 37–43 (2023)
13. Mukha, V.S.: Decision of the open assignment problem by standard simplex method. Doklady BGUIR **6**(84), 104–107 (2014)
14. Afraimovich, L.G., Tyuntyaev, A.S., Tyuntyaeva, L.A.: Study of a combined solution to the three-index assignment problem. Syst. Adm. **5**(198), 84–87 (2019)
15. Salehi, K.: An approach for solving multi-objective assignment problem with interval parameters. Manage. Sci. Let. **4**, 2155–2160 (2017)
16. Shraideh, A., Camus, H., Yim, P.: Two stages optimization problem: New variant of Bin Packing Problem for decision making. In: Proceedings of the International Multiconference on Computer Science and Information Technology (IMCSIT), pp. 921–925. IEEE, Poland (2008)

Comparison of Machine Learning Based Anomaly Detection Methods for ADS-B System

Nurşah Çevik[1,2(✉)] and Sedat Akleylek[3,4]

[1] HAVELSAN, Ankara, Türkiye
nursah.kaya@bil.omu.edu.tr
[2] Computational Sciences, Ondokuz Mayis University, Samsun, Türkiye
[3] University of Tartu, Institute of Computer Science, Tartu, Estonia
sedat.akleylek@bil.omu.edu.tr
[4] Department of Computer Engineering, Istinye University, Istanbul, Türkiye

Abstract. This paper introduces an anomaly/intrusion detection system utilizing machine learning techniques for detecting attacks in the Automatic Detection System-Broadcast (ADS-B). Real ADS-B messages between Türkiye's coordinates are collected to train and test machine learning models. After data collection and pre-processing steps, the authors generate the attack datasets by using real ADS-B data to simulate two attack scenarios, which are constant velocity increase/decrease and gradually velocity increase or decrease attacks. The efficacy of five machine learning algorithms, including decision trees, extra trees, gaussian naive bayes, k-nearest neighbors, and logistic regression, is evaluated across different attack types. This paper demonstrates that tree-based algorithms consistently exhibit superior performance across a spectrum of attack scenarios. Moreover, the research underscores the significance of anomaly or intrusion detection mechanisms for ADS-B systems, highlights the practical viability of employing tree-based algorithms in air traffic management, and suggests avenues for enhancing safety protocols and mitigating potential risks in the airspace domain.

Keywords: ADS-B · Anomaly Detection System · Intrusion Detection System · IDS · Machine Learning · Avionics Security · Cyber Security

1 Introduction

Modern aircraft are equipped with an Automatic Dependent Surveillance–Broadcast (ADS-B) system to enable continuous monitoring by air traffic control units and other aircraft. This system allows aircraft to periodically broadcast essential information such as their position, speed, altitude, and other relevant data. Using ADS-B out devices has become mandatory in civil aviation as of 2020 [1]. A report published by the Federal Aviation Administration (FAA) in 2010 highlighted potential security challenges associated with ADS-B [2]. The report warned that various organizations could eavesdrop on ADS-B broadcasts and even interrupt signals through injection attacks, posing a potential threat to aircraft [2]. This situation underscores a critical issue that needs consideration regarding the cybersecurity of ADS-B. The aviation industry is actively working to develop practical solutions to address these security challenges and make this technology more secure [3].

It is acknowledged in the literature that three primary methods are employed for the secure transmission of ADS-B messages: encryption-based, location-based, and anomaly detection-based. Encryption-based methods involve encrypting ADS-B messages using cryptographic techniques to ensure confidentiality and integrity during transmission by protecting the information from unauthorized access. There are two approaches: asymmetric and symmetric encryption. The asymmetric encryption methods use ECC-based methods for data authentication or signature verification [3–5]. On the other hand, symmetric encryption methods are more efficient in encrypting data [6]. However, they have been required to share symmetric keys. Therefore, a hybrid solution is proposed for this problem [7].

Location-based methods leverage geographical information to authenticate the transmitted data by verifying that the reported location aligns with expected positions. One of the most common location-based methods is multilateration technology based on the signal arrival time difference (TDOA) [8–10]. Another one is group authentication, which is based on authenticating a group aircraft to each other. Many methods exist, such as Kalman filtering and distance bounding [11].

Anomaly-based detection methods involve identifying abnormal patterns or deviations from the expected behavior within ADS-B messages. These methods can detect unusual or suspicious patterns that may indicate a security threat. Anomaly detection methods can detect jamming [12] and spoofing [13–16] attacks. With the increased reliability of machine learning and deep learning methods in recent years, the practical application of these methods has expanded, and their usability has also improved. With these improvements, machine learning and deep learning-based methods have become an option for flight systems. Deep learning methods have recently been commonly preferred for anomaly detection [13–26].

ADS-B security solutions enhance the cybersecurity of ADS-B systems, each addressing specific aspects of security concerns in transmitting ADS-B messages. The comparisons of different security approaches are presented in [11].

1.1 Motivation and Contribution

According to recent research on the security of ADS-B systems, generating an attack on ADS-B systems has a low cost since the attacker can easily reach the open information related to flight. Therefore, there is a need for an applicable security solution for ADS-B systems. While comparing ADS-B security solutions, anomaly-based detection systems are more practical and low-cost security solutions for ADS-B systems. For these reasons, the paper focused on anomaly-based detection systems for the security of the ADS-B system.

- We introduce a novel anomaly detection system tailored for ADS-B devices. This system could utilize advanced machine learning techniques to effectively identify unusual patterns or behaviors in the ADS-B velocity data.

- We could comprehensively evaluate the proposed anomaly detection system's performance. Evaluation metrics for our system include detection accuracy, recall, precision, and F1 score.
- We generate attack datasets for ADS-B systems. Understanding potential threats and weaknesses is crucial for designing effective anomaly detection mechanisms.

1.2 Organization

This study is organized as follows: Sect. 2 provides an overview of the ADS-B system, including its technical infrastructure and the machine learning techniques used. Section 3 outlines the system model and the corresponding threat model. Section 4 explains the proposed anomaly detection system and details the data processing steps. The model results are presented and analyzed in Sect. 5. Finally, Sect. 6 concludes the study with a discussion of the findings and their implications.

2 Preliminaries

2.1 Introduction to ADS-B

ADS-B devices periodically broadcast real-time flight and position information obtained from satellite navigation systems to the nearest ground station and other aircraft within the airspace. It enables visibility on the radars of other aircraft and air traffic controllers [21, 22].The ADS-B packet contains 112 bits; the packet content is shown in Table 2 (Table 1).

Table 1. Comparison of Security Solution for ADS-B Device

Method	Protocol	Sensor or Input
Encryption-based Methods	There is a need for encryption mechanisms, key exchange mechanisms and new message format	There is a need for certificate authority for key exchange
Location-based Methods	There is a need for a new protocol format	There is a need for additional sensors and input values
Anomaly-based Detection Methods	There is no change in protocol	There is no need for additional sensors or input values

Automatic Dependent Surveillance-Broadcast (ADS–B) is an advanced surveillance technology with two types: ADS-B Out and ADS-B In. ADS-B Out, situated within the aircraft, transmits exact positional data, including the aircraft's GPS location, altitude, and ground speed, to ground controllers and other aircraft. Its accuracy surpasses that of traditional radar surveillance. ADS-B In is positioned on the aircraft or at ground stations, acquiring data from ADS-B Out devices within its range. Initially deployed in ground and air traffic control systems due to its faster position transmission than other radar systems, it is currently being integrated into aircraft for air traffic monitoring purposes [21, 22].

Table 2. ADS-B Message Format

Context	Definition
Preamble	A predefined sequence of bits to mark the start of a message
DF Format	Downlink Format, indicating the type of message
Capability	Aircraft capabilities and equipment codes
ICAO	A unique 24-bit identifier assigned to each aircraft
CRC	Cyclic Redundancy Check, used for error detection
ADS-B	Various fields containing position, velocity, and other information

2.2 Machine Learning Algorithms

In this paper, we focus on classification-based algorithms. Here are some commonly used ML algorithms in the context of ADS-B anomaly detection:

1. **Logistic Regression (LR):** A popular classification algorithm can be adapted for anomaly detection in ADS-B data. LR models the probability that an instance belongs to a particular class, such as flight or attack data. In the context of ADS-B anomaly detection, it can be used to estimate the probability of an instance being normal or abnormal based on its features.
2. **Decision Tree (DT):** This algorithm is used to handle both categorical and numerical data. They are utilized in various applications, including classification, regression, and anomaly detection. A decision tree is a practical algorithm to understand which features contribute to anomaly detection.
3. **Gaussian Naive Bayes (GNB):** ADS-B data often includes continuous features such as altitude, speed, and heading. Gaussian Naive Bayes offers a practical and efficient approach to ADS-B anomaly detection, leveraging its ability to model the Gaussian distribution of continuous features in aviation data.
4. **K-Nearest Neighbors Classifier (KNN):** KNN is a non-parametric, instance-based algorithm that classifies data points based on the majority class of their k-nearest neighbors. In the context of ADS-B anomaly detection, KNN can be applied to identify abnormal flight behavior, deviations from typical flight patterns, or potential security threats in aviation data.
5. **Extra Tree Classifier:** The Extra Trees Classifier, an extension of Random Forests, builds multiple decision trees and combines their outputs to improve overall performance. In the context of ADS-B anomaly detection, Extra Trees can reduce overfitting, enhance generalization, and effectively capture complex relationships and variations in the data.

The choice of algorithm depends on three factors: 1. The characteristics of the ADS-B data, 2. The types of anomalies, 3. The computational resources. Experimentation and evaluation are crucial to determine which algorithm or combination of algorithms works best for the ADS-B anomaly detection system.

2.3 Evaluation Metrics

Evaluation metrics were employed to assess the anomaly detection models. During the evaluation process, parameters such as accuracy and recall are computed and used as benchmarks for model comparison. This section provides a comprehensive overview of commonly utilized performance metrics in the literature.

A confusion matrix, given in Fig. 1, is a table used to evaluate a classification model's performance. It is beneficial in machine learning and statistics, where it helps assess the accuracy and effectiveness of a predictive model, especially in binary classification problems [23].

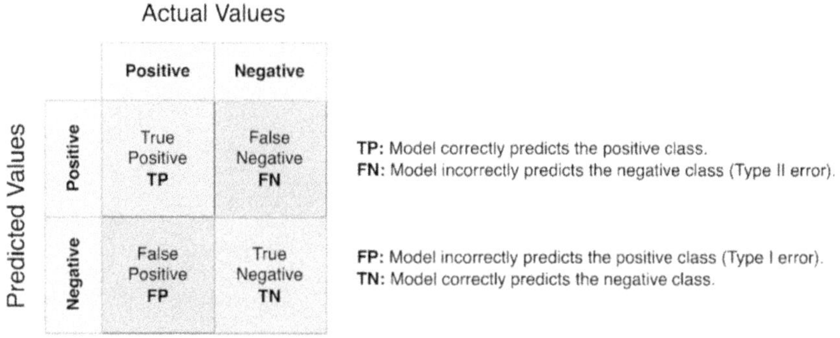

Fig. 1. Confusion Matrix

The confusion matrix represents the model's performance, allowing for calculating various evaluation metrics such as accuracy, precision, recall (sensitivity), and F1 score.

$$Accuracy = \frac{TN + TP}{TN + TP + FP + FN}, \qquad (1)$$

$$Precision = \frac{TP}{TP + FP}, \qquad (2)$$

$$Recall = \frac{TP}{TP + FN}, \qquad (3)$$

$$F1Score = 2\frac{Precision * Recall}{Precision + Recall}. \qquad (4)$$

3 Problem Statement

In this section, we explain the anomaly detection system model based on machine learning models and present our threat model by defining attack scenarios.

3.1 System Model

The ADS-B system is recognized for broadcasting an aircraft's positional information and enabling air traffic monitoring. The system comprises two primary components: ADS-B IN, which receives ADS-B messages from aircraft within a 200 km range, and ADS-B OUT, which transmits the positional coordinates and relevant data of an aircraft to other aircraft and ground terminals within a 200 km range [21, 22].

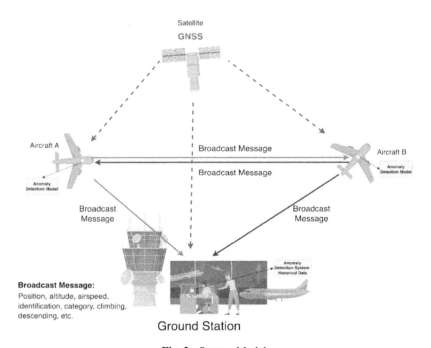

Fig. 2. System Model

As illustrated in Fig. 2, our system model comprises general ADS-B components ADS-B IN and ADS-B OUT. Aircrafts broadcast positional information periodically through ADS-B OUT. Equipped with ADS-B receivers, surrounding aircraft and ground stations can receive the messages through ADS-B In. Ground stations have a crucial role in surveillance by presenting aircraft positions on the monitor based on received location information. The assurance of on-ground communication security can be presumed due to the interconnectivity of civil aviation intranet networks. In our design, the anomaly detection model, deployed on both aircraft and ground stations, processes ADS-B data and detects anomalies in air traffic. In this paper, the authors aim to develop a regular anomaly-based attack detection system against some security vulnerabilities of the ADS-B device. The system architecture includes classic machine learning steps. In addition to these classic steps, this study extensively investigated potential attack scenarios. Since there was no data labeled as anomaly/attack within ADS-B data, attack scenarios were simulated using real ADS-B data.

3.2 Threat Model

ADS-B messages are transmitted in an unencrypted form, and the receiver doesn't authenticate the message source. Therefore, an attacker can easily listen to the channel and inject messages into an ADS-B packet to spoof the receiver. In our threat model, given in Fig. 3, there are two approaches to attack scenarios: active and passive. Active attack scenarios consist of an injection, which includes adding or deleting something to ADS-B messages. Passive attack scenarios consist of eavesdropping broadcasted ADS-B messages to collect information related to flight routes.

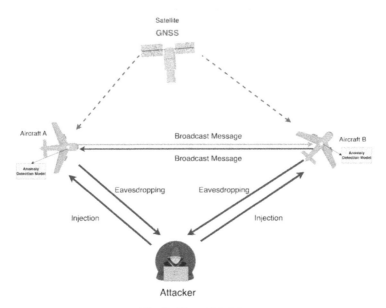

Fig. 3. Threat Model

For the training of the models, attack scenarios were simulated: velocity increase/decrease attacks. Changes were made to the velocity values within the scope of the velocity increase/decrease attacks, creating a new dataset. In the velocity increase/decrease attack, the parameter representing the aircraft's velocity in the real ADS-B message is either increased or decreased. The attacker has passive and active capabilities, such as eavesdropping, injecting, and modifying messages within the network. In this type of attack, velocity is manipulated in two ways: increasing it with a constant value or gradually increasing it over time [24].

4 ADS-B Anomaly Detection System

As demonstrated in the threat model in Sect. 3, we mainly focus on discussing two types of potential threats, namely constant and gradual velocity increase/decrease attack. This section presents our solution to resist these malicious attacks as follows.

4.1 Evaluation Metrics

In this study, the most critical step is the data collection phase. Upon reviewing the literature, it is evident that the accuracy rate of machine learning models is directly dependent on the dataset. First, real flight data collected from the OpenSky Network [25] has been used in this study. The dataset includes features such as time, icao24, lat, lon, velocity, heading, vertrate, callsign, onground, alert, spi, squawk, baroaltitude, geoaltitude, lastposupdate, and lastcontact. It is essential to define a specific route between cities for aircraft and establish a baseline scenario [26]. Within this dataset, flights that fall within Türkiye's latitude and longitude coordinates were selected. The method of filling rows containing NaN elements with the mean, which may be suitable for some datasets, is not applicable for features determining flight routes. Therefore, rows containing NaN values were removed from the dataset.

In the raw data, the time is stored as a Unix timestamp. This value has been separated into date and time. However, the usage of Unix time was preferred in the initial stages of the model. ICAO numbers have been replaced with integer values and were not used within the model due to their categorical nature. These values have been labeled as 1 since they represent normal flight data, given in Fig. 4.

Subsequently, attack data was required to train the models based on different attack scenarios defined in Sect. 3. Attack data was generated according to these scenarios, utilizing historical flight data. When generating attack data, the scenario was where the attacker modifies previously obtained flight data and broadcasts it [24]. The attack data produced under this scenario was labeled as 0, given in Fig. 4.

	icao24	time	lat	lon	heading	velocity	baroaltitude	geoaltitude	label
686	1	1638831610	41.104835	33.857616	118.004857	256.374859	11277.60	11407.14	1
918	2	1638831610	39.401442	29.834106	162.540797	200.615352	11879.58	11986.26	1
1581	3	1638831610	38.938248	28.090966	130.236358	227.777307	11894.82	11917.68	1
2206	4	1638831610	39.578247	27.127812	136.620247	218.711315	11277.60	11292.84	1
2237	5	1638831610	39.723404	33.359209	129.641353	229.813714	11879.58	12016.74	1
	icao24	time	lat	lon	heading	velocity	baroaltitude	geoaltitude	label
686	1	1638831610	41.054835	33.857816	128.004857	236.374859	22555.20	11427.14	0
918	2	1638831610	39.361442	29.842106	266.540797	293.615352	11727.58	10450.26	0
1581	3	1638831610	38.888248	28.091166	140.236358	207.777307	23789.64	11937.68	0
2206	4	1638831610	39.588247	27.124812	126.620247	318.711315	11227.60	11192.84	0
2237	5	1638831610	39.673404	33.359409	139.641353	209.813714	23759.16	12036.74	0

Fig. 4. Real Flight Data and Modified Attack Data

Finally, real flight and attack data were combined to create a dataset. Critical data from the ADS-B device is broadcast every two seconds, and less critical data is broadcast every five seconds to reduce the transmission load. Therefore, missing values occur in the data. Additionally, they may arise in the data due to transmission faults. When examining studies in the literature, it has been observed that filling in these missing data negatively affects the model's accuracy. Therefore, the method of generally deleting rows with missing data was preferred. Data transformation processes were carried out before feature extraction. The data was initially normalized, then values with a deviation above a certain threshold were removed. After the data preprocessing steps are completed, it is necessary to determine the parameters that most affect the performance of the models. In this step, studies in the literature were considered.

5 Results

In the velocity increase/decrease attack, the velocity modification was applied in two ways: Fig. 5 represents a constant increase/decrease, and Fig. 6 shows a gradual increase/decrease. The value 1 in the figures represents normal behavior, while 0 represents anomalies.

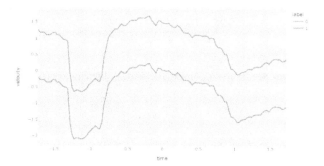

Fig. 5. Constant Velocity Increase/Decrease Attack

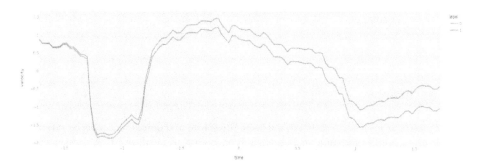

Fig. 6. Gradually Velocity Increase/Decrease Attack

When looking at the model results for constant velocity increase/decrease given in Table 3, it is observed that models other than logistic regression and gaussian naive bayes cannot detect the attack quickly. Similarly, when examining the model results for gradual velocity increase/decrease in Table 4, it is observed that models other than logistic regression and gaussian naive bayes can quickly detect the attack, like the constant velocity increase/decrease attack. The model performance rates are observed to be lower compared to the constant increase/decrease. When examining the correlations between the parameters of the data, it is observed that there is a high correlation between velocity and geoaltitude as well as baroaltitude parameters. For this reason, logistic regression and naive bayes models are generally thought to have low performance values.

Table 3. Results for Constant Velocity Increase/decrease

Algorithm	Acc	Pre	Recall	F1 score	ROC AUC
Logistic Regression	0.56	0.57	0.53	0.55	0.56
Decision Tree Classifier	0.94	0.95	0.94	0.94	0.94
Gaussian Naive Bayes	0.56	0.57	0.47	0.51	0.56
K-Nearest Neighbors	0.97	0.97	0.96	0.97	0.97
Extra Trees Classifier	0.95	0.96	0.93	0.95	0.95

Table 4. Results for Gradually Velocity Increase/decrease

Algorithm	Acc	Pre	Recall	F1 score	ROC AUC
Logistic Regression	0.58	0.57	0.60	0.59	0.58
Decision Tree Classifier	0.91	0.90	0.91	0.91	0.91
Gaussian Naive Bayes	0.56	0.55	0.59	0.57	0.56
K-Nearest Neighbors	0.90	0.90	0.90	0.90	0.90
Extra Trees Classifier	0.87	0.88	0.86	0.87	0.87

6 Conclusions

The objective is to detect intrusion, message injection, and deletion attacks by ensuring reliable data communication in the ADS-B system and verifying that messages are transmitted from the correct source. The system must implement protocol changes to achieve authenticity, integrity, and confidentiality using cryptographic methods, which require hardware and software changes. However, it is known that ADS-B messages do not contain confidential data. Therefore, within the scope of this study, practical constraints have been considered, and have been decided not to include data encryption in our design goals. It is expected that the developed anomaly detection system can detect changes in ADS-B velocity data.

For future studies, different models and scenarios can be defined for various flight stages, such as takeoff, landing, and taxi. If specific parameters need to be added for each flight stage, these parameters can be identified and integrated into the system. In addition, the challenge of anomaly-based systems is distinguishing between normal and abnormal data or defining normal behavior. At this stage, obtaining expert opinions for labeling normal and abnormal situations is expected to enhance the performance of the models.

References

1. Olive, X., Basora, L.: Detection and identification of significant events in historical aircraft trajectory data. Transp. Res. Part C: Emerg. Technol. **119**, 102737 (2020)
2. Federal Aviation Administration. Automatic dependent surveillance-broadcast (ADS-B) out performance requirements to support air traffic control (ATC) service - final rule. Federal Aviation Administration (FAA) (2010)
3. Çevik, N., Akleylek, S.: Cyber Security in Aviation Systems Nobel Academic (2022). https://dergipark.org.tr/en/pub/ijiss/page/13335
4. Wesson, K.D., Humphreys, T.E., Evans, B.L.: Can cryptography secure next-generation air traffic surveillance? Res. UT Austin, Univ. Texas Austin, Austin, TX, USA, Tech. Rep, Radionavigation Secur (2014)
5. Feng, Z., Pan, W., Wang, Y.: A data authentication solution of ADS-B system based on X. 509 certificate. In: Proceedings of the 27th International Council of the Aeronautical Sciences (ICAS), pp. 1–6 (2010)
6. Samuelson, K., Valovage, E., Hall, D.: Enhanced ADS-B research. In: Proceedings of the IEEE Aerospace Conference p. 7 (2006)
7. Baek, J., Hableel, E., Byon, Y.-J., Wong, D.S., Jang, K., Yeo, H.: How to protect ADS-B: confidentiality framework and efficient realization based on staged identity-based encryption. IEEE Trans. Intell. Transp. Syst. **18**(3), 690–700 (2017)
8. Kim, Y., Jo, J.-Y., Lee, S.: ADS-B vulnerabilities and a security solution with a timestamp. IEEE Aerosp. Electron. Syst. Mag. **32**(11), 52–61 (2017)
9. Kaune, R., Steffes, C., Rau, S., Konle, W., Pagel, J.: Wide area multilateration using ADS-B transponder signals. In: Proceeding of the 15th International Conference on Information Fusion, pp. 727–734 (2012)
10. Johnson, J., Neufeldt, H., Beyer, J.: Wide area multilateration and ADS-B proves resilient in Afghanistan. In: Proceedings af the ICNS, pp. A6-1–A6-8 (2012)
11. Wu, Z., Shang, T., Guo, A.: Security issues in Automatic Dependent Surveillance - Broadcast (ADS-B): a survey. IEEE Access **8**, 122147–122167 (2020). https://doi.org/10.1109/ACCESS.2020.3007182
12. Manesh, M.R., Velashani, M.S., Ghribi, E., Kaabouch, N.: Performance comparison of machine learning algorithms in detecting jamming attacks on ADS-B devices. In: 2019 IEEE International Conference on Electro Information Technology (EIT), pp. 200–206 (2019). https://doi.org/10.1109/EIT.2019.8833789
13. Manesh, M.R., Kenney, J., Hu, W.C., Devabhaktuni, V.K., Kaabouch, N.: Detection of GPS spoofing attacks on unmanned aerial systems. In: 2019 16th IEEE Annual Consumer Communications & Networking Conference (CCNC) pp. 1–6 (2019). https://doi.org/10.1109/CCNC.2019.8651804
14. Chen, S., Zheng, S., Yang, L., Yang, X.: Deep learning for large-scale real-world ACARS and ADS-B radio signal classification. IEEE Access (2019). https://doi.org/10.1109/ACCESS.2019.2925569
15. Ying, X., Mazer, J., Bernieri, G., Conti, M., Bushnell, L., Poovendran, R.: Detecting ADS-B spoofing attacks using deep neural networks. In: IEEE Conference on Communications and Network Security (CNS) (2019). https://doi.org/10.1109/CNS.2019.8802732

16. Wang, J., Zou, Y., Ding, J.: ADS-B spoofing attack detection method based on LSTM. EURASIP J. Wireless Commun. Network. **2020**(1) (2020). https://doi.org/10.1186/s13638-020-01756-8
17. Fried, A., Last, M.: Facing airborne attacks on ADS-B data with autoencoders. Comput. Secur. **109** (2021). https://doi.org/10.1016/j.cose.2021.102405
18. Habler, E., Shabtai, A.: Using LSTM encoder-decoder algorithm for detecting anomalous ADS-B messages. Comput. Secur. **78** (2017). https://doi.org/10.1016/j.cose.2018.07.004
19. Chevrot, A., Vernotte, A., Legeard, B.: CAE: contextual autoencoder for multivariate time-series anomaly detection in air transportation. Comput. Secur. **116**, 102652 (2022). https://doi.org/10.1016/j.cose.2022.102652
20. Yi, J., Lin, L., Nisi, L., Wang, J.: ADS-B anomaly detection algorithm based on LSTM-ED and SVDD. In: Proceedings of the 10th Chinese Society of Aeronautics and Astronautics Youth Forum, pp. 245–257. Springer Nature Singapore (2023). https://doi.org/10.1007/978-981-19-7652-0
21. Radio Tech. Commiss. Aeronautics (RTCA) Inc. Minimum aviation system performance standards for automatic dependent surveillance broadcast (ADS-B). Washington, DC, USA, DO-242A (including Change 1), December 2006 (2006)
22. Radio Tech. Commiss. Aeronautics (RTCA) Inc. Minimum operational performance standards for 1090 MHz extended squitter automatic dependent surveillance Broadcast (ADS-B) and traffic information services broadcast (TIS-B). Washington, DC, USA, DO-260B with Corrigendum 1, December 2011 (2011)
23. Ting, K.M.: Confusion matrix. In: Sammut, C., Webb, G.I. (eds.) Encyclopedia of Machine Learning. Springer, Boston, MA (2011). https://doi.org/10.1007/978-0-387-30164-8_15
24. Clay, C., Khan, M., Bajracharya, B.: A look into the vulnerabilities of automatic dependent surveillance-broadcast. In: Proceedings of the 2023 IEEE 13th Annual Computing and Communication Workshop and Conference (CCWC), pp. 933–938, ELECTR NETWORK, March 8–11. IEEE; SMART; IEEE Reg 1, IEEE USA, Institute of Engineering & Management, University of Engineering & Management (2023). https://doi.org/10.1109/CCWC57344.2023.10099369. ISBN: 979-8-3503-3286-5
25. OpenSkyNetwork. The OpenSkyNetwork-FreeADS-BandMode S Data (2015). https://opensky-network.org/
26. Habler, E., Shabtai, A.: Analyzing sequences of airspace states to detect anomalous traffic conditions. IEEE Trans. Aerospace Electron. Syst. **58**(3), 1843–1857 (2022). https://doi.org/10.1109/TAES.2021.3124199

Image Processing Based Wood Defect Detection

Merve Özkan[1(✉)] and Caner Özcan[2]

[1] Computer Engineering, Institute of Postgraduate Education,
Karabuk University, Karabuk, Turkey
`merveozkan867@gmail.com`
[2] Software Engineering, Faculty of Engineering, Karabuk University, Karabuk, Turkey

Abstract. Detection of defects in wooden structures in the forestry industry has become a crucial area of research. Existing studies have focused on specific categories of wood defects, failing to provide a comprehensive classification for high-quality wood. Trained human operators currently perform a variety of wood quality in wood processing facilities. However, this human-dependent process leads to time and performance losses and inaccurate type. This study aims to address all these challenges in future intelligent production systems by targeting the detection of the fungus in oak wood, one of the wood defect classes. The algorithm created based on image processing utilizes median filtering, Canny edge detection, and masking technologies using the HSV color space. The algorithm then calculates the fungal area ratio to the wooden piece's surface area on the masked image to reach the final result. While existing studies in the literature are primarily based on deep learning methods, there has been limited focus on fungus detection. The novelty of this study, conducted on oak wood, lies in its use of a specific dataset, fungal detection, and image processing. An algorithm has been developed and presented in the literature that can be used in the software of future intelligent production systems in the forestry industry.

Keywords: Image Processing · Wood Material · Object Detection · Canny Edge Detection · HSV

1 Introduction

Detecting defects in wooden structures has become a significant research focus in forestry in recent years [1]. Existing studies have predominantly concentrated on specific types of defects, compromising the ability to meet the quality requirements of high-quality wood. Additionally, trained human operators conduct visual quality control inspections in industrial forest production facilities, making it a repetitive and tedious task with a high probability of errors. The execution of this task can also be hindered by varying light conditions. Although algorithms capable of performing duties similar to those conducted by humans with high-resolution cameras are being developed, solutions for real-time industrial applications are not sufficiently fast and accurate. In a study conducted by Wang et al. in 2018, defect detection and quality assessment in hardwood logs were addressed [2]. The most effective quality determinants for the internal integrity of hardwood logs are acoustic parameters such as acoustic velocity, time central, damping ratio, combined time, and frequency domain parameters. Wang and colleagues integrated acoustic parameters with high-resolution laser scanning results and made inferences regarding log dimensions, shape, surface defects, and integrity levels. The findings suggested combining high-quality auditory assessment and low visual quality often indicates a robust structure but may include many knots. Therefore, an integrated system was proposed to examine log quality and recovery [2].

In 2019, a defect detection study covering cracks, knots, stains, and core classes was conducted using a faster region-based convolutional neural network (Faster R-CNN) for faster region-based defect detection. Various neural network models, such as AlexNet, VGG16, BNInception, and ResNet152, were employed to enhance results. The dataset was synthetically augmented, and the ResNet152 model achieved the highest accuracy with an average precision (mAP) value of 80.6% [3]. In a study by Co and Sc in 2019, defect detection was conducted on American red oak, yellow poplar, and maple wood types. Pareto analysis determined that usage defects, particularly dents and knife marks, were the most common. The Spaeth/Richman Contrast Sensitivity test was identified as suitable for detecting low-contrast defects. The study also recommended periodic eye examinations for factory workers [4]. Another study in 2019 focused on knot detection by Aleksi and colleagues [5]. The proposed method included image accumulation, image preprocessing, feature extraction, calculation of vector distance length, and classification of wood as defective or defect-free based on a specified threshold.

In 2020, YOLOv3 convolutional neural network architecture was utilized for knot detection [6]. It was trained on a specific dataset using Darknet-53 and demonstrated fast and accurate identification of small, medium, and large knots, achieving an mAP value of 80%. The potential application of this algorithm in lumber factories as both workstations and mobile device applications was suggested upon reaching a sufficient number of images [6].

In 2021, a study was conducted on wood defects consisting of dead knots, live knots, decay, mold, cracks, and wormhole structures, using the original Single Shot MultiBox Detector (SSD) algorithm with ResNet instead of VGG16. TensorFlow was used for training, and the SSD network structure was further developed. The model achieved an accuracy rate of 89.7%, with an average detection time of 90 ms [7].

In 2022, Wu et al. extracted directed gradient histograms from wood images, followed by kernel principal component analysis to reduce feature dimensions. A gray-level co-occurrence matrix was obtained from photos, and Support Vector Machines were employed as classifiers. The study achieved an accuracy value of 91.26% [1]. In another study in the same year, reinforced learning was applied in the wood image processing module, resulting in an efficiency ratio of 90.19%, a 20% improvement compared to traditional wood quality assessment systems [8]. Urtans and colleagues detected knots in oak woods in 2022, achieving a 90% accuracy [9]. In a study by Han et al., wood defects such as live knots, resin, cracked knots, dead knots, cracks, and pith were addressed. The proposed STC-YOLOv5 demonstrated a 3.1% increase in mAP compared to YOLOv5 [10]. Uniquely, in 2022, defect detection in Faster R-CNN was made real-time using the MobileNetV3 backbone, achieving a 99% accuracy rate [11].

In 2023, Mohsin and colleagues included region and crack objects in wood defect classes, achieving a global pixel accuracy of 96%, distinct from their 2022 study [12]. Ji et al. worked on images of birch woods, focusing on wood defect classes, including dead knots, cracks, minerals, and live knots, and achieving an average detection accuracy of 89.69% on ten different types [13]. They proposed that their work could be a foundation for future innovative manufacturing systems in the wood industry [13]. Meng and Yuan used a Semi-Global Aggregation-based YOLOv5 model for knot detection in a different approach. The mapped value was determined to be 86.4%, showing a 7.1% improvement over the SSD algorithm and a 13.6% improvement over Faster R-CNN [14]. Another study in 2023 covered wood defects such as live knots, dead knots, resin, cracks, and cracked knots. YOLOv5m and YOLOv5n were optimized, and the introduction of the SimAM attention model, the replacement of the learning rate decay strategy with CosLR, and the use of ghost convolution to minimize model parameters led to a 1.5% increase in mAP for YOLOv5n and a 1.6% increase for YOLOv5m [15]. In the same year, Ge and colleagues designed a Detection Transformer-based detection line for wood defects, focusing on dead knots, live knots, and wormholes. Although the proposed method was 6% lower in speed than the best model, it was twice as fast. Experiments indicated that the proposed plan did not balance detection accuracy and speed [16]. Another study by Wang and colleagues in 2023 addressed eight wood defect classes. The presented Multi-Dimensional Dynamic Convolution Coordinate Attention-based YOLO model outperformed the YOLOv7 algorithm by 9.1% and exhibited enhanced detection accuracy for all seven defect types compared to the YOLOx algorithm [17].

The existing studies have primarily focused on specific classes of wood defects. To establish intelligent workstations in production lines of companies operating in the forestry sector, detecting all defect classes in high-quality wood is essential. This study concentrates on a different type of wood defect, namely the fungal class, which has not been extensively addressed in the literature. The uniqueness of this work lies in proposing an image processing-based algorithm for detecting fungal surfaces in oak wood. This study aims to contribute to intelligent manufacturing systems in the forest industry by addressing a distinct defect class in timber. While numerous wood defect classes have been identified in previous studies, detecting fungal surfaces has not been extensively explored. This study's originality lies in seeing fungal characters in oak wood, presented as a specialized dataset, through an image processing-based algorithm.

2 Matrials and Methods

2.1 Data

Images were acquired through human intervention from a factory's solid panel production line specializing in industrial forest product manufacturing. The Fujifilm X-S1 12MP camera was employed for image capture. A total of 1861 images depicting wooden pieces containing fungal were collected.

The image acquisition environment is depicted in Fig. 1. Two projectors with 20 and 30 watts were utilized to prevent shadowing on wood. Additionally, the 30-W projector served as a brightness source in low-light situations. A five-sided object, shaped like a rectangular prism and covered with paper on all sides, was created on the table. The

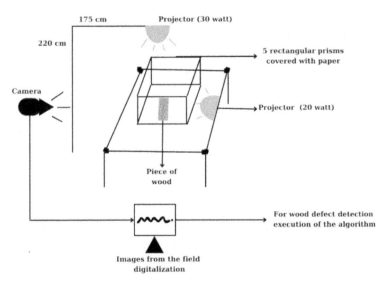

Fig. 1. Image acquisition environment

paper covering mechanism prevents the dispersion of rays from the camera into the environment. The 30-W projector source has parts measuring 220 cm in height and 175 cm in width. These dimensions ensure that the projector beams fall vertically onto the wooden piece. Images taken with a digital camera from the experimental setup were transferred to the computer environment and prepared for input to the wood defect detection algorithm.

2.2 Brightness Ratio

The analog image, represented by the function I(x, y), where 'I' is the unit of light intensity and x, y are variables describing the horizontal and vertical coordinates of the image, is sampled to create a digital image with M columns and N rows. Reducing an image's brightness implies altering each pixel's brightness value.

$$I_k(x, y) + g, \text{ if } I_k(x, y) + g < 256 \\ 255, \quad \text{ if } I_k(x, y) + g > 255 \quad (1)$$

If a coefficient is added or subtracted from the pixel value of the image, the brightness is increased or decreased. Increasing the brightness means that the pixel value approaches 255. Equation 1 shows the brightness enhancement process with the help of the transfer function. Figure 2 shows the graph of the transfer function of Eq. 1.

2.3 Median Filtering

In the filtering processes performed in spatial environments, the statistical results of the neighborhood relations and brightness values of the pixels on the image are taken as

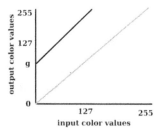

Fig. 2. Graph of the transfer function

basis [18]. The median filter is used to reduce or remove noise in an image. Neighboring pixels are used to calculate new pixel values. An odd number of sample values are sorted to calculate the output, and the middle value is used as the filter output. Since the target pixel is not affected by the window's lowest or highest pixel values, noise is removed while preserving important feature information, such as edge information.

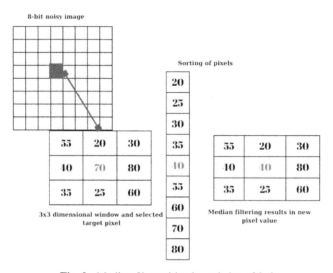

Fig. 3. Median filter with a kernel size of 3x3

Figure 3 shows how a median filter with a kernel size 3x3 is implemented. The median filter type successfully reduces salt and pepper noise compared to other filters [19]. Since there are small pieces of woods on the table in the collected images and these noises in the image matrix need to be removed, the median filter was preferred before canny edge detection. The kernel size was gradually increased, and the best value was 13x13 for this study. It was observed that when the kernel size increases, the blurring rate in the images will increase, causing loss of information.

2.4 Canny Edge Detection

It was developed by John Canny in 1986 [20]. The original algorithm aimed to satisfy three performance criteria. The first criterion is the ability to perform good detection. The probability of not marking actual edge points and the probability of observing non-edge points should be low. Since both of these probabilities are decreasing functions of the output signal-to-noise ratio, this vital property maximizes the signal-to-noise ratio.

$$SNR = \frac{|\int_{-w}^{+w} G(-x)h(x)dx|}{\sigma \sqrt{\int_{-w}^{+w} h^2(x)dx}} \qquad (2)$$

Equation 2 shows the mathematical formula for this process. In the equation, $G(x)$ is the edge function. $H(x)$ is the impulse response of the filter of width w. σ is the mean square deviation of Gaussian noise.

As the second criterion, the algorithm should be able to perform the positioning process well. The points marked as edges should be very close to the center of the actual border.

$$L = \frac{|\int_{-w}^{+w} G(-x)h'(x)dx|}{\sigma \sqrt{\int_{-w}^{+w} h'^2(x)dx}} \qquad (3)$$

The mathematical formula for the second criterion is given in Eq. 3. In this equation, $G'(x)$ represents the derivative of $G(x)$, and $h'(x)$ represents the derivative of $h(x)$.

The third criterion addresses the issue of not answering a single margin. This was addressed indirectly in the first item. When two answers are given to the same side, one of these answers should be considered wrong. The mathematical formula of the first item did not capture the multiple response requirement, and it was observed that this should be addressed at other stages.

$$D(f\prime) = \pi \sqrt{\frac{\int_{-w}^{+w} h\prime 2(x)dx}{\int_{-w}^{+w} h\prime\prime 2(x)dx}} \qquad (4)$$

The formula for the third criterion is given in Eq. 4. Here, $h(x)$ represents the second derivative of $h(x)$. $D(f')$ is the average distance of the zero crossing point of the result of the impulse response of the edge detection algorithm.

The Canny algorithm consists of four steps [20];

1. The image is smoothed using a Gaussian filter. The smoothed image is called $I(x, y)$.
2. The horizontal gradient $G_X(x, y)$ and vertical gradient $G_y(x, y)$ of each pixel are calculated by convolving the image $I(x, y)$ with the partial derivatives of the two-dimensional Gaussian filter.
3. Non-maximal suppression (NMS) is applied to the thin pixels. In this step, the gradient's magnitude and direction are considered. A comparison is made between the analyzed pixel and the relevant neighbors determined by the law of the slope. Suppose there is a value of ten pixels more significant than the threshold value. Among these

pixels, the one with the maximum value should be selected, and the others should be eliminated. For this reason, the gradient direction is needed. For the edge direction 90^0 the importance of pixels in the same row and neighboring columns are compared. If the value of the current pixel is higher than both neighboring pixels to its right and left, it is considered an edge [21].
4. There are two different threshold values, upper and lower. If the gradient magnitude of a pixel is greater than the lower threshold value, that pixel is an edge. It is not an edge if it is less than the lower threshold. However, if the gradient value is between these two thresholds, it is potentially an edge. Any of its corresponding neighbors is an edge. This pixel will be accepted as an edge. Otherwise, it will be discarded.

2.5 HSV Color Space

The Hue, Saturation, and Value (HSV) color space maps to the definitions of color, saturation, and brightness, respectively. The concept of saturation defines the vividness of the color, while brightness deals with the brightness of the color. The reason for using HSV color space compared to Red, Green, and Blue (RGB) space is that it consists of a structure close to the human eye mechanism. HSV color Space is used for color discrimination on objects.

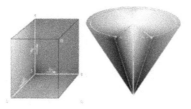

Fig. 4. RGB and HSV color spaces [22] (Color figure online)

In RGB color space, all three components affect brightness, while in HSV color space, only Value affects intelligence. Figure 4 shows an image of RGB and HSV color spaces.

3 Experimental Studies and Results

3.1 Proposed Algorithm

Images are collected from the field by the manpower. After being converted to digital form, the first step of the algorithm is to convert the image to a grey level. After the image is converted into grey form, the algorithm steps begin. Pictures taken in real life necessarily contain noise. Small wood particles falling on the plane constitute noise in this data set. This example is reminiscent of salt and pepper noise. The filter that gives the best solution for salt and pepper noise is the median filter. When applying the median filter, the wood parts are not destroyed sufficiently when the kernel size is small. When chosen too large, the canny edge detection algorithm has difficulty finding the edges. Starting with a kernel size of 5×5, it was observed by trial and error that the kernel size that gave the best result over the data set was 13×13.

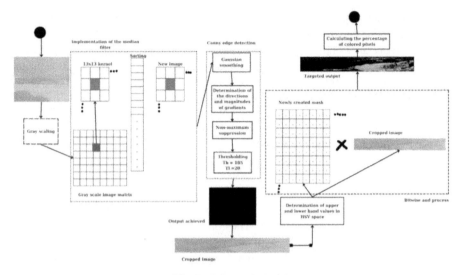

Fig. 5. Proposed Model

The study preferred the Canny edge detection algorithm since it produces successful results. This algorithm must determine a threshold value for threshold edge detection. There is no generalizable threshold value for all data sets. The user should determine it according to the characteristics of the data. The lower threshold value found for the canny edge detection algorithm by trial and error was determined to be 20, and the upper threshold value was defined as 185. After the process, the edges were cropped according to the coordinates returned from the algorithm.

```
def crop(edges)
    if edges > 0
        pts = edges
        min_y,min_x = pts.min()
        max_y,max_x = pts.max()
        crop = img[min_y:max_y, min_x:max_x]
    return crop
```

Fig. 6. Pseudocode of the cropping process according to the array information returned from edge detection.

Figure 6 shows the pseudo-code of the cropping process. The region's lower and upper band values to be detected in HSV space were determined by trial and error method. For Hue, Saturation, and Value values, the lower threshold was set as 23, 70, and 170, and the upper hall was set as 50, 255, and 255, respectively. As a result of these determined values, a mask was created. The show makes the value of that index 0 in the image matrix for areas that do not contain fungus. The front and cropped images are subjected to the multiplication process, and the regions with fungal are obtained. According to the newly formed image matrix, the pixel-based percentage fungal rate is calculated.

3.2 Results

Five samples were selected from the dataset, and the results obtained from the proposed algorithm are presented in Table 1.

Table 1. Results produced by the algorithm on the data set of oak wood species belonging to the fungal defect class

Image	Fungal Percentage of Wood
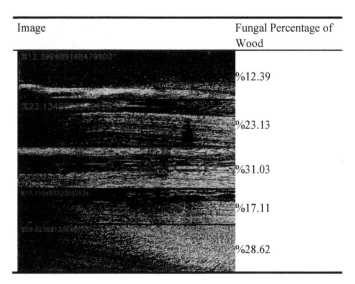	%12.39
	%23.13
	%31.03
	%17.11
	%28.62

4 Discussion and Conclusion

Detecting defects in wooden structures is a subject that needs more studies today. The studies carried out have concentrated on the detection of certain defects. This is not sufficient in the process of categorizing high-quality wood into quality classes. In companies working in industrial forest products, the quality control system is mainly carried out with the help of human hands. While this situation causes losses in terms of time and performance in production, errors are frequently encountered. For all these reasons, the method and algorithm to be used in designing intelligent production systems in the future has become necessary. This study uses an image processing-based algorithm to detect a fungal class of wood defects. The fact that there is not much focus on fungal area detection in the literature and the proposal of an image processing-based algorithm constitutes the unique aspect of the study. The wood type used in the experiments is oak.

1861 images were collected from the production line by Fujifilm X-S1 12 MP camera. Brightness enhancement was performed on the pictures to make the fungal area more prominent. A median filter with a kernel size of 13 × 13 was applied to eliminate the noise in the images. At this stage, it was observed that if the wood pieces falling on the ground were very large, the median filter had difficulty removing this structure. When the kernel size was increased to eliminate the system, it was observed that errors occurred in the cropping stage after the canny edge detection algorithm. To achieve high success in cropping only the wooden structure from the image environment, there should not be a large piece of wood in the image frame. When designing the machine, dust falling into the frame should be cleaned at short intervals using a structure. Cropping is performed with the help of minimum and maximum values from the edges returned from the Canny edge detection algorithm. After cropping, a mask is created by obtaining the upper and lower band values in HSV space. For Hue, Saturation, and Value values, the lower threshold is set as 23, 70, 170, and the upper point is set as 50, 255, and 255, respectively. The cropped image with the created mask is subjected to bitwise operation, and the matrix value of the fungal-free areas in the image matrix becomes 0. The mushroom percentage of the wooden plate is calculated, and the algorithm is terminated.

After Canny edge detection, images containing more than the wooden plate in the cropping stage cause the value obtained in the percentage calculation of the fungal area to be incorrect. To overcome this problem, the code block given in Fig. 5 can be developed. During image acquisition, light should be uniformly distributed over the wooden plate. Otherwise, the lower and upper values determined in HSV space for the changing amount of illuminance will not be valid. Future studies are planned to produce a new solution to these situations and to develop a method for another defect class among wood surface defects.

Acknowledgment. Images were obtained with special permission from Kaswood and used in the study. For academic collaborations and partnerships, the full data set can be shared if desired. Project codes and partial datasets have been shared [23].

References

1. Wu, C., Zou, X., Yu, Z.: A detection method for wood surface defect based on feature fusion. In: 2022 4th International Conference on Frontiers Technology of Information and Computer, ICFTIC 2022, vol. 1, no. 2, pp. 876–880 (20220. https://doi.org/10.1109/ICFTIC57696.2022.10075158
2. Wang, X., Thomas, E., Xu, F., Liu, Y., Brashaw, B.K., Ross, R.J.: Defect detection and quality assessment of hardwood logs: part 2-combined acoustic and laser scanning system. Wood Fiber Sci. **50**(3), 310–322 (2018). https://doi.org/10.22382/wfs-2018-030
3. Urbonas, A., Raudonis, V., Maskeliunas, R., Damaševičius, R.: Automated identification of wood veneer surface defects using a faster region-based convolutional neural network with data augmentation and transfer learning. Appl. Sci. **9**(22) (2019). https://doi.org/10.3390/app9224898
4. Tan, C.O., Ng, S.C.: Wood veneer surface manufacturing defects-prevalence in Malaysian industry and human baseline defect detection performance. J. Trop. For. Sci. **31**(4), 384–397 (2019). https://doi.org/10.26525/jtfs2019.31.4.384
5. Aleksi, I., Susac, F., Matic, T.: Features extraction and texture defect detection of sawn wooden board images. In: 27th Telecommunication Forum, TELFOR 2019, pp. 1–4 (2019). https://doi.org/10.1109/TELFOR48224.2019.8971381
6. Lopes, D.J.V., Bobadilha, G.D.S., Grebner, K.M.: A fast and robust artificial intelligence technique for wood knot detection. BioResources **15**(4), 9351–9361 (2020). https://doi.org/10.15376/biores.15.4.9351-9361

7. Yang, Y., Wang, H., Jiang, D., Hu, Z.: Surface detection of solid wood defects based on ssd improved with resnet. Forests **12**(10), 2–11 (2021). https://doi.org/10.3390/f12101419
8. Sun, P.: A wood quality defect detection system based on deep learning and multicriterion framework. J. Sensors **2022** (2022). https://doi.org/10.1155/2022/3234148
9. Urtans, E., et al.: Detection of knots in oak wood planks: instance versus semantic segmentation. In:5th International Conference on Artificial Intelligence BDAI 2022, pp. 163–168 (2022). https://doi.org/10.1109/BDAI56143.2022.9862633
10. Han, S., Jiang, X., Wu, Z.: An improved YOLOv5 algorithm for wood defect detection based on attention. IEEE Access **11**, 71800–71810 (2022). https://doi.org/10.1109/ACCESS.2023.3293864
11. Mohsin, M., Balogun, O.S., Haataja, K., Toivanen, P.: Real-time defect detection and classification on wood surfaces using deep learning. IS&T International Symposium on Electronic Imaging Science and Technology, vol. 34, no. 10 (2022). https://doi.org/10.2352/EI.2022.34.10.IPAS-382
12. Mohsin, M., Balogun, O.S., Haataja, K., Toivanen, P.: Convolutional neural networks for real-time wood plank detection and defect segmentation. F1000Research, vol. 12, no. March, p. 319 (2023). https://doi.org/10.12688/f1000research.131905.1
13. Ji, M., Zhang, W., Diao, X., Wang, G., Miao, H.: Intelligent automation manufacturing for betula solid timber based on machine vision detection and optimization grading system applied to building materials. Forests **14**(7) (2023). https://doi.org/10.3390/f14071510
14. Meng, W., Yuan, Y.: SGN-YOLO : detecting wood defects with improved YOLOv5 based on semi-global network, pp. 1–20 (2023)
15. Xu, J., Yang, H., Wan, Z., Mu, H., Qi, D., Han, S.: Wood surface defects detection based on the improved YOLOv5-C3Ghost with SimAm module. IEEE Access 1 (2023). https://doi.org/10.1109/ACCESS.2023.3303890
16. Ge, Y., Jiang, D., Sun, L.: Wood veneer defect detection based on multiscale DETR with position encoder net. Sensors **23**(10) (2023). https://doi.org/10.3390/s23104837
17. Wang, R., Liang, F., Wang, B., Mou, X.: ODCA-YOLO: an omni-dynamic convolution coordinate attention-based YOLO for wood defect detection. Forests **14**(9), 1885 (2023). https://doi.org/10.3390/f14091885
18. Oğuzhanoğlu, S., Kapucuoğlu, İ., Sunar, F.: Landsat 8 VeSentinel 2Görüntülerinde MedyanFiltre Ve Curvelet DönüşümüIle GürültüGiderme. no. November (2022). https://doi.org/10.15659/uzalcbs2022.12789
19. Sharmila, B.S., Kaulgud, N.: Comparison of time complexity in median filtering on multi-core architecture. In: Proceedings - 2017 3rd International Conference on Advances in Computing, Communication and AutomationICACCA 2017, vol. 2018-Janua, pp. 1–4 (2018). https://doi.org/10.1109/ICACCAF.2017.8344734
20. Canny, J.: A computational approach to edge detection. IEEE Trans. Pattern Anal. Mach. Intell. PAMI-8(6), 679–698 (1986). https://doi.org/10.1109/TPAMI.1986.4767851
21. Xu, Q., Chakrabarti, C., Karam, L.J.: A distributed Canny edge detector and its implementation on FPGA. In: 2011 Digital Signal Processing and Signal Processing Education Meeting DSP/SPE 2011 - Proceedings, pp. 500–505 (2011). https://doi.org/10.1109/DSP-SPE.2011.5739265
22. Kuncan, M., et al.: Görüntü İşleme Tabanlı Zeytin Ayıklama Makinesi Mekatronik Mühendisliği Bölümü, no. September 2013, pp. 26–28 (2013)
23. Wood Defect Detection Based Image Procesing. https://github.com/MRV-1/wood_defect_detection_based_image_processing. Accessed 24 Jan 2024

Anomaly Detection with Machine Learning Models Using API Calls

Varol Sahin[1]([✉]) , Hami Satilmis[1] , Bilge Kagan Yazar[1] ,
and Sedat Akleylek[2,3]

[1] Ondokuz Mayıs University, Samsun, Türkiye
{varol.sahin,hami.satilmis,kagan.yazar}@bil.omu.edu.tr
[2] Department of Computer Engineering, İstinye University, Istanbul, Türkiye
[3] University of Tartu, Tartu, Estonia

Abstract. Malware is malicious code developed to damage telecommunications and computer systems. Many malware causes anomaly events, such as occupying the systems' resources, such as CPU and memory, or preventing their use. Malware causing these events can hide their destructive activities. Therefore, monitoring their behavior to detect and block such malicious software is necessary. In other words, the anomalies they cause are detected and intervened by monitoring the behaviors exhibited by malware. Various features such as application programming interface (API) calls or system calls, registry modification, and network activities constitute malware behavior. API calls and various statistical information of these calls, extracted by dynamic analysis, are considered one of the most representative features of behavior-based detection systems. Each API call in the sequences is associated with previous or subsequent API calls. Such relationships may contain patterns of destructive functions of malware. Many intrusion/anomaly detection systems are proposed, including machine and deep learning models, in which various information about API/system calls are used as features. This paper aims to evaluate the effect of various statistical information of API calls on the models in detecting anomaly events and classification performances. The anomaly detection performances of various machine learning (ML) models with known effects in the literature are examined using a dataset containing API calls. As a result of the experiments, it is seen that the models using statistical features of API calls have reached high performance in terms of precision, recall, f1-score, and accuracy metrics.

Keywords: Anomaly Detection · API Call · Machine Learning · Deep Learning · Comparative Analysis

1 Introduction

Malware is malicious code developed to damage telecommunications and computer systems [1]. Many malware causes anomaly events such as occupying the systems' resources such as CPU and memory or preventing their use. On the other hand, newly developed malware includes many different characteristics than traditional malware, such as zero-day and obfuscation [1, 2]. Malware with these features can hide their destructive activities. Therefore, it is necessary to monitor their behavior to detect and block such malicious software. In other words, the anomalies they cause are detected and intervened by monitoring the behaviors exhibited by malware.

Behavior-based intrusion/anomaly detection systems are being developed to detect anomalies. In these detection systems, malware's behavior (activity) is monitored. Even if a monitored malware contains zero-day or obfuscation features, it can be detected because it will cause anomaly events in the system due to its behavior during its operation [1].

Various features such as API calls or system calls, registry modification, and network activities constitute the behavior of malware [3]. API calls and various statistical information of these calls, which are extracted by dynamic analysis, are considered one of the most representative features of behavior-based detection systems [4]. On the other hand, API call sequences are created to represent the relationships between API calls. Each API call in the sequences is associated with previous or subsequent API calls [5]. Such relationships may contain patterns of destructive functions of malware [6].

Many intrusion/anomaly detection systems are proposed, including machine/deep learning models, in which various information about API/system calls are used as features. Supervised or unsupervised machine/deep learning models such as logistic regression (LR), support vector machine (SVM), random forest (RF), k-means, and deep neural network (DNN) are widely used in detection systems.

1.1 Related Works

The method using API call sequences to detect Windows malware was suggested in [6]. Malware was detected using clustering models based on Markov chains. The method tested with the dataset created by mixing various datasets reached 0.01 false positive rate (FPR) and 0.99 precision values.

In [7], a method including long short-term memory based (LSTM-based) autoencoders and probability models was introduced to understand the behavior of system processes and detect anomalies. The method was tested on a dataset with information about system calls from an autonomous aircraft application. As a result of the test operations, the method achieved 1.00 true negative rate (TNR), and true positive rate (TPR) values up to 0.90 for benign and malicious samples, respectively.

In [5], a bidirectional LSTM-based method was proposed to extract and combine more meaningful features of API sequences. To test the performance of this method, which included DNN models, data samples from various sources such as VirusShare were used. As a result of the test, the method obtained 0.973 accuracy (ACC) and 0.972 f1-score (F1) values.

In [8], the influence of model parameters on the classification of malware was researched, focusing on using API call information as a feature in machine learning models. In this research, various models such as SVM, RF, and naive Bayes (NB) were trained and tested with data samples from multiple sources such as VirusShare and VirusTotal. As a result of the research, the RF model with optimum parameters reached 0.991 ACC and became the model with the best performance.

In [9], a host-based intrusion detection system (HIDS) was proposed using the ADFA-LD [10] and ADFA-WD [11] datasets, which contain system call sequences for Linux and Windows operating systems, respectively. In this HIDS, SVM, neural network (NN), and decision tree (DT) models were employed to detect abnormal system processes. In experiments conducted on the ADFA-LD dataset, SVM achieved false positive rates (FPR) of 3.34% and 9.12% in binary and multiclass classification, respectively. On the ADFA-WD dataset, NN exhibited FPR values of 8.63% and 15.11% in binary and multiclass classification, respectively.

In [12], WaveNet [13], LSTM [14], and convolutional neural network/recurrent neural network (CNN/RNN) models [15] were used to represent the sequence-to-sequence behaviors of system calls. These models were evaluated with the ADFA-LD and PLAID [12] datasets, achieving area under the curve (AUC) values of 99% on both datasets.

In [16], RF, J48, RIPPER, NB, SVM, and k-nearest neighbors (KNN) models using system call sequences were developed. These models were trained and tested on ADFA-LD and VMM [17] datasets. In the experiments, 100% ACC and 0% false alarm rate (FAR) values were achieved.

In [18], an ensemble-based HIDS was developed, incorporating LSTM, gated recurrent unit (GRU), and fully connected neural network (FCNN) models. This HIDS, detecting abnormal system calls, achieved ACC rates of 91.1% and 68.7% in binary and multiclass classification on the ADFA-WD dataset, respectively.

In [19], an intrusion detection framework was proposed to analyze system call sequences. This framework included a hybrid model combining the LSTM model and the frequency-based anomaly detection method. In experiments conducted on the ADFA-LD dataset, the framework performed with a 97.2% ACC.

1.2 Motivation and Contribution

When the studies in the literature are examined, it is concluded that various features of API calls and API call sequences increase the performance of behavior-based intrusion/anomaly detection systems consisting of machine/deep learning models. As a result of this deduction, this paper it is aimed to evaluate the effect of various statistical information of API calls on the models in terms of detecting anomaly events and classification performances. Within the scope of this aim, this paper makes the following contributions:

- As far as is known, the dataset [20], which has not been used in any scientific paper before and contains various statistical features related to API calls, is used in developing and evaluating models.
- LR, SVM, NB, RF, KNN, and DNN models are developed as machine/deep learning models in the paper.
- In the experiments performed on the developed models, the performance values of the models are observed. As a result of the observation, it is seen that the models using statistical features of API calls have reached high performance values.

1.3 Organization

The remainder of this article is organized as follows. Machine/deep learning models and assessment metrics are described in Sect. 2. Details of the dataset and preprocessing, training and testing phases of models are mentioned in Sect. 3. In Section 4, the results of the experiments on the developed models are given and evaluated. Eventually, the results of this paper and future work are outlined in Sect. 5.

2 Background

Machine learning (ML) methods involve techniques for making predictions (regression) or classifications based on data [21]. These methods have the ability to identify non-linear relationships between independent and dependent variables. Machine learning techniques are split into two categories: supervised and unsupervised [21, 22]. Supervised learning depends on utilizing pre-labeled data for training. Classification is the most prevalent process in supervised learning, but the data must be labeled manually. If there is not enough labeled data, it may prevent supervised learning methods from achieving successful results. Unsupervised learning entails extracting features from unlabeled data, and detection performance typically falls short compared to supervised learning methods. The study primarily focused on supervised learning models. Subsequently, this section will provide brief information on the supervised learning methods used in the study, along with the metrics used to assess the performances of these methods.

2.1 Logistic Regression (LR)

The LR model is an ML method for addressing two-class classification problems. Input variables may encompass one or more features. In binary logistic regression, outcomes are typically represented as either 0 or 1. Logistic regression employs the sigmoid function to compute the probability values and performs classification tasks. The sigmoid function yields outputs within the range of 0 to 1. Instances with values below 0.5 are designated as belonging to the negative class, while those equal to or exceeding 0.5 are assigned to the positive class [23].

2.2 Support Vector Machine (SVM)

SVM is a commonly utilized supervised learning model for binary and multi-class classification problems. SVM functions by mapping input data points into an n-dimensional space and constructing an n-1 dimensional hyperplane to separate the data into distinct groups [23]. The objective of the SVM algorithm is to delineate these points into two separate clusters via a hyperplane while maximizing the margin between the two groups. Moreover, in cases where linear separation is unattainable, a technique referred to as the "kernel trick" is employed [22]. Subsequently, a linear hyperplane is delineated in this transformed space and projected back into the original feature space. Various types of kernels, such as Gaussian, radial basis function, and polynomial kernels, are commonly utilized in practice. The most significant advantages of SVM are that it avoids over-fitting and is not probabilistic [22].

2.3 Naive Bayes (NB)

NB classifiers are simple probabilistic classifiers [22]. The term "naive" refers to the presumption that there are no correlation among any of the input features and that they are all independent of one another. The naive Bayes algorithm is fundamentally grounded in Bayes' theorem. This method gives the probability of the input features belonging to a particular class [24].

2.4 Random Forest (RF)

RF is an ensemble learning algorithm comprised of a set of tree-structured classifiers. Within the model, each tree employs a decision tree algorithm to choose a subset of features. Once the forest is constructed using the RF method, data is presented to each tree to classify a new sample. Subsequently, each tree votes for a specific class, signifying its decision. The forest selects the class with the most votes for the given instance. The primary advantages of the RF method include its resilience to noise and its reduced susceptibility to over-fitting compared to other models [25].

2.5 K-Nearest Neighbour (KNN)

The KNN algorithm, one of the simplest supervised ML methods, classifies input data into discrete results [22]. Due to its not probabilistic and not parametric nature, it is used in classification problems when there is no prior knowledge about the data distribution. The classification process relies on computing a similarity measure (distance) between inputs. Subsequently, any unknown sample classified based on the majority vote of its k closest neighbors. However, as data size grows, so does complexity. Dimensionality reduction techniques are often applied to mitigate the effects of dimensionality before employing KNN [26].

2.6 Deep Neural Network (DNN)

DNN is one of the most frequently employed methods in machine learning. In contrast to traditional ML techniques, it demonstrates superior performance when handling large datasets. A notable characteristic of DNN models is their deep architecture comprising multiple hidden layers [27]. The design of a DNN reflects the working logic of the human brain and typically consists of layers for input, several hidden, and output. Consequently, DNN models encompass many units, making them suitable for classifying non-linear and complex data. However, training DNN models requires more time than other methods due to their complex model structures and sizes [27].

2.7 Evaluation Metrics

The classification performances of ML algorithms are assessed using diverse metrics. The confusion matrix is an essential method for scrutinizing the effectiveness of a classification algorithm. True positive (TP), true negative (TN), false positive (FP), and false negative (FN) values obtained from confusion matrices are important in computing the metrics under examination [27]. Here, we outline the metrics employed in this study:

- **Accuracy (ACC):** Accuracy represents the proportion of correctly classified samples to the total number of samples. This metric is suitable when dealing with a balanced dataset. However, real-world datasets frequently exhibit imbalances. Consequently, it becomes essential to explore alternative metrics alongside accuracy. The computation of accuracy is as follows:

$$ACC = \frac{TP + TN}{TP + FP + FN + TN} \quad (1)$$

- **Precision (PRC):** Precision is the ratio of true positive samples to predicted positive samples and is calculated as follows:

$$PRC = \frac{TP}{TP + FP} \quad (2)$$

- **Recall (REC):** Recall is the ratio of true positive samples to total positive samples and is calculated as follows:

$$REC = \frac{TP}{TP + FN} \quad (3)$$

- **F1-score (F1):** F1-score is defined as the harmonic average of the precision and the recall. It is often preferred over other metrics because it provides more reliable results when data sets are unbalanced. F1-score is calculated as follows:

$$F1 = \frac{2 * PRC * REC}{PRC + REC} \quad (4)$$

3 Approach

In this section, the dataset used in the training and testing phases of machine/deep learning models is mentioned. Following, the preprocessing phase performed on the dataset is explained. Then, the training and testing phases of the models are detailed. The relationship between the preprocessing, training, and testing phases is illustrated in Fig. 1.

3.1 Dataset

Long-running system or application API calls can represent the behavior of the system or application. Therefore, with various features extracted from API calls, whether there is an anomaly in the system or application behavior can be determined.

In this paper, a file of the dataset in [20], which contains various API call features, is used to develop and test the machine/deep learning models. The dataset contains four files, two in "CSV" and two in "JSON" formats. Among these files, the "supervised dataset.csv" file is used as a dataset to develop and test models. In the "supervised dataset.csv" file, there are a total of 1699 samples, of which 1106 are normal, and 592 are anomalies. However, definitions of ten features of API calls representing each sample are given in Table 1.

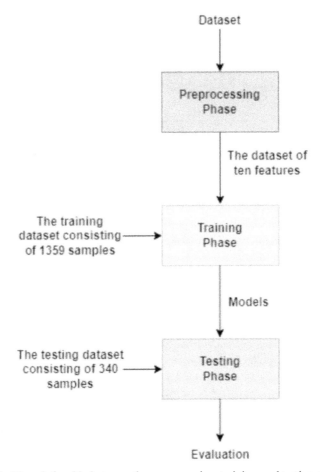

Fig. 1. The relationship between the preprocessing, training, and testing phases.

3.2 Preprocessing Phase

In the preprocessing stage, the features of API calls that will be used as inputs in the machine/deep learning models are determined. In this direction, when the dataset is examined, it is seen that the "ip type" property has a "default" value of 91%. On the other hand, the proportion of samples with an 'E' value for the "source" feature is 88%. Therefore, it is concluded that these two features do not have any distinguishing features during the training and testing phases. For this reason, while these two features are not used in the training and testing of the models, all the other features are used.

3.3 Training Phase

In the training phase of each model, the same dataset, which has gone through the preprocessing phase, is used. Before the training phase, the dataset is randomly divided into 80% and 20% partitions. The larger size of these partitions is used as the training

Table 1. Description of features.

Feature	Description	Sample Value
inter_api_access_duration	The median time interval between consecutive API calls in a session	0.011626192826587695
api_access_uniqueness	The ratio of the number of different APIs in a session to the total number of API calls	0.029285949325436
sequence_length	The average total number of API calls in a session	60.78
vsession_duration	The duration, in minutes, of a session within a window	4240
ip_type	The IP type of access	default
num_sessions	The number of sessions	107.0
num_users	The number of users with the same API call sequences	100.0
num_unique_apis	The number of different API calls	178.0
source	The source of data	E
classification	The class label	normal

dataset for training the models. The training dataset contains a total of 1359 samples, with 877 normal samples and 482 anomaly samples. On the other hand, the way the training phase is carried out varies according to the models. The parameters used in the training stages of the models and how they are applied are explained respectively below.

1. **LR:** In the training phase of this model, all parameters are used with their "default" values in the Scikit-learn library.
2. **SVM:** In the training phase of this model, the type of the "kernel" parameter is chosen as "linear". Except for the "kernel" parameter, all parameters are left with the "default" values in the Scikit-learn library.
3. **NB:** In the training of this model, which uses the Gaussian distribution, all parameters are set with their "default" values in the Scikit-learn library.
4. **RF:** During the training phase of this model, the value of "n_estimators" parameter is set to "10". In the "criterion" parameter, the "entropy" quality metric is selected. Except for these parameters, all parameters are assigned "default" values in the Scikit-learn library.
5. **KNN:** In the training phase of this model, "5" and "2" values are assigned to "n_neighbors" and "p" parameters, respectively. In the "metric" parameter, the "Minkowski" distance metric is set. All parameters except these are given their default values in the Scikit-learn library.
6. **DNN:** This model is created with three layers: input, hidden, and output. While the input and hidden layers contain 16 nodes, there is only one node in the output layer. The "activation" parameter of the input and hidden are set to the "relu" activation function, while this of the output layer is set to the "sigmoid" activation function.

In the training phase of the DNN model, the "RMSprop" optimization algorithm is used in the "optimizer" parameter, with "learning_rate = 0.001". In addition, the "binary_crossentropy" loss function is chosen for the "loss" parameter. On the other hand, before beginning the training, the values of "20" and "256" are given to the parameters "epochs" and "batch_size". All parameters other than these are used to the training with the "default" values in the Tensorflow library.

3.4 Testing Phase

In the testing phase, the performance of the models created as a result of the training phase in detecting anomaly samples is measured. In addition, their ability to classify normal samples and anomaly samples is evaluated. At this phase, 20% of the dataset is used as the test dataset. The test dataset consists of a total of 340 samples, of which 229 are normal samples, and 111 are anomaly samples.

4 Experimental Results

In this section, firstly, the characteristics of the environment in which the models are developed and tested are mentioned. Then, the models are evaluated based on the performance values they have reached as a result of the experiments.

4.1 Experimental Environment

The training and testing phases of the models are carried out on a computer equipped with Intel Core i7-10750H 2.60 GHz and 16 GB RAM. The models are developed using the Python 3.8.5 programming language and the Scikit-learn 0.23.2 and TensorFlow 2.7.0 libraries.

4.2 Results and Discussion

Confusion matrices representing the classification performances of the models as a result of the experiments performed are shown in Fig. 2. In confusion matrices, normal samples are represented by the label "Positive", while the anomaly samples are represented by "Negative".

Anomaly detection performances of the models calculated based on the confusion matrices in Fig. 2 are given in Table 2.

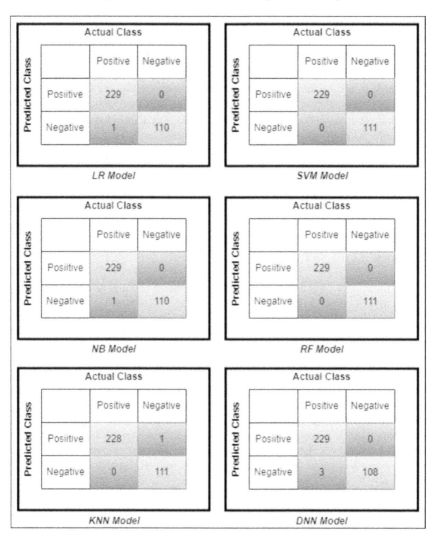

Fig. 2. The confusion matrices of the models

Table 2. Experimental results.

Model	PRC	REC	F1	ACC
LR	0.99	1.00	1.00	1.00
SVM	1.00	1.00	1.00	1.00
NB	0.99	1.00	1.00	1.00
RF	1.00	1.00	1.00	1.00
KNN	1.00	0.99	1.00	1.00
DNN	0.97	1.00	0.99	0.99

5 Conclusion and Future Work

In this paper, machine/deep learning models that detect and classify anomaly events are developed using a dataset that features based on API calls. When the developed models are examined, it is seen that they detect anomaly events with very high performance. While the DNN model reaches an accuracy of 0.99, all other models detect anomaly events with an accuracy of 1.00. The fact that the dataset does not contain missing data and is smooth causes the models to have such high detection rates. Another reason is that various statistical features of API calls are effective in distinguishing normal and anomaly examples of models. As a result, in this study, it is concluded that statistical features of API calls are distinctive features in detecting and classifying anomaly examples of models. Future works aim to develop different deep learning models that use API call sequences as features and detect anomaly events.

References

1. Singh, J., Singh, J.: A survey on machine learning-based malware detection in executable files. J. Syst. Architect. **112**, 101861 (2021)
2. Muzammal, S.M., Shah, M.A., Zhang, S.J., Yang, H.J.: Conceivable security risks and authentication techniques for smart devices: a comparative evaluation of security practices. Int. J. Autom. Comput. **13**(4), 350–363 (2016)
3. Qiao, Y., Yang, Y., He, J., Tang, C., Liu, Z.: CBM: free, automatic malware analysis framework using API call sequences. In: Sun, F., Li, T., Li, H. (eds.) Knowledge Engineering and Management. Advances in Intelligent Systems and Computing, vol. 214. Springer, Berlin, Heidelberg (2014). https://doi.org/10.1007/978-3-642-37832-4_21
4. Elhadi, A.A.E., Maarof, M.A., Barry, B.I.: Improving the detection of malware behaviour using simplified data dependent API call graph. Int. J. Secur. Appl. **7**(5), 29–42 (2013)
5. Li, C., Lv, Q., Li, N., Wang, Y., Sun, D., Qiao, Y.: A novel deep framework for dynamic malware detection based on API sequence intrinsic features. Comput. Secur. **116**, 102686 (2022)
6. Amer, E., Zelinka, I.: A dynamic windows malware detection and prediction method based on contextual understanding of API call sequence. Comput. Secur. **92**, 101760 (2020)
7. Ezeme, O.M., Mahmoud, Q., Azim, A.: A framework for anomaly detection in time-driven and event-driven processes using kernel traces. IEEE Trans. Knowl. Data Eng. (2020)
8. Singh, J., Singh, J.: Assessment of supervised machine learning algorithms using dynamic API calls for malware detection. Int. J. Comput. Appl. **44**(3), 270–277 (2022)
9. Subba, B., Gupta, P.: A tfidfvectorizer and singular value decomposition based host intrusion detection system framework for detecting anomalous system processes. Comput. Secur. **100**, 102084 (2021)
10. Creech, G., Hu, J.: Generation of a new ids test dataset: time to retire the KDD collection. In: 2013 IEEE Wireless Communications and Networking Conference (WCNC), pp. 4487–4492. IEEE (2013)
11. Creech, G.: Developing a high-accuracy cross platform Host-Based Intrusion Detection System capable of reliably detecting zero-day attacks. Ph.D. thesis, UNSW Sydney (2014)
12. Ring, J.H., IV., Van Oort, C.M., Durst, S., White, V., Near, J.P., Skalka, C.: Methods for host-based intrusion detection with deep learning. Digital Threats: Res. Pract. **2**(4), 1–29 (2021)
13. Oord, A.V.D., et al.: Wavenet: A generative model for raw audio. arXiv preprint arXiv:1609.03499 (2016)

14. Kim, G., Yi, H., Lee, J., Paek, Y., Yoon, S.: LSTM-based system-call language modeling and robust ensemble method for designing host-based intrusion detection systems. arXiv preprint arXiv:1611.01726 (2016)
15. Chawla, A., Lee, B., Fallon, S., Jacob, P.: Host based intrusion detection system with combined CNN/RNN model. In: Alzate, C., et al. ECML PKDD 2018 Workshops. ECML PKDD 2018. LNCS, vol. 11329. Springer, Cham (2019). https://doi.org/10.1007/978-3-030-13453-2_12
16. Melvin, A.A.R., Kathrine, G.J.W., Pasupathi, S., Shanmuganathan, V., Naganathan, R.: An AI powered system call analysis with bag of word approaches for the detection of intrusions and malware in australian defence force academy and virtual machine monitor malware attack data set. Expert Syst. e13029 (2022)
17. Melvin, A.A.R., et al.: Dynamic malware attack dataset leveraging virtual machine monitor audit data for the detection of intrusions in cloud. Trans. Emerg. Telecommun. Technol. **33**(4), e4287 (2022)
18. Kumar, Y., Subba, B.: Stacking ensemble-based hids framework for detecting anomalous system processes in windows based operating systems using multiple word embedding. Comput. Secur. **125**, 102961 (2023)
19. Chaudhari, A., Gohil, B., Rao, U.P.: A novel hybrid framework for cloud intrusion detection system using system call sequence analysis. Cluster Comput. pp. 1–17 (2023)
20. Guntur, R.: API security: access behavior anomaly dataset. Kaggle, https://www.kaggle.com/datasets/tangodelta/api-access-behaviour-anomaly-dataset
21. Buskirk, T.D., Kirchner, A., Eck, A., Signorino, C.S.: An introduction to machine learning methods for survey researchers. Surv. Pract. **11**(1) (2018)
22. Muhammad, I., Yan, Z.: Supervised machine learning approaches: a survey. ICTACT J. Soft Comput. **5**(3) (2015)
23. Shanthamallu, U.S., Spanias, A., Tepedelenlioglu, C., Stanley, M.: A brief survey of machine learning methods and their sensor and IOT applications. In: 2017 8th International Conference on Information, Intelligence, Systems & Applications (IISA), pp. 1–8. IEEE (2017)
24. Rish, I., et al.: An empirical study of the naive bayes classifier. In: IJCAI 2001 Workshop on Empirical Methods in Artificial Intelligence, vol. 3, pp. 41–46 (2001)
25. Min, E., Long, J., Liu, Q., Cui, J., Chen, W.: Tr-ids: anomaly-based intrusion detection through text-convolutional neural network and random forest. Security and Communication Networks 2018 (2018)
26. Beyer, K., Goldstein, J., Ramakrishnan, R., Shaft, U.: When is "nearest neighbor" meaningful? In: Database Theory—ICDT 1999: 7th International Conference Jerusalem, Israel, January 10–12, 1999 Proceedings 7. pp. 217–235. Springer (1999)
27. Liu, H., Lang, B.: Machine learning and deep learning methods for intrusion detection systems: a survey. Appl. Sci. **9**(20), 4396 (2019)

Information Technology in Decision Making

Finding a Minimum Distance Between Two Smooth Curves in 3D Space

Majid Abbasov[✉] [iD] and Lyudmila Polyakova [iD]

St. Petersburg State University, SPbSU, 7/9, Universitetskaya nab., St. Petersburg 199034, Russia
{m.abbasov,l.polyakova}@spbu.ru

Abstract. This study investigates the problem of determining the minimum distance between two smooth 3D curves. We propose its solution on the basis of the newly formulated idea of Charged Balls Method. The main idea behind this method is the physical model which tends to the solution of the original problem. We can derive equations of motion for the system and apply the difference scheme for the solution of the obtained ordinary differential equations. This is the way we get an iterative algorithm for the initial problem. We employ Lyapunov theory of stability to prove the convergence of the proposed method. More specifically, the Barbashin-Krasovskii theorem on asymptotic stability is employed. The convergence rate of the algorithm is analyzed and the corresponding results are presented. We provide a number of numerical examples to demonstrate its work. These examples acknowledge the effectiveness of the method and stresses that the choice of the parameters one may significantly affect the convergence rate.

Keywords: Minimal Distance · Charged Balls Method · Mathematical Programming · Lyapunov's Second Method · LaSalle's Invariance Principle · Computational Geometry

1 Introduction

To create new optimization algorithms, researchers often use the transition from the original problem to a dynamic system which tends to an equilibrium point that coincides with the solution of the original problem. The equations of motion of a given system describe a continuous version of the method, and the convergence of this method is equivalent to the stability of the equilibrium position. After applying the difference scheme to the equations of motion, an iterative algorithm is obtained. This common approach is based on the application of physical analogies (see [1–10]). There are many optimization methods that have been derived in this way. One of them is the Charged Balls Method, which allows us to solve certain computational geometry problems [11].

Let the problem of orthogonal projection of the origin onto a set

$$X = \{x \in \mathbb{R}^n | f(x) \leq 0\}$$

is given. The function $f : \mathbb{R}^n \to \mathbb{R}$ is convex, continuously differentiable up to the second order inclusive and $0 \notin X$. The problem is formalized as

$$\begin{cases} \|x\| \to \min, \\ x \in X. \end{cases} \qquad (1)$$

We can assume that the gradient $\nabla f(x) \neq 0$ for any $x \in bdX$. Here bdX stands for the boundary of the set X. We place a negative charge q in the origin and fix it. Select an arbitrary point $\hat{x}$ in the interior of X and place a positively charged ball with charge q and mass m in it. We assume that the Coulomb force and viscous friction are the only active forces here. The ball starts its movement along the line towards the origin, till its collision with the boundary of X. The coordinate of the collision point is found by solving the system

$$\begin{cases} f(x) = 0, \\ x = \lambda \hat{x}, \, \lambda \in (0, \infty), \end{cases} \qquad (2)$$

and in case when the solution is not unique, choosing the one which is closer to the origin. As a result, we get $x_0 \in bdX$. Note that for the search of an initial point x_0 one can also use some easy heuristic procedures [12].

Then the ball moves along the set bdX, oscillating around the equilibrium position, which is equal to the solution of (1). In this point the normal reaction force of bdX is counterbalanced by the Coulomb force. At any other point, the tangential component of the Coulomb force is nonzero and directed to the equilibrium position. For the sake of getting damping oscillations, we introduce viscous friction force, which is proportional to the velocity and directed opposite to it. This provides the convergence of the process to the desired solution. Deriving equations of motion and passing to the difference scheme of their solution, we come to an iterative method of solving the problem (1).

Denote by $x(t)$ the coordinate of the ball, where t is a time. Coulomb and Normal forces can be represented as

$$F(t) = -\frac{c_1 q^2}{\|x\|^3} x, \ N(t) = -\frac{\nabla f(x)}{\|\nabla f(x)\|^2} \langle F(t), \nabla f(x) \rangle - m \frac{\langle H(x) \dot{x}, \dot{x} \rangle}{\|\nabla f(x)\|^2} \nabla f(x),$$

respectively, where c_1 is the electric constant, m is the mass of the ball, $H(x)$ is the second-order derivatives matrix of the function f at a point x. The first summand in the expression defining N balances the Coulomb's force normal component, whereas the second one balances centrifugal force. Define viscous friction as $R(t) = -c_2 \dot{x}$, where c_2 is a coefficient of viscosity.

Using Newton's second low we get

$$m\ddot{x}(t) = F(t) + N(t) + R(t). \qquad (3)$$

Let us move from (3) to a first-order system by initiating an n-dimensional vector of fictitious variables $z(t)$:

$$\begin{cases} \dot{x} = z, \\ \dot{z} = p_1 \psi(x) - p_2 z - \chi(x, z). \end{cases} \qquad (4)$$

where

$$\psi(x) = -\frac{1}{\|x\|^3}x + \frac{\langle x, \nabla f(x)\rangle}{\|x\|^3\|\nabla f(x)\|^2}\nabla f(x), \quad \chi(x,z) = \frac{\langle H(x)z, z\rangle}{\|\nabla f(x)\|^2}\nabla f(x),$$

and p_1, p_2 are parameters depending on m, q, c_1, c_2.

Note that if a point x_* is the solution of the problem (1) then $(x_*, 0)^T$ is the equilibrium position for the system (4).

Let a trajectory $x(t)$ is a solution of the system (4) with an initial conditions $x(0)$, $z(0)$, where $f(x(0)) = 0$ and the inner product $\langle \nabla f(x(0)), z(0)\rangle = 0$. It can be shown that $f(x(t)) \equiv 0$, that is $x(t) \in bdX$ for all $t \geq 0$.

The system (4) can be solved by Euler method. Choose $x(0) = x_0$, $z(0) = 0$, a positive δ as a step length. Let x_{k-1}, z_{k-1} be given, then

$$\begin{cases} x_k = x_{k-1} + \delta z_{k-1}, \\ z_k = z_{k-1} + \delta(\psi(x_{k-1}) - p_2 z_{k-1} - \chi(x, z_{k-1})). \end{cases} \quad (5)$$

We can take the condition

$$\|\psi(x_k)\| \leq \varepsilon$$

as a stopping criteria.

2 Finding the Minimum Distance Between Two Curves

The Charged Ball Method concept is quite universal and can be used to solve more complex problems. Let's take two curves in $\mathbb{R}^3$

$$\Gamma_1 : r_1(u) = (x_1(u), y_1(u), z_1(u)),$$

$$\Gamma_2 : r_2(v) = (x_2(v), y_2(v), z_2(v)),$$

which have empty intersection and $r_1, r_2 : \mathbb{R} \to \mathbb{R}^3$ are vector functions of the scalar parameters u and v respectively and continuously differentiable up to the second order. We seek points

$$\tilde{r}_1 = (x_1(\tilde{u}), y_1(\tilde{u}), z_1(\tilde{u})),$$

$$\tilde{r}_2 = (x_2(\tilde{v}), y_2(\tilde{v}), z_2(\tilde{v})),$$

on which the minimum distance between these curves is attained. The problem is stated as follows

$$\begin{cases} \|r_1 - r_2\| \to \inf, \\ r_1 \in \Gamma_1, \\ r_2 \in \Gamma_2. \end{cases} \quad (6)$$

To find a solution of this problem using the Charged Ball Method, we place a positively charged ball on one of the curves and a negatively charged ball on the other. These balls can only move along their curves. We assume that their movement is due to the Coulomb force and the force of viscous friction. The latter is proportional to the speed of balls and directed against it, which leads to a decrease in the energy of the system and the movement of the balls to positions $\tilde{r}_1$ and $\tilde{r}_2$ as time increases. To describe the motion of the balls, we compose equations of motion and solve them numerically using a difference scheme. This gives us an iterative method for solving the original problem. At each moment of time, the position of the balls on the curves is determined by the values of the parameters u and v. Thus, the coordinates of the balls become functions. $r_1(u(t))$, $r_2(v(t))$ of time, which are determined by the curves on which they move. The normal component of the Coulomb force impacting the balls is compensated by the normal reactions of the corresponding curves. In this regard, the equations of motion take into account only viscous friction, centrifugal force and the tangential component of the Coulomb force. The latter is obtained by projecting onto the tangential direction. This direction for the first curve is determined by a vector

$$\tau_1(u) = \frac{dr_1(u)}{du} / \|\frac{dr_1(u)}{du}\|,$$

while for the second one by the vector

$$\tau_2(v) = \frac{dr_2(v)}{dv} / \|\frac{dr_2(v)}{dv}\|.$$

The Coulomb forces impacting the first and second balls are defined by expressions

$$\frac{r_2 - r_1}{\|r_2 - r_1\|^3}, \quad -\frac{r_2 - r_1}{\|r_2 - r_1\|^3}.$$

The centrifugal forces impacting the first and second balls are defined by expressions

$$\frac{1}{\rho_1(u)}\left(\frac{dr_1}{dt}\right)^2 n_1(u), \quad \frac{1}{\rho_2(v)}\left(\frac{dr_2}{dt}\right)^2 n_2(v),$$

where $\rho_1(u)$, $\rho_2(v)$ are radiuses of curvature of curves $r_1(u)$ and $r_2(v)$ respectively, and the normal vectors $n_1(u)$, $n_2(v)$ are given by expressions

$$n_1(u) = \frac{d\tau_1(u)}{du} / \|\frac{d\tau_1(u)}{du}\|, \quad n_2(v) = \frac{d\tau_2(v)}{dv} / \|\frac{d\tau_2(v)}{dv}\|.$$

By means of Newton's second law we derive the equations of motion in the form

$$\begin{cases} \frac{d^2 r_1}{dt^2} = p_1 \langle \frac{r_2 - r_1}{\|r_2 - r_1\|^3}, \tau_1 \rangle \tau_1 - p_2 \frac{dr_1}{dt} + \frac{1}{\rho_1}\left(\frac{dr_1}{dt}\right)^2 n_1, \\ \frac{d^2 r_2}{dt^2} = p_1 \langle \frac{r_1 - r_2}{\|r_2 - r_1\|^3}, \tau_2 \rangle \tau_2 - p_2 \frac{dr_2}{dt} + \frac{1}{\rho_2}\left(\frac{dr_2}{dt}\right)^2 n_2, \end{cases} \quad (7)$$

where p_1, p_2 are some constants.

Thereby, the continuous method for problem (6) was derived. An auxiliary results are required in order to analyze the convergence of the method as well as obtain estimates for the convergence rate.

Proceed to an autonomous system of differential equations

$$\frac{dx_i}{dt} = X_i(x_1, \ldots, x_n), i = 1, \ldots, n, \qquad (8)$$

where functions $X_i(x)$ are Lipschitz continuous in some domain D of the phase space so that $0 \in D$. Let $X_i(0) = 0$, then we have theorems (see [13, 14])

Theorem 1. If there exists a differentiable function V in D for the system (8) so that

1. $V(x) \geq 0$ for all $x \in D$ and $V(x) = 0$ iff $x = 0$,
2. derivative of the function V with respect to the system satisfy the condition

$$\dot{V} = \frac{dV}{dt} = \sum_{i=1}^{n} \frac{\partial V}{\partial x_i} X_i \leq 0 \forall x \in D,$$

then the equilibrium position 0 is stable in the sense of Lyapunov.

Theorem 2 [Lassalle's Invariance Principle (Barbashin-Krasovskii Theorem on Asymptotic Stability)]. If there exists a differentiable function V in D so that

1. $V(x) \geq 0$ for all $x \in D$ and $V(x) = 0$ iff $x = 0$,
2. The derivative of function V with respect to the system satisfy the condition

$$\dot{V} = \frac{dV}{dt} = \sum_{i=1}^{n} \frac{\partial V}{\partial x_i} X_i \leq 0, \forall x \in D,$$

3. the set $\Lambda = \{x \in D | \dot{V}(x) = 0\}$ does not contain whole trajectories except the point 0,

then the equilibrium position 0 is asymptotic stable in the sense of Lyapunov.

Now we can formulate and prove a theorem on the convergence of the method (7).

Theorem 3. Assume that the problem (6) has a finite number of stationary points and $[\tilde{r}_1, \tilde{r}_2]$ is the one of solutions of (6). Then there is $\delta > 0$ so that for all $[r_1^0, \dot{r}_1^0] \in B_\delta([\tilde{r}_1, 0]) \cap \mathcal{W}_1, [r_2^0, \dot{r}_2^0] \in B_\delta([\tilde{r}_2, 0]) \cap \mathcal{W}_2$, where

$$\mathcal{W}_1 = \left\{ w \in \mathbb{R}^6 | w = [r(u), \dot{r}(u)], r(u) \in \Gamma_1, \exists \lambda \in \mathbb{R} : \dot{r}(u) = \lambda \tau_1(u) \right\},$$

$$\mathcal{W}_2 = \left\{ w \in \mathbb{R}^6 | w = [r(v), \dot{r}(v)], r(v) \in \Gamma_2, \exists \lambda \in \mathbb{R} : \dot{r}(v) = \lambda \tau_2(v) \right\},$$

the method (7) with an initial condition $[r_1^0, \dot{r}_1^0, r_2^0, \dot{r}_2^0]$ converge to $[\tilde{r}_1, 0, \tilde{r}_2, 0]$. Here $[r, \dot{r}]$ is a vector in $\mathbb{R}^{2n}$, obtained by concatenating of three-dimensional vectors r and $\dot{r}$. $B_\delta([r, 0])$ is the ball with the center $[r, 0]$ and radius δ.

Proof. Let $[\tilde{r}_1, \tilde{r}_2]$ be a solution of the problem (6). At this point the necessary condition for the minimum of the function $\|r_1(u) - r_2(v)\|^2$ should be satisfied, whence

$$\langle \tilde{r}_1 - \tilde{r}_2, \frac{d\tilde{r}_1}{du} \rangle = 0, \langle \tilde{r}_1 - \tilde{r}_2, \frac{d\tilde{r}_2}{dv} \rangle = 0.$$

Therefore $[\tilde{r}_1, 0, \tilde{r}_2, 0]$ is the equilibrium point of (7). The opposite is true as well, i.e. any equilibrium $[\tilde{r}_1, 0, \tilde{r}_2, 0]$ of the system (7) corresponds to some stationary point $[\tilde{r}_1, \tilde{r}_2]$ of the problem (6).

Since there are a finite number of stationary points for the problem (6), there exists $\varepsilon > 0$ for which $B_\varepsilon([\tilde{r}_1, 0]) \times B_\varepsilon([\tilde{r}_2, 0])$ does not contain any other equilibrium positions except $[\tilde{r}_1, 0, \tilde{r}_2, 0]$. The right side of (7) is continuously differentiable on $B_\varepsilon([\tilde{r}_1, 0]) \times B_\varepsilon([\tilde{r}_2, 0])$ which means that it is Lipschitz continuous on the indicated set. With no loss of generality, we assume that for all $[r_1, 0, r_2, 0] \in D$, where

$$D = (B_\varepsilon([\tilde{r}_1, 0]) \cap W_1) \times (B_\varepsilon([\tilde{r}_2, 0]) \cap W_2),$$

the inequality $\|\tilde{r}_1 - \tilde{r}_2\| \leq \|r_1 - r_2\|$ is true.

Consider the Lyapunov function
$V(r_1, r_2, \dot{r}_1, \dot{r}_2) = \frac{p_1}{\|\tilde{r}_2 - \tilde{r}_1\|} - \frac{p_1}{\|r_2 - r_1\|} + \frac{1}{2}(\|\dot{r}_1\|^2 + \|\dot{r}_2\|^2).$

The derivative of this function by virtue of the system (7) is written in a following form

$$\dot{V}(r_1, r_2, \dot{r}_1, \dot{r}_2) = p_1 \langle \frac{r_2 - r_1}{\|r_2 - r_1\|^3}, \frac{dr_2}{dv} \dot{v} - \frac{dr_1}{du} \dot{u} \rangle + \langle \dot{r}_1, \ddot{r}_1 \rangle + \langle \dot{r}_2, \ddot{r}_2 \rangle =$$

$$p_1 \langle \frac{r_2 - r_1}{\|r_2 - r_1\|^3}, \frac{dr_2}{dv} \dot{v} - \frac{dr_1}{du} \dot{u} \rangle + p_1 \langle \frac{r_2 - r_1}{\|r_2 - r_1\|^3}, \tau_1 \rangle \langle \tau_1, \dot{r}_1 \rangle -$$

$$p_2 \|\dot{r}_1\|^2 + p_1 \langle \frac{r_1 - r_2}{\|r_2 - r_1\|^3}, \tau_2 \rangle \langle \tau_2, \dot{r}_2 \rangle - p_2 \|\dot{r}_2\|^2 =$$

$$p_1 \langle \frac{r_2 - r_1}{\|r_2 - r_1\|^3}, \frac{dr_2}{dv} \dot{v} - \frac{dr_1}{du} \dot{u} \rangle + p_1 \langle \frac{r_2 - r_1}{\|r_2 - r_1\|^3}, \frac{dr_1}{du} \dot{u} \rangle -$$

$$p_2 \|\dot{r}_1\|^2 + p_1 \langle \frac{r_1 - r_2}{\|r_2 - r_1\|^3}, \frac{dr_2}{dv} \dot{v} \rangle - p_2 \|\dot{r}_2\|^2.$$

Hence

$$\dot{V}(r_1, r_2, \dot{r}_1, \dot{r}_2) = -p_2 \left(\|\dot{r}_1\|^2 + \|\dot{r}_2\|^2 \right) \leq 0.$$

By Theorem 2 we conclude that the equilibrium position $[\tilde{r}_1, 0, \tilde{r}_2, 0]$ of the system (7) is Lyapunov stable. Thus, there exists $\delta > 0$, for which all trajectories starting from $(B_\delta([\tilde{r}_1, 0]) \cap W_1) \times (B_\delta([\tilde{r}_2, 0]) \cap W_2)$ do not abandon D. Evidently, any trajectory entirely lying in the set

$$\Lambda = \{[r_1, r_2, \dot{r}_1, \dot{r}_2] \in D | \dot{V}(r_1, r_2, \dot{r}_1, \dot{r}_2) = 0\}$$

is an equilibrium position. But there are no other equilibrium positions in D except $[\tilde{r}_1, 0, \tilde{r}_2, 0]$. Whence all conditions of Theorem 2 are satisfied and all trajectories that start from $(B_\delta([\tilde{r}_1, 0]) \cap W_1) \times (B_\delta([\tilde{r}_2, 0]) \cap W_2)$ tend to $[\tilde{r}_1, 0, \tilde{r}_2, 0]$.

Refering to the proof above we get that the point $[r_1, r_2]$ is stationary point for the problem (6) if and only if

$$\psi_1(r_1, r_2, \tau_1) = \langle \frac{r_2 - r_1}{\|r_2 - r_1\|^3}, \tau_1 \rangle \tau_1 = 0,$$

$$\psi_2(r_1, r_2, \tau_2) = \langle \frac{r_1 - r_2}{\|r_2 - r_1\|^3}, \tau_2 \rangle \tau_2 = 0.$$

Therefore we can choose the condition

$$\|\psi_1(r_1, r_2, \tau_1)\|^2 + \|\psi_2(r_1, r_2, \tau_2)\|^2 \le \varepsilon \qquad (9)$$

as a stopping criterion for some small positive ε.

Theorem 4 [see [15]]. Assume that $V(x)$ is a function bounded from below, that is $V(x) \ge v_*$ and let there is a continuous function $\omega(x)$ such that $-\dot{V}(x) \ge \omega(x) \ge 0$ for all x. Then

$$\min_{0 \le t \le T} \omega(x(t)) \le \frac{V(x(0)) - v_*}{T}.$$

Let us assume that the conditions of the Theorem 3 are satisfied and $[\tilde{r}_1, 0, \tilde{r}_2, 0]$ is an equilibrium position of the system (7) that corresponds to the solution $[\tilde{r}_1, \tilde{r}_2]$ of (6). Then there exists some $\varepsilon > 0$ such that $B_\varepsilon([\tilde{r}_1, 0, \tilde{r}_2, 0])$ does not include any other equilibrium positions except $[\tilde{r}_1, 0, \tilde{r}_2, 0]$. Via Theorem 3 we get that there is $\delta > 0$ such that for any initial point $[r_1^0, \dot{r}_1^0, r_2^0, \dot{r}_2^0]$, $[r_1^0, \dot{r}_1^0] \in B_\delta([\tilde{r}_1, 0]) \cap \mathcal{W}_1$, $[r_2^0, \dot{r}_2^0] \in B_\delta([\tilde{r}_2, 0]) \cap \mathcal{W}_2$, trajectories $[r_1(u(t)), \dot{r}_1(u(t))]$, $[r_2(v(t)), \dot{r}_2(v(t))]$ of the system (7) do not leave sets $B_\varepsilon([\tilde{r}_1, 0]) \cap \mathcal{W}_1$, $B_\varepsilon([\tilde{r}_2, 0]) \cap \mathcal{W}_2$ and tend to $[\tilde{r}_1, 0]$, $[\tilde{r}_2, 0]$ respectively.

Now we can propose and prove a theorem that estimates the convergence rate of the method.

Theorem 5. Let p_1, p_2 be parameters of the method and Jacobians $\psi_{1'}(r_1, r_2, \tau_1)$, $\psi_{2'}(r_1, r_2, \tau_1)$ satisfy inequalities

$$\langle \psi_{1'}(r_1, r_2, \tau_1)[\dot{r}_1, \dot{r}_2, \dot{\tau}_1], \dot{r}_1 \rangle + \langle \psi_{2'}(r_1, r_2, \tau_1)[\dot{r}_1, \dot{r}_2, \dot{\tau}_1], \dot{r}_2 \rangle \ge \frac{p_2^2}{p_1}\left(\|\dot{r}_1\|^2 + \|\dot{r}_2\|^2\right)$$

for all $[r_1, \dot{r}_1, r_2, \dot{r}_2] \in (B_\varepsilon([\tilde{r}_1, 0]) \cap \mathcal{W}_1) \times (B_\varepsilon([\tilde{r}_2, 0]) \cap \mathcal{W}_2)$. For any initial points $[r_1^0, \dot{r}_1^0, r_2^0, \dot{r}_2^0] \in (B_\delta([\tilde{r}_1, 0]) \cap \mathcal{W}_1) \times (B_\delta([\tilde{r}_2, 0]) \cap \mathcal{W}_2)$ for the method (7) the following inequality

$$\min_{0 \le t \le T}(\|\psi_1(r_1, r_2, \tau_1)\|^2 + \|\psi_2(r_1, r_2, \tau_2)\|^2) \le \frac{V(r_1^0, \dot{r}_1^0, r_2^0, \dot{r}_2^0) - v_*}{p_1 T}$$

is true. Here

$$V(r_1, \dot{r}_1, r_2, \dot{r}_2) = -\langle \psi_1(r_1, r_2, \tau_1), \dot{r}_1 \rangle - \langle \psi_2(r_1, r_2, \tau_2), \dot{r}_2 \rangle - \frac{p_2}{2p_1}\left(\|\dot{r}_1\|^2 + \|\dot{r}_2\|^2\right),$$

$$\psi_1(r_1, r_2, \tau_1) = \langle \frac{r_2 - r_1}{\|r_2 - r_1\|^3}, \tau_1 \rangle \tau_1, \quad \psi_2(r_1, r_2, \tau_2) = \langle \frac{r_1 - r_2}{\|r_2 - r_1\|^3}, \tau_2 \rangle \tau_2,$$

$$v_* = \min_{[r_1, \dot{r}_1] \in B_\varepsilon([\tilde{r}_1, 0]) \cap \mathcal{W}_1; [r_2, \dot{r}_2] \in B_\varepsilon([\tilde{r}_2, 0]) \cap \mathcal{W}_2} V(r_1, \dot{r}_2, r_2, \dot{r}_2).$$

Proof. Provided the conditions of the theorem, we have the following chain od equalities for the derivative of the function V by virtue of the system (7)

$$\dot{V}(r_1, \dot{r}_2, r_2, \dot{r}_2) =$$
$$-\langle \psi_{1\prime}(r_1, r_2, \tau_1)[\dot{r}_1, \dot{r}_2, \tau_1], \dot{r}_1 \rangle - \langle \psi_1(r_1, r_2, \tau_1), p_1\psi_1(r_1, r_2, \tau_1) - p_2\dot{r}_1 \rangle -$$
$$-\langle \psi_{2\prime}(r_1, r_2, \tau_2)[\dot{r}_1, \dot{r}_2, \tau_2], \dot{r}_2 \rangle - \langle \psi_2(r_1, r_2, \tau_2), p_1\psi_2(r_1, r_2, \tau_2) - p_2\dot{r}_2 \rangle -$$
$$-\frac{p_2}{p_1}(\langle \dot{r}_1, p_1\psi_1(r_1, r_2, \tau_1) - p_2\dot{r}_1 \rangle + \langle \dot{r}_2, p_1\psi_2(r_1, r_2, \tau_2) - p_2\dot{r}_2 \rangle) \leq$$
$$-p_1(\|\psi_1(r_1, r_2, \tau_1)\|^2 + \|\psi_2(r_1, r_2, \tau_2)\|^2).$$

Whence using Theorem 4 we finish the proof.
Considering that

$$\frac{d^2 r_1}{dt^2} = \frac{d^2 r_1}{du^2}\left(\frac{du}{dt}\right)^2 + \frac{d^2 u}{dt^2}\frac{dr_1}{du},$$

$$\frac{d^2 r_2}{dt^2} = \frac{d^2 r_2}{dv^2}\left(\frac{dv}{dt}\right)^2 + \frac{d^2 v}{dt^2}\frac{dr_2}{dv}.$$

We can rewrite (7) in the form

$$\begin{cases} \ddot{u}\frac{dr_1}{du} + \dot{u}^2\frac{d^2 r_1}{du^2} + p_2\dot{u}\frac{dr_1}{du} = p_1\langle \frac{r_2-r_1}{\|r_2-r_1\|^3}, \tau_1\rangle\tau_1 + \frac{1}{\rho_1}\left(\frac{dr_1}{dt}\right)^2 n_1, \\ \ddot{v}\frac{dr_2}{dv} + \dot{v}^2\frac{d^2 r_2}{dv^2} + p_2\dot{v}\frac{dr_2}{dv} = p_1\langle \frac{r_1-r_2}{\|r_2-r_1\|^3}, \tau_2\rangle\tau_2 + \frac{1}{\rho_2}\left(\frac{dr_2}{dt}\right)^2 n_2. \end{cases}$$

Multiplying the left an the right side of these equations by the corresponding tangential directions, along which a movement occurs, we get

$$\begin{cases} \ddot{u}\|\frac{dr_1}{du}\| + \dot{u}^2\langle \tau_1, \frac{d^2 r_1}{du^2}\rangle + p_2\dot{u}\|\frac{dr_1}{du}\| = p_1\langle \frac{r_2-r_1}{\|r_2-r_1\|^3}, \tau_1\rangle, \\ \ddot{v}\|\frac{dr_2}{dv}\| + \dot{v}^2\langle \tau_2, \frac{d^2 r_2}{dv^2}\rangle + p_2\dot{v}\|\frac{dr_2}{du}\| = p_1\langle \frac{r_1-r_2}{\|r_2-r_1\|^3}, \tau_2\rangle. \end{cases} \quad (10)$$

Introducing dummy variables $\xi = \dot{u}, \eta = \dot{v}$ we downgrade the order of the system (10)

$$\begin{cases} \dot{u} = \xi \\ \dot{\xi} = p_1\langle \frac{r_2-r_1}{\|r_2-r_1\|^3}, \tau_1\rangle\|\frac{dr_1}{du}\|^{-1} - p_2\xi - \langle \tau_1, \frac{d^2 r_1}{du^2}\rangle\|\frac{dr_1}{du}\|^{-1}\xi^2, \\ \dot{v} = \eta \\ \dot{\eta} = p_1\langle \frac{r_1-r_2}{\|r_2-r_1\|^3}, \tau_2\rangle\|\frac{dr_2}{dv}\|^{-1} - p_2\eta - \langle \tau_2, \frac{d^2 r_2}{dv^2}\rangle\|\frac{dr_2}{dv}\|^{-1}\eta^2. \end{cases} \quad (11)$$

Now we assume that u_k, ξ_k, v_k, η_k are given and apply the forward Euler method to the system (11) with step δ. By doing this we obtain an algorithm for solving the initial problem

$$\begin{cases} u_{k+1} = u_k + \delta \xi_k, \\ \xi_{k+1} = \xi_k + \delta \left(p_1 \phi_1(u_k, v_k) \| \frac{dr_1}{du} \|^{-1} - p_2 \xi_k - \zeta_1(u_k) \xi_k^2 \| \frac{dr_1}{du} \|^{-1} \right), \\ v_{k+1} = v_k + \delta \eta_k \\ \eta_{k+1} = \eta_k + \delta \left(p_1 \phi_2(u_k, v_k) \| \frac{dr_2}{dv} \|^{-1} - p_2 \eta_k - \zeta_2(v_k) \eta_k^2 \| \frac{dr_2}{dv} \|^{-1} \right). \end{cases}$$

where

$$\phi_1(u, v) = \langle \frac{r_2(u) - r_1(v)}{\|r_2(v) - r_1(u)\|^3}, \tau_1(u) \rangle, \phi_2(v, v) = \langle \frac{r_1(v) - r_2(u)}{\|r_2(v) - r_1(u)\|^3}, \tau_2(v) \rangle,$$

$$\zeta_1(u) = \langle \frac{d^2 r_1(u)}{du^2}, \tau_1(u) \rangle, \zeta_2(v) = \langle \frac{d^2 r_2(v)}{dv^2}, \tau_2(v) \rangle.$$

Obviousely, the tangential component of the Coulomb forces should be zero in the points r_1^*, r_2^*. Therefore we can take a condition

$$\phi_1^2(u_k, v_k) + \phi_2^2(u_k, v_k) \le \varepsilon, \qquad (12)$$

as the stopping criterion for some $\varepsilon > 0$. Evidently, this condition is equivalent to (9).

3 Numerical Experiments

In this section we provide numerical examples for solution of problems using the derived algorithm. Calculations were made in Matlab 2013a mathematical package, via a computer with Core i5-3450 3.1GHz processor, 8GB of DDR3 RAM, working under Windows 10 Pro x64 operating system.

3.1 Example 1

Find a distance between two curves (see Fig. 1)

Table 1 shows the results for different values of the parameters of the method.

3.2 Example 2

Find the minimum distance between two flat curves (see Fig. 2).

$$r_1(u) = \left(u, 5e^{-u^2/0.1}, 0 \right), r_2(v) = \left(v, -\cos\frac{v}{2} + 7, 0 \right).$$

It is obvious that the points $r_1^* = (0, 5)$, $r_2^* = (0, 6)$ are the global solutions. Curve $r_1(u)$ have a high curvature at the point r_1^*. This leads to some problems in case of inappropriate choice of p_1, p_2.

Table 2 shows the results for different values of the parameters of our method.

Table 1. The solution of the problem in Example 1 for $\varepsilon = 10^{-8}$, $u_0 = 1$, $v_0 = 1$, $\xi_0 = 0$, $\eta_0 = 0$ and different parameters of the method. [a] The value of the expression (12), [b] number of iterations, [c] distance to the true solution, [d] time of calculations, s.

p_1	p_2	δ	criteria[a]	N[b]	$\|r_1 - r_1^*\|$[c]	$\|r_2 - r_2^*\|$[c]	t[d]
100	10	0.1	$0.4 \cdot 10^{-9}$	63	$1 \cdot 10^{-4}$	$1 \cdot 10^{-4}$	0.003
100	20	0.1	$9.7 \cdot 10^{-9}$	125	$6 \cdot 10^{-4}$	$3 \cdot 10^{-4}$	0.004
100	20	0.01	$9.9 \cdot 10^{-9}$	1277	$6 \cdot 10^{-4}$	$3 \cdot 10^{-4}$	0.041
10	20	0.01	$10 \cdot 10^{-9}$	1 3127	$6 \cdot 10^{-4}$	$3 \cdot 10^{-4}$	0.395

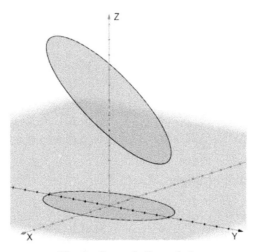

Fig. 1. Curves in Example 1.

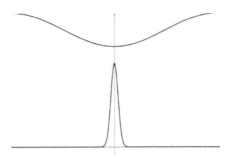

Fig. 2. The curves in Example 2.

Table 2. The solution of the problem in Example 2 for $\varepsilon = 10^{-6}$, $\delta = 0.1$, $u_0 = -0.4\pi$, $v_0 = -0.4\pi$, $\xi_0 = 0$, $\eta_0 = 0$ and different parameters of the method. [a] The value of the expression (12), [b] number of iterations, [c] distance to the true solution, [d] time of calculations, s.

p_1	p_2	δ	criteria[a]	N[b]	$\|r_1 - r_1^*\|$[c]	$\|r_2 - r_2^*\|$[c]	t[d]
0.5	10	0.1	$1 \cdot 10^{-6}$	3254	$8 \cdot 10^{-6}$	$8 \cdot 10^{-4}$	0.012
1	10	0.1	$0.9 \cdot 10^{-7}$	1920	$0.3 \cdot 10^{-5}$	$1.96 \cdot 10^{-5}$	0.009
2	10	0.01	$1 \cdot 10^{-7}$	8561	38.04	37.7	0.02
5	10	0.01	$1 \cdot 10^{-7}$	5205	100.67	100.54	0.015

4 Conclusion

In the paper we propose Charged Balls Method for solving the problem of finding the minimum distance between two smooth curves in $\mathbb{R}^3$. Convergence of the proposed method was proved and the estimate for the rate of convergence obtained. The effectiveness of the method is confirmed by the results of numerical experiments, from which it is also clear that due to the choice of the parameters p_1, p_2, δ we may significantly affect the convergence rate.

Generally speaking, the algorithm guarantees obtaining local solutions only. So the results depend on the choice of initial points and parameters. To get a global solution one can consider various multi-start (from various initial points) modifications of the algorithm or use appropriate values for p_1 and p_2 to provide "jumping" through local solutions (this is possible due to the inertness property of the balls). Example 2 shows that the right choice of parameters is critically important. Despite the fact that the stopping criteria is fulfilled, we get points far from the global solution of the problem in some cases. However, the same situation may happen if one try to minimize function $f(u, v) = \|r_1(u) - r_2(v)\|^2$, because the norm of the gradient of this function equals zero iff $\phi_1^2(u, v) + \phi_2^2(u, v) = 0$.

Acknowledgements. The research was supported by the Russian Science Foundation (RSF), project No 23-21-00027, https://rscf.ru/project/23-21-00027/.

References

1. Diener, I.: Trajectory methods in global optimization. In: Handbook of Global Optimization. Nonconvex Optimization and Its Applications, pp. 649–668. Kluwer Acad. Publ, Dordrecht (1995)
2. Ramm, A.G.: Dynamical systems method for solving operator equations. Elsevier, Amsterdam (2007)
3. Treccani, G.: A new strategy for global optimization. In: Dixon, L.C.W., Szegio, G.P. (eds.) Towards Global Optimisation. Kluwer Acad. Publ.,North-Holland, Amsterdam (1975)
4. Snyman, J.A., Fatti, L.P.: A multistart global minimization algorithm with dynamic search trajectories. J. Optim. Theory Appl. **54**(1), 121–141 (1987)

5. Incerti, S., Parisi, V., Zirilli, F.: A new method for solving nonlinear simultaneous equations. SIAM J. Numer. Anal. **16**(5), 779–789 (1979)
6. Snyman, J.A.: A new dynamic method for unconstrained minimization. Appl. Math. Model. **6**(6), 449–462 (1982)
7. Griewank, A.O.: Generalized descent for global optimization. J. Optim. Theory Appl. **34**(1), 11–39 (1981)
8. Aluffi-Pentini, F., Parisi, V., Zirilli, F.: A global optimization algorithm using stochastic differential equations. ACM Trans. Math. Software **14**(4), 345–365 (1988)
9. Antipin, A.S.: On finite convergence of processes to a sharp minimum and to a smooth minimum with a sharp derivative. Differen. Equ. **30**(11), 1703–1713 (1994)
10. Zhidkov, N., Shchdrin, B.: On the search of minimum of a function of several variables. Comput. Meth. Program. **10**, 203–210 (1978)
11. Abbasov, M.E.: Charged balls method for solving some computational geometry problems. Vestnik SPbSU. Math. Mechan. Astron. **4**(62), 359–369 (2017). https://doi.org/10.21638/11701/spbu01.2017.301
12. Abbasov, M.E.: Randomized heuristic algorithms for orthogonal projection of a point onto a set. Commun. Statist. – Simul. Comput. **48**(10), 2866–2876 (2019). https://doi.org/10.1080/03610918.2018.1469764
13. LaSalle, J.P., Lefschetz, S.: Stability by Liapunov's direct method. Academic Press, New York-London (1961)
14. Barbashin, E.A.: Introduction to the theory of stability. Nauka, Moscow (1967). [in Russian]
15. Polyak, B.T.: Optimization and asymptotic stability. Int. J. Control (2016). https://doi.org/10.1080/00207179.2016.1257154

On Numerical Solution of Block-Structured Discrete Systems

Jamila Asadova[✉]

Azerbaijan University of Architecture and Construction, Institute of Control Systems, Baku, Azerbaijan
asadova.jamilya64@gmail.com

Abstract. The paper presents a numerical methodology for solving high-dimensional discrete dynamical systems characterized by a block structure with boundary conditions that are non-separated between blocks. Direct application of classical methods of transferring boundary conditions is inefficient, since the block structure of the conditions, as for many other classes of problems, allows to speed up their solution considerably.

In this approach the decomposition of complex processes into simpler sub-processes with known mathematical models is used, and the decomposition of a complex process is performed in such a way that the intermediate states of sub-processes do not affect each other, i.e. they are independent, and the connection between sub-processes is implemented only through their input and output states. In this case, as a rule, the conditions determining the links between sub-processes are characterized by poorly occupied Jacobi matrices.

In the paper formulas that perform transferring of non-separated boundary conditions are presented and practical results demonstrating the efficacy of the proposed approach on a test problem are provided.

Keywords: Discrete Dynamic Models · Decomposition of Complex Objects · Technique for Transferring Boundary Conditions · Poorly Occupied Jacobi Matrices · Discrete Systems with Concentrated or Distributed Parameters

1 Introduction

The focus of the article centers on numerically solving large-dimensional discrete dynamical systems characterized by a block structure with boundary conditions undivided between blocks.

Such problems arise in the numerical study of dynamic processes in complex multi-branch systems of network structure, the functioning of each link of which is described by partial differential equations. In this case, it is impossible to measure the value of each process parameter separately at the junction points of separate sections, but the physical regularities are known, which the parameters must satisfy for the node as a whole, which leads to non-separability of the boundary conditions specification. In real life, such problems arise in mathematical modelling of electric power transfer processes in complex power transmission systems, unsteady motion of liquid and gas in pipeline systems of looped structure, processes of plasma oscillations, sorption and desorption of gases, calculation of oscillations of complex structures during earthquakes, etc.

If we take, for example, the problem of calculation of a branched electric circuit, then for the nodes and circuits of the electric circuit, equations according to the first and second Kirchhoff's laws regarding currents and voltage drops are made; in the calculation of complex network pipelines, which is also made with the use of electrical analogues of Kirchhoff's laws, for each node a balance of flow rates is made, and for each ring (circuit) - a balance of the pressures (Fig. 1).

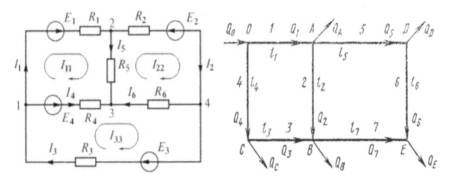

Fig. 1. a) Schematic of a branched electrical circuit; b) schematic of a complex circular pipeline.

Since even the numerical solution of a system of partial differential equations is a very complicated problem, a numerical approach based on the application of the straight-line method to the original problem involving a system of partial differential equations and its reduction to a boundary value problem for a discrete system of ordinary differential equations with unseparated boundary conditions is proposed. Further, taking into account the specificity of the problem under consideration, i.e. the fact that the functions involved in the equations of the initial system are independent of each other, a variant of the boundary condition transfer method is introduced, which consists in transferring the boundary conditions step by step, one by one for each function. Thus, a sequence of auxiliary problems with Cauchy initial conditions and their subsequent solution is constructed.

The direct application of boundary condition transfer methods is deemed inefficient due to the block-oriented nature of conditions. This block structure, common to various problem classes, enables a notable acceleration in the solution process.

Many mathematical models for discrete dynamic systems representing complex processes encountered in practical applications have been derived through the decomposition of intricate entities into simpler subcomponents with established mathematical models. This decomposition, as documented in various sources [1, 5–8, 10], involves breaking down complex objects with respect to spatial and/or temporal variables. It's important to highlight that this decomposition ensures the independence of intermediate states among subcomponents; in other words, these states are not mutually influenced. The interconnection between subcomponents occurs solely through input and output states [5–7]. Additionally, in real-world scenarios, subcomponents typically exhibit associations with a small, arbitrary number of other subcomponents. Consequently, the conditions governing the relationships between subcomponents are characterized by sparsely filled Jacobi matrices [1, 5].

In this work, a numerical method for resolving discrete systems characterized by a block structure was suggested, featuring weak and arbitrary interconnections between subsystems. The methodology is rooted in the concept of employing boundary condition transfer methods [1–3, 5]. Building upon the principles outlined in reference [3], which explored the numerical solution of a system governed by independent three-point discrete equations with non-separated boundary conditions, this paper extends those ideas to discrete systems of block structure with conditions not shared between blocks. The article not only derives the formulas for implementing condition transfers but also presents the outcomes of numerical experiments conducted in this context.

There have been explored solutions to the challenges that arise when employing finite difference approximation methods [9] for systems of equations involving partial derivatives of hyperbolic type.

2 Problem Formulation

Examine the set of equations characterizing a complex discrete process (the object), comprising discrete sub-processes that are mutually independent. Each sub-process is defined by a system of linear algebraic equations.

$$x^{\nu i+1} = A^{\nu i} x^{\nu i} + B^{\nu i}, \quad i = 1, ..., N_\nu - 1, \quad x^{\nu i} \in R^{n_\nu} \quad \nu = 1, ..., L. \tag{1}$$

Here, we have a n_ν-dimensional vector $x^{\nu i} = \left(x_1^{\nu i}, ..., x_{n_\nu}^{\nu i}\right)^*$ denoting the state of the $\nu-$ th process at the $i-$ th discrete time instant. The $A^{\nu i}$ and $B^{\nu i}$ are appropriately n_ν-dimensional square matrix and vector associated with this representation, $rang A^{\nu i} = n_\nu$, $i = 1, ..., N_\nu$; N_ν is the variable represents the duration of the $\nu-$ th process, $\nu = 1, ..., L$; and * denotes the transpose operation.

Additionally, we introduce the notation:

$$n = \sum_{\nu=1}^{L} n_\nu, \quad M = \sum_{\nu=1}^{L} n_\nu N_\nu \quad x^{\nu k} = \left(x_1^{\nu k}, ..., x_{n_\nu}^{\nu k}\right)^* \in R^{n_\nu},$$

$$x^1 = \left(x^{1,1}, x^{2,1}, ..., x^{L,1}\right)^* \in R^n, \quad x^N = \left(x^{1,N_1}, x^{2,N_2}, ..., x^{L,N_L}\right)^* \in R^n.$$

Here, M represents the overall dimension of the entire system comprising subsystems (1). The variables $x^1 \in R^n$ and $x^N \in R^n$ correspondingly denote the states of all subprocesses at the initial and final instants of time, with each subprocess having its individual final state. The sub-processes under consideration are interconnected through the initial and final states, and this interconnection is expressed as unshared boundary conditions. These conditions are written in the following form:

$$Gy^1 + Qy^N = R, \tag{2}$$

Here, the matrices $G = ((g^{ij})), Q = ((q^{ij}))$ are of $n \times n$ dimensions, and $R = (R^1, ..., R^n)^*$ is a predefined n-dimensional vector.

We suppose that the augmented matrix (G, Q) has a rank equal to n, i.e., $rang(G, Q) = n$, indicating that the system of Eqs. (1), (2) generally possesses a unique solution. The conditions outlined in (2) are expressed in vector form, which will be employed in subsequent discussions.

$$\sum_{s=1}^{L} g^{is*} x^{s1} + \sum_{s=1}^{L} q^{is*} x^{sN_s} = R^i, i = 1, ..., n, \quad (3)$$

where $g^{is} = (g_1^{is}, ..., g_{n_s}^{is})^*, q^{is} = (q_1^{is}, ..., q_{n_s}^{is})^*$.

Equations (1) and (2) serve as mathematical representations for numerous intricate objects and processes with lumped or distributed parameters, that operate discretely [1–4]. Simultaneously, the mathematical modeling of these objects and processes involves the utilization of the decomposition method by temporal and/or spatial variables. This method involves dividing the entire object into distinct parts, each functioning independently. The connection between these parts is established through their input and output states, as defined by conditions (2).

Boundary value problems, characterized by systems of differential equations involving ordinary or partial derivatives and solvable by grid methods, can also be transformed into problems expressed in the form of Eqs. (1) and (2) [1–6]. In this scenario, the systems of equations are composed of distinct independent subsystems, interconnected solely through initial and/or boundary conditions.

Specifically, when addressing the simulation of the oscillatory movement of intricate systems during seismic events or modeling fluid flow in intricate interconnected pipeline systems, the challenge can be redefined as a system depicted by Eqs. (1) and (2). The inherent wave motion process at each segment can be defined by a hyperbolic system consisting of two first-order partial differential equations [1–4]. Mathematical representations of numerous large real-world objects with intricate structures exhibit distinct characteristics: 1) an extensive array of sub-entities L; 2) a small size of vectors describing the state of each component of the object, i.e. of vectors $n_\nu, \nu = 1, ..., L$; 3) prolonged operational durations $N_\nu, \nu = 1, ..., L$; 4) feeble and random connections among the sub-processes, denoting weak and random matrix and vector fillings, represented as G, Q and $g^{is}, q^{is}, s = 1, ..., L$ respectively.

Attributes 1) and 3) for actual objects result in the algebraic system's order (represented by (1) and (3)), being equivalent to M, which may surpass several thousand or even tens of thousands. This makes it impractical to apply established numerical methods for solving algebraic equation systems. The characteristic 4) is leading to non-separated boundary conditions, necessitating the utilization of methods involving the transfer of these boundary conditions.

A specificity of feature 2) facilitates the derivation of relationships equivalent to (1), but in the opposite order of calculation.

$$x^{\nu,i} = \bar{A}^{\nu,i+1} x^{\nu,i+1} + \bar{B}^{\nu,i+1} \quad i = N_\nu - 1, ..., 1 \quad \nu = 1, ..., L, \quad (4)$$

$$\bar{A}^{s,i} = \left(A^{s,i+1}\right)^{-1} \quad \bar{B}^{s,i} = \left(A^{s,i+1}\right)^{-1} B^{s,i} \quad i = 1, ..., N_s - 1, s = 1, ..., L.$$

This study aims to formulate a proficient numerical approach to solve the system of discrete Eqs. (1) with non-separable boundary conditions (2) and (3), while considering the aforementioned features. The method utilizes the concept of transferring conditions and involves solving a sequence of specially constructed discrete Cauchy problems for individual subsystems within the overarching system (1).

3 Numerical Problem Solution

The devised method for solving the given problem relies on relocating boundary conditions (3) to either the left or right extremity. This implies replacing relations (2) or (3) by equivalent expressions in which the vector x^1 will be absent when the conditions are transferred to the right end:

$$\tilde{Q}x^N = \tilde{R} \tag{5}$$

or

$$\sum_{s=1}^{L} \tilde{q}^{is*} x^{sN_s} = \tilde{R}^i, i = 1, ..., n, \tag{6}$$

and the vector x^N will be absent when conditions are passed to the left end

$$\tilde{G}x^1 = \tilde{R} \tag{7}$$

or

$$\sum_{s=1}^{L} \tilde{g}^{is*} x^{s1} = \tilde{R}^i, i = 1, ..., n. \tag{8}$$

Upon relocating conditions to a single extremity, we will derive systems (5), (6) or (7), (8), which are systems of n algebraic equations with n unknowns: x^1 or x^N. By resolving these systems and determining x^1 or x^N, the overall problem solution is accomplished through straightforward computations utilizing explicit recursive formulas (Cauchy problems). This is carried out in relation to separate subsystems of the discrete Eqs. (1) (during leftward transference) or subsystems of (4) (when shifting conditions to the right).

The choice of transferring conditions (2) and (3) hinges on the extent of filling the matrices G and Q. Specifically, if matrix G is less filled than matrix, then conditions should be shifted to the right. Conversely, if matrix G is more densely populated than matrix Q, conditions ought to be moved to the left. This guideline will become evident when we give the following explanation of the transfer procedure. The transfer of conditions (2), or more precisely (3), will be executed independently for each i- th condition, $i = 1, ..., n$.

Hence, let's examine an arbitrary i- th condition in (3), which, after being transferred to the right side, will take the form of (6), wherein the $\tilde{q}^{is*}$, $\tilde{R}^i$ represent new coefficient values yet to be determined. The process of deriving the conditions in the form (6) will proceed in stages.

Suppose that not all vector coefficients $g^{ij}, j = 1, ..., L$, are zero; otherwise, there would be no necessity to shift the $i-$ th condition to the right, given that this condition solely involves the values x^N. Let the initial non-zero coefficient be g^{iv}, that is, $g^{iv} \neq 0_{n_v}$, $g^{ij} = 0_{n_j}, j < v$, $(0_{n_v} - n_v$-dimensional vector with all components equal to 0).

In this scenario, it can be stated that n_v- dimensional vectors $\alpha^k \in R^{n_v}$ and scalars $\beta^k, k = 1, ..., N_v$ facilitate the transfer of the $i-$ th condition (3) concerning the $v-$ th vector of unknowns x^{vk} to the right. This occurs when, for all vectors x^{vk} satisfying the $v-$ th subsystem (1), the following equalities are met:

$$\alpha^{k*}x^{vk} + \sum_{s=v+1}^{L} g^{is*}x^{s1} + \sum_{s=1}^{L} q^{is*}x^{sN_s} = \beta^k, k = 1, ..., N_v. \quad (9)$$

It is evident that, given the condition $k = 1$, the following equalities must be satisfied:

$$\alpha^1 = g^{iv}, \quad \beta^1 = R^i. \quad (10)$$

Vectors α^k and scalars $\beta^k, k = 1, ..., N_v$ that satisfy (9) and (10) will be referred to as transfer coefficients. By substituting the values of these transfer coefficients into (10) with $k = N_v$, we derive a new condition.

$$\sum_{s=v+1}^{L} g^{is*}x^{s1} + \sum_{s=1}^{L} \tilde{q}^{is*}x^{sN_s} = \tilde{R}^i,$$

In this, the following notation is introduced:

$$\tilde{q}^{iv} = q^{iv} + \alpha^{N_v}, \tilde{q}^{ij} = q^{ij}, j = 1, ..., L, j \neq v, \tilde{R}^i = \beta^{N_v}.$$

The transfer coefficients α^k, β^k, which fulfil the rightward transfer of condition (3), can be obtained by various methods. One of such methods is presented in the following theorem.

Theorem 1. Suppose that n_v- dimensional vectors α^k and scalars $\beta^k, k = 1, ..., N_v$ are specified through the following recurrence relationships (discrete Cauchy problems):

$$\alpha^{k+1} = \bar{A}^{vk}\alpha^k, \quad \alpha^1 = g^{iv} \quad k = 1, ..., N_v$$

$$\beta^{k+1} = \beta^k + \alpha^{k+1*}B^{vk}, \quad \beta^1 = R^i \quad k = 1, ..., N_v \quad (11)$$

Hence, α^k, β^k serve as the transfer coefficients, facilitating the transfer of the $i-$ th condition in (3) to the right with respect to the x^v- th solution of the $v-$ th subsystem (1).

Proof. In accordance with (10), the condition (9) is identical to the $i-$ th condition in (3). Let's assume that α^k, β^k and $\alpha^{k+1}, \beta^{k+1}$ at $k > 1$ satisfy the condition for transferring the $i-$ th condition with respect to the x^v- th solution of the $v-$ th subsystem (1).

As a result, this implies:

$$\alpha^{k*}x^{vk} + \left[\sum_{j=v+1}^{L} g^{ij*}x^{j1} + \sum_{j=1}^{L} q^{ij*}x^{jN_j}\right] = \beta^k, k = 1, ..., N_v, \quad (12)$$

$$\alpha^{k+1*}x^{\nu,k+1} + \left[\sum_{j=\nu+1}^{L} g^{ij*}x^{j1} + \sum_{j=1}^{L} q^{ij*}x^{jN_j}\right] = \beta^{k+1}. \tag{13}$$

We consider the $\nu-$ th subsystem of (1) in (13):

$$\alpha^{k+1*}\left[A^{\nu k}x^{\nu k} + B^{\nu k}\right] + \left[\sum_{j=\nu+1}^{L} g^{ij*}x^{j1} + \sum_{j=1}^{L} q^{ij*}x^{jN_j}\right] = \beta^{k+1}.$$

Upon subtracting Eq. (12) from this equation, and subsequently grouping the resulting expression, we get:

$$[\alpha^{k+1*}A^{\nu k} - \alpha^{k*}]x^{\nu k} + [-\beta^{k+1} + \beta^{k} + \alpha^{k+1*}B^{\nu k}] = 0.$$

Considering that this equation must be valid for all potential solutions of the $\nu-$ th subsystem (1), we demand that the expressions within square brackets involving coefficients $\alpha^k, \beta^k, \alpha^{k+1}, \beta^{k+1}$, be equal to zero.

Bearing in mind (4), we derive the essential relationships for transfer coefficients in the form (11). Following the process of substituting the values of the $\nu-$ th vector $x^{\nu 1}$ in the $i-$ th condition with the value $x^{\nu N_\nu}$ with a new coefficient $\tilde{q}^{\nu N_\nu}$, we derive a new condition that is equivalent to the previous one.

Under this condition, no value for $x^{\nu 1}$ exists. Subsequently, we move on to the next non-zero coefficient $g^{ij}, j > \nu$, until we reach the condition $g^{ij} = 0_{n_j}, j = 1, ..., L$. This indicates the successful transfer of the $i-$ th condition entirely to the right. Further the entire specified procedure is then repeated for the $(i+1)$-th condition. If $(i+1) > n$, then all conditions (3) have been successfully shifted to the right, resulting in conditions in the form of (5) or (6), which are equivalent to the initial conditions (3).

Conditions (5) and (6) constitute a system of n linear algebraic equations in relation to the $n-$ dimensional vector x^N. Solving this system results in determining the vector x^N, and subsequently, the sought-after solution $x = (x^1, ..., x^N)^*$ to the problem is obtained in accordance with (1).

In a manner analogous to the previously described process of passing conditions to the right, a successive passing of conditions to the left is performed to obtain conditions (7) or (8), which are equivalent to the initial conditions (3).

Let's assume that in the $i-$ th condition, among the vectors $q^{ij}, j = 1, ..., L$, the first non-zero vector is $q^{i\nu}$, that can be denoted as $q^{ij} = 0_{n_j}, j < \nu$, $q^{i\nu} \neq 0_{n_j}$.

We will assume that $n_\nu-$ dimensional vectors α^k and scalars $\beta^k, k = 1, ..., N_\nu$ implement the transfer of the $i-$ th condition (3) to the left concerning the vectors $x^{\nu k}$ being solutions of the $\nu-$ th subsystem (1) if the following equalities are satisfied:

$$\alpha^{k*}x^{\nu k} + \sum_{s=1}^{L} g^{is*}x^{s1} + \sum_{s=\nu+1}^{L} q^{is*}x^{sN_s} = \beta^k, \quad k = 1, ..., N_\nu, \tag{14}$$

$$\alpha^{N_\nu} = q^{i\nu}, \beta^{N_\nu} = R^i. \tag{15}$$

It is evident that (14) aligns with the $i-$ th condition (3) when $k = N_\nu$.

In the case when $\alpha^k, \beta^k, k = 1, ..., N_\nu$, are sweep coefficients, the equality (14) under $k = 1$ yields a new condition:

$$\sum_{s=1}^{L} \tilde{g}^{is*}x^{s1} + \sum_{s=\nu+1}^{L} q^{is*}x^{sN_s} = \tilde{R}^i,$$

that is identical to the i- th condition, wherein the notation has been introduced:

$$\tilde{g}^{iv} = g^{iv} + \alpha^1, \tilde{g}^{ij} = g^{ij}, j = 1, ..., N_v, j \neq v.$$

This condition deviates from the condition (3) in that its i- th component lacks the term with x^{vN_v}. Subsequently, this process is reiterated until there is at least one coefficient q^{is} that is not zero. Following that, the transfer is executed for the next $(i+1)$- th condition in the case $i + 1 \leq n$. The coefficients for leftward transfer, responsible for shifting the i- th condition to the left, can be obtained using the next theorem.

Theorem 2. Suppose that n_v- dimensional vectors α^k and scalars $\beta^k, k = 1, ..., N_v$ are specified through the following recurrent relationships (discrete Cauchy problems):

$$\alpha^k = A^{vk*}\alpha^{k+1}, \quad \alpha^{N_v} = q^{iN_v}, \quad k = N_v - 1, N_v - 2, ..., 1,$$

$$\beta^k = \beta^{k+1} - B^{vk*}\alpha^{k+1}, \quad \beta^{N_v} = R^i, \quad k = N_v - 1, N_v - 2, ..., 1. \quad (16)$$

Then α^k, β^k serve as the sweep coefficients for transferring the i- th condition in (3) to the right concerning the x^v- th solution of the v- th subsystem (1).

The proof is analogous to the previous proof of Theorem 1. The overall process of transforming all conditions (2), (3) into (5), (6) by transferring values to the left is analogous to the procedure outlined earlier for transferring conditions to the right. Completion of the transfer process results in a system of n algebraic equations with respect to the n- dimensional vector x^1. Subsequently, after solving this system, we proceed with a required solutions of the subsystem x^{vk} of the system (1) from left to right, $k = 1, ..., N_v, v = 1, ..., L$.

4 Numerical Experimental Results

Examine the subsequent system of discrete equations, comprising five subsystems. $(L = 5, n_v = 2, N_v = 201, v = 1, ..., 5)$:

$$x_1^{1,k+1} = x_1^{1,k} + 0{,}005x_2^{1,k}, \quad x_2^{1,k+1} = x_1^{1,k} + 1{,}01x_2^{1,k} - 0{,}01e^{0{,}005k} + 0{,}0099,$$
$$k = 1, ..., 200,$$

$$x_1^{2,k+1} = x_1^{2,k} + 0{,}005x_2^{2,k}, x_2^{2,k+1} = x_1^{2,k} + 0{,}085x_2^{2,k} - 0{,}0175e^{0{,}0025k} -$$
$$- 0{,}005\cos(0{,}005k) - 0.015\sin(0{,}005k) + 0{,}045,$$

$$x_1^{3,k+1} = x_1^{3,k} + 0{,}005x_2^{3,k},$$
$$x_2^{3,k+1} = x_1^{3,k} + 1{,}0025x_2^{3,k} - 0{,}0025,$$
$$x_1^{4,k+1} = x_1^{4,k} + 0{,}005x_2^{4,k}, x_2^{4,k+1} = x_1^{4,k} + 1{,}005x_2^{4,k} - 0{,}0025e^{0{,}0025k} -$$
$$- 0{,}25 \times 10^{-4}k + 0{,}005, \quad (17)$$

$$x_1^{5,k+1} = x_1^{5,k} + 0{,}005x_2^{5,k}, x_2^{5,k+1} = x_1^{5,k} + 1{,}005x_2^{5,k} - 0{,}0025e^{0{,}0025k} -$$
$$- 1{,}25 \times 10^{-7}k^2 + 0{,}5 \times 10^{-4}k$$

with the subsequent set of ten non-separable conditions, encompassing states at the initial and terminal instants:

$$x_1^{1,1} + x_1^{2,1} + x_1^{3,1} = 0, \tag{18}$$

$$x_2^{1,1} - x_2^{3,1} = 0, \tag{19}$$

$$x_2^{2,1} - x_2^{3,1} = 0, \tag{20}$$

$$x_1^{4,1} = 4, \tag{21}$$

$$x_1^{5,1} = -1, \tag{22}$$

$$x_1^{3,N_3} + x_1^{4,N_4} + x_1^{5,N_5} = 4\sqrt{e} - 1/4, \tag{23}$$

$$x_2^{3,N_3} - x_2^{5,N_5} = 0, \tag{24}$$

$$x_2^{4,N_4} - x_2^{5,N_5} = 0, \tag{25}$$

$$x_1^{1,N_1} = -1 + 3e, \tag{26}$$

$$x_1^{2,N_2} = 3 - 2\sqrt{e} + sin(1). \tag{27}$$

It can be convinced that the solution to the problem (17)–(27) with precision up to 10^{-8} is expressed by the following vectors with components defined for $k = 1, ..., 201$:

$x_1^{1,k} = 0{,}25 \times 10^{-4}(k-1)^2 + 2e^{0{,}005(k-1)} - 2$, $x_2^{1,k} = 0{,}01(k-1) + 2e^{0{,}005(k-1)}$,
$x_1^{2,k} = 0{,}015(k-1) - 2e^{0{,}0025(k-1)} + \cos(0{,}005(k-1))$, $x_2^{2,k} = 3 - e^{0{,}0025(k-1)}$
$- \sin(0{,}005(k-1))$, $x_1^{3,k} = 0{,}005(k-1) + 2e^{0{,}0025(k-1)} - 1$, $x_2^{3,k} = 1 + e^{0{,}0025(k-1)}$,
$x_1^{4,k} = 0{,}125 \times 10^{-4}(k-1)^2 + 2e^{0{,}0025(k-1)} + 2$, $x_2^{4,k} = 0{,}005(k-1) + e^{0{,}0025(k-1)}$,
$x_1^{5,k} = 0{,}125 \times 10^{-6}(k-1)^3 + 2e^{0{,}0025(k-1)} - 3$, $x_2^{5,k} = 0{,}25 \times 10^{-4}(k-1)^2$
$+ e^{0{,}0025(k-1)}$.

.

It's worth mentioning that the system of Eqs. (17) and conditions (18)–(27) is derived through simulating the finite-difference approximation of a system of differential equations of hyperbolic type with partial derivatives, specifically describing fluid movement within a complex looped network structure. Another illustrative example could be the oscillatory motion of objects with intricate structures during earthquakes.

System (17) determines the traffic mode at the first time layer at discrete points of the network section. All sections have equal length and are divided into 200 parts.

Conditions (18), (23) determine the law of material balance at the nodal points of the network, conditions (19), (20), (24), (25) - characterize the conditions of flow continuity, conditions (21), (22), (26), (27) determine the modes of operation of external sources.

Considering an equal count of non-zero coefficients under $x^{\nu 1}$ and $x^{\nu N_\nu}$ in the conditions (18)–(27), the direction of transferring conditions becomes irrelevant.

Following the transfer of conditions (18)–(22) to the right, the conditions were transformed into the shape of an algebraic system (5), featuring a ten-dimensional matrix $\tilde{Q}$ as outlined in Table 1, and a right-side vector $\tilde{R}$ structured as:

$$\tilde{R}^* = [0.00353 \; 9.72566 \; 0.0 \; 0.0 \; 0.005009 \; -.07064 \; 4.43656 \; 0.24286 \; 0.03244 \; -.00821]$$

By employing the Gaussian method and selecting the main element, the solution to the resulting system of equations yielded the vector:

$$x^N = (4.39986 \; 4.43656 \; 0.24032 \; 0.24286 \; 3.27706 \; 3.29026 \; 5.78862 \; 5.80182 \; 0.62037 \; 0.63357)^*.$$

Utilizing this vector, recursive computations were executed to determine $x^{\nu k}$, $k = \overline{201,1}$, $\nu = \overline{1,5}$, from the subsystems of system (17). The precision of the obtained results did not surpass the threshold.
$$\max_{k}\max_{\nu}\max_{s} |\Delta x_s^{\nu k}| \leq 10^{-6}.$$

Table 1. Elements of matrix $\tilde{Q}$

j / i	1	2	3	4	5	6	7	8	9	10
1	-0.5601	0.5638	-8.1103	8.1040	-1	0.9836	0	0	0	0
2	0.2331	-0.2331	0	0	-1	1	0	0	0	0
3	0	0	23.1173	-23.1173	-1	1	0	0	0	0
4	0	0	0	0	0	0	1	-0.9821	0	0
5	0	0	0	0	0	0	0	0	1	-0.9821
6	0	0	0	0	0	1	0	1	0	1
7	0	0	0	0	-1	1	0	0	1	-1
8	0	0	0	0	0	0	-1	1	1	-1
9	0	1	0	0	0	0	0	0	0	0
10	0	0	0	1	0	0	0	0	0	0

5 Conclusion

This study explores the numerical solution of large-dimensional discrete equation systems characterized by a block structure and featuring "weak" and arbitrary connections between subsystems. Such systems necessitate repeated solving, especially in the optimization of parameters for complex structures or the discretization of optimal control problems of processes represented by equations with ordinary and partial derivatives. The proposed schemes and corresponding formulas are grounded in the concept of carrying over boundary conditions, considering the features of the system's Jacobian and the "poor" occupying of the Jacobian matrix related to connection conditions between subsystems.

The outcomes of numerical experiments derived from solving a test problem encountered in hydraulic calculations within intricate network structures, employing the implicit scheme of the grid method, are provided.

References

1. Aida-zade, K.R.: Investigation of nonlinear optimization problems of network structure. Autom. Remote Control **2**, 3–14 (1990). (in Russian)
2. Aida-zade, K.R., Abdullaev, V.M.: On the solution of boundary value problems with non-separated multipoint and integral conditions. Differ. Equ. **49**(9), 1114–1125 (2013)
3. Asadova, J.A.: Numerical solution of a system of independent three point discrete equations with non-separated boundary conditions. Proc. IAM **4**, N1, 58–69 (2015)
4. Aida-zade, K.R., Asadova, D.A.: The study of transients in pipelines. Autom. Remote. Control. **12**, 156–172 (2011)
5. JuergenGeiser: Decomposition methods for differential equations: theory and applications, p. 304. CRC Press (2010)
6. Qingying, Q., Bing, L., Peien, F., Gao, Y.: Decomposition method of complex optimization model based on global sensitivity analysis. Chin. J. Mech. Eng. **27**(4), 722–729 (2014)
7. Zbrishchak, S.G., Zvyagin, L.S.: Methods of system composition and decomposition as a tool for efficient processing of large amounts of experimental data. J. Phys.: Conf. Ser. **1703**, 012015 (2020)
8. Wang, W., Jiang, M.: Generalized decomposition method for complex systems. Annual Symposium Reliability and Maintainability, - RAMS, pp. 12–17, Los Angeles, CA, USA (2004) https://doi.org/10.1109/RAMS.2004.1285416
9. Samarskii, A.A., Nashed, Z., Taft, E.: The Theory of Difference Schemes, Edition 1, Taylor & Francis, pp. 788 (2001)
10. Dzurkov, V.I.: The decomposition in problems of large dimensionality, p. 352. Nauka, Moscow (1981). (in Russian)

Detection of Leaks in the Water Distribution Network

Kamil Aida-zade[1,2] and Yegana Ashrafova[1,3(✉)]

[1] Institute of Control Systems, Baku, Azerbaijan
ashrafova.yegana@gmail.com
[2] Azerbaijan University of Architecture and Construction, Baku, Azerbaijan
[3] Baku State University, Baku, Azerbaijan

Abstract. Numerical methods are employed to address an inverse problem associated with a water distribution network characterized by a complex loopback structure. The problem is to ascertain the locations and magnitudes of leaks based on measurements of unsteady flow characteristics at certain points in the pipeline. Key aspects of the problem include the involvement of impulse functions within a system of hyperbolic differential equations, the lack of traditional initial conditions, and the specification of nonseparated boundary conditions between states at the endpoints of adjacent pipeline segments. The problem is transformed into a parametric optimal control problem, devoid of initial conditions but featuring nonseparated boundary conditions. The latter problem is tackled using first-order optimization methods. The paper presents the outcomes of numerical experiments. Notably, this research distinguishes itself from others by addressing the inverse problem of determining leak locations and magnitudes within an unsteady flow scenario in a water distribution network with a complex (loopback) structure, as opposed to studies focusing on steady flow or transient flow in simpler pipeline configurations.

Keywords: water distribution network · locations of water leaks · amounts of water leaks · an inverse problem · impulse functions · hyperbolic differential equations · nonseparated boundary conditions

1 Introduction

This study focuses on the inverse problem, specifically addressing the determination of both the power and, in contrast to numerous other investigations, the precise locations of point sources (external and internal) influencing the underlying process or object. The process itself is characterized by a system of numerous first-order hyperbolic-type differential equations, interconnected solely through boundary conditions. Notably, in this particular problem, the extended duration of the process renders precise information about the initial conditions unavailable. Instead, there exists a multitude of potential values for the initial states. To resolve the inverse problem, additional observations of the process state at specific points are considered.

© The Author(s), under exclusive license to Springer Nature Switzerland AG 2025
G. Mammadova et al. (Eds.): ITTA 2024, CCIS 2226, pp. 336–347, 2025.
https://doi.org/10.1007/978-3-031-73420-5_28

Instances of such problems are readily encountered in the realms of ecology, geophysics, and underground hydrodynamics. This study extensively explores this matter, focusing specifically on the problem of identifying both the volumes and precise locations of water leaks in the transportation process through a pipeline characterized by a complex looped structure. It is worth noting that, to address the problem of monitoring the condition of pipeline sections, much attention has been devoted to the development of methods and technical devices for detecting leak locations [1–9]. For underground and underwater segments of pipeline networks, where visual inspection and the use of such technical devices pose challenges, mathematical methods can be employed to determine leak locations based on monitoring the movement modes of the substance at specific points in the pipeline network. Currently, there is increased interest in such approaches, given the growing level of development of numerical methods and computational and measuring technology [1–11].

This research diverges from many others, which predominantly focused on addressing the problem of identifying the location and volume of leaks in either a complex-structured pipeline under steady-state flow conditions or during unsteady flow on a pipeline consisting of a singular linear segment [10, 11]. Here, the study delves into numerical solutions for the inverse problem of determining both the locations and volumes of leaks during unsteady water flow in a pipeline network characterized by a complex looped structure. The problem is characterized by a system comprising numerous subsystems, each involving two hyperbolic-type partial differential equations with impulse inputs at points where leaks may occur in the pipeline network segments, equal to the number of segments in the pipeline network.

Another characteristic of the considered problem formulation lies in the assumption that due to the extended duration of the transportation process, precise information about the initial state of the modes at the start of monitoring is unavailable, and promptly measuring the modes at all points in the pipeline network is not feasible. Instead, information is available about a multitude of possible mode values across all pipeline segments at some initial moment in time, and measurements of the modes at specific points in the pipeline network have been taken from that moment onward.

Another distinctive aspect of the problem is the particularity of the boundary conditions. These conditions are characterized as integral relationships between the modes (states) at the ends of adjacent segments of the pipeline network, established through analogs of Kirchhoff's first law and the continuity of movement.

The specified formulation of the inverse problem, utilizing the variational approach, falls within the category of optimal control problems for objects with distributed parameters [12–14]. To address this, the study proposes the application of numerical methods involving first-order optimal control, resulting in the development of formulas for the components of the gradient of the minimized objective functional concerning the identifiable parameters.

2 Problem Statement

Within the general formulation of the inverse problem, let's consider the issue of determining the locations of leaks and the volume of water during its transportation through a pipeline network with a complex structure.

We examine a pipeline network comprising m segments (blocks) that are interconnected at their ends in a random manner. The structure of this complex network is conveniently depicted as a directed graph.

The collection of all vertices in the graph is denoted as V, and the set of links (i,j) with a length represented as $l^{i,j}$ and starting at vertex $i \in V$ and ending at vertex $j \in V$ is expressed as $E = \{(i,j) : i, j \in V\}$. Here, $|V| = N$ indicates the number of elements in the set V, and $|E| = K$, |E| = K represents the number of links in set E.

Let the sets of links $E_i^+ = \{(j,i) : j \in V_i^+\}$, $E_i^- = \{(i,j) : j \in V_i^-\}$ — respectively entering and leaving the i-th vertex, V_i^+ and V_i^- — the sets of vertices adjacent to the i-th vertex, which correspond to the endpoints and starting points of links from the set E_i, $E_i = E_i^+ \cup E_i^-$, $V_i = V_i^+ \cup V_i^-$. Let us denote

$$\left|E_i^+\right| = |V_i^+| = \bar{n}_i, \left|E_i^-\right| = |V_i^-| = \underline{n}_i, \bar{n}_i + \underline{n}_i = n_i, i \in V.$$

It's evident that

$$\sum_{i \in V} \underline{n}_i + \sum_{i \in V} \bar{n}_i = K, \sum_{i \in V} n_i = 2K.$$

In practical applications, typically, the condition $n_i \ll N$, where $i \in V$, holds true. This means that the number of vertices adjacent to any given vertex is significantly less than the overall total number of vertices.

Each link in the graph is associated with an independent subobject (block). Let the state of each links $(i,j) \in E$, with $j \in V_i^-$, $i \in V\}$ be described by a system of two partial differential equations of hyperbolic type.

We will assume that the flow regime of the droplet liquid in the network is isothermal, unsteady, and laminar [15, 16].

The state of the liquid movement process along each arc (link, linear segment) is described by the following system of hyperbolic-type differential equations [10, 13, 15, 16]:

$$-\frac{\partial p^{ij}(x,t)}{\partial x} = \frac{\partial q^{ij}(x,t)}{\partial t} + 2a^{ij}q^{ij}(x,t), x \in \left(0, l^{ij}\right), (i,j) \in E,$$

$$-\frac{\partial p^{ij}(x,t)}{\partial t} = c^2 \frac{\partial q^{ij}(x,t)}{\partial x} + c^2 v^{ij}(t)\delta\left(x - \xi^{ij}\right), x \in (0, l^{ij}), (i,j) \in E$$

$$v^{ij}(t) = 0, x \in (0, l^{ij}), (i,j) \in E \setminus E^{loss} \tag{1}$$

Here: $t \in (t_0, T]$; $q^{ij} = \frac{\rho}{S^{ij}} q^{ij}(x,t)$; $p^{ij}(x,t), q^{ij}(x,t), (i,j) \in E, j \in V_i^-, i \in V$ — pressure and flow rate of water at time, at the point $x \in (0, l^{ij})$ of the (i,j)-th segment of the pipeline network; c – is the speed of sound in a water; the area of the internal cross-section of the (i,j)-th segment of the pipeline is denoted as S^{ij}. The coefficient of resistance on the (i,j)-th segment, denoted as a^{ij} can be considered as $2a^{ij} = \frac{32\gamma}{(d^{ij})^2} = const$, assuming laminar flow where the coefficient of kinematic viscosity γ — is considered independent of pressure. The values of $q^{ij}(x,t)$ can be either positive or negative. A positive $q^{ij}(x,t)$ signifies that the actual flow on the (i,j)-th segment is directed from vertex i to vertex, while a negative value indicates that the flow is from vertex j to vertex i. Clearly, each segment acts as an inflow for one vertex and an outflow for another vertex. Considering that $l^{ji} = l^{ij}$, the following relationships hold for $j \in V_i^-, i \in V$:

$$q^{ij}(x,t) = -q^{ji}\left(l^{ij} - x, t\right), p^{ij}(x,t) = p^{ji}\left(l^{ij} - x, t\right), x \in (0, l^{ij})$$

This implies that the flow rate from j to i is equal in magnitude but opposite in direction to the flow rate from i to j.

The system (1) comprises of $2M$ equations, encompassing M subsystems, each consisting of two equations corresponding to all segments of the network. Specifically, each segment aligns with only one subsystem. Considering that each segment serves as an outflow for some vertex of the network, the fluid movement process in the network can be expressed in terms of outflow segments instead of (1), resulting in a similar system with indices taking values only for outflow segments. Alternatively, the system can be expressed in a mixed form, considering both inflow and outflow segments.

For the system (1), it is necessary to specify $2M$ boundary conditions. Material balance conditions are applied for internal vertices of the network.

$$\sum_{j \in V_i^+} q^{ij}(l^{ij}, t) - \sum_{j \in V_i^-} q^{ij}(0, t) = \tilde{q}^i(t), i \in V^{int}, t \in [t_0, T], \quad (2)$$

and continuity of flow:

$$p^{si}\left(l^{si}, t\right) = p^{ik}(0, t), s \in V_i^+, k \in V_i^-, i \in V^{int}, t \in [t_0, T] \quad (3)$$

here $\tilde{q}^i(t)$ – represents the specified external inflow ($\tilde{q}^i(t) > 0$) or outflow ($\tilde{q}^i(t) < 0$) for the i-th internal vertex.

The relationships (2) include N^{int} conditions. In relationships (3), for each i-th internal vertex from the set V^{int}, the number of independent conditions is $(N_i - 1)$, where each internal segment of the network is linked to two boundary conditions (3). For all N_f segments, where one end belongs to V^f, the overall number of conditions is N_f. Consequently, the total number of boundary conditions in (2), (3) is

$$N^{int} + \sum_{i \in V^{int}} (N_i - 1) = 2M - N^f$$

In each vertex that is not an internal vertex of the network, i.e., from the set V^f, we will assign a pressure value (the set of such vertices will be denoted as $V_p^f \subset V^f$) or a flow value (the set $V_q^f \subset V^f$). Then, the following conditions will be added to the N^f conditions (2), (3):

$$\begin{cases} p^i(t) = p^{is}(0, t) = \tilde{p}^i(t), & s \in V_i^+, \text{ if } V_i^- = \emptyset, \\ p^i(t) = p^{si}(l^{si}, t) = \tilde{p}^i(t), & s \in V_i^-, \text{ if } V_i^+ = \emptyset, \end{cases} i \in V_p^f, t \in [t_0, T], \quad (4)$$

$$\begin{cases} q^i(t) = q^{is}(0, t) = \tilde{q}^i(t), & s \in V_i^+, \text{ if } V_i^- = \emptyset, \\ q^i(t) = q^{si}(l^{si}, t) = \tilde{q}^i(t), & s \in V_i^-, \text{ if } V_i^+ = \emptyset, \end{cases} i \in V_q^f, t \in [t_0, T], \quad (5)$$

Moreover, it is important to ensure that the conditions are satisfied: $V^f = V_q^f \cup V_p^f$, $V_q^f \cap V_p^f = \varnothing$, $V_p^f \neq \varnothing$. Here $\tilde{q}^i(t)$, $\tilde{p}^i(t)$, $i \in V_q^f$, $j \in V_p^f$ – are the specified functions that define the operating modes of the sources.

We will assume that the initial conditions for the process (1) are not precisely defined at the initial moment in time t_0. However, the initial conditions belonging to the set U^0 can be parameterized as follows:

$$\varphi^{ij}(x, \gamma) = (q^{ij}(x, t_0; \gamma), p^{ij}(x, t_0; \gamma)), \ (i,j) \in E, \ \gamma \in \Gamma \in R^\nu \qquad (6)$$

with the corresponding probability density functions $\rho_\Gamma(\gamma)$. The set of possible initial conditions can be defined by a finite number N_{U^0} of specified functions or parameters that determine them.

Therefore, to compute the liquid flow regimes $(p^{ij}(x, t), q^{ij}(x, t))$, $(i, j) \in E$, across all sections of the water supply network, considering the locations and volumes of leaks, it is imperative to solve the system of $2M$ differential equations (1). This entails determining specific $2M$ initial conditions at $t = t_0$ that satisfy (6), along with $2M$ boundary conditions (2)–(3). The solution obtained for the problem at $t > \tau$ will correspond to the actual current regimes [17].

Let's assume that the volumes $v^{ij}(t)$ and the leakage locations ξ^{ij} on pre-defined N^{loss} sections of the set $E^{loss} \subset E$ are unknown and must be determined utilizing additional information about the flow regimes.

To identify the unknown locations and volumes of leaks $(\xi^{ij}, v^{ij}(t))$, $(i, j) \in E^{loss}$, we posit that there are observations of pressure and/or flow rates at certain points on various sections of the network. The number of these points exceeds $2N^{loss}$-the number of identifiable parameters. It is reasonable to assume that the sought-after leak locations do not coincide with the points of measurements. For simplicity and to avoid introducing new indices and notations, let's assume that the points of additional measurements are once again the vertices of the inlet and outlet pipes, i.e., from V^f.

Furthermore, if pressure measurements are employed for the vertices of the set V_p^f in boundary conditions (4), (5), and flow rate measurements are utilized for the vertices V_q^f, then the additional information will entail pressure measurement results at certain vertices of the subset V_{qp}^f of the set V_q^f, i.e., $V_{qp}^f \subset V_q^f$, and flow rate measurements at vertices of some subset V_{pq}^f of the set V_p^f, i.e., $V_{pq}^f \subset V_p^f$:

$$\begin{cases} p^i(t) = p^{is}(0, t) = p^i_{mes}(t), & s \in V_i^+, \text{ if } V_i^- = \varnothing, \\ p^j(t) = p^{sj}(l^{sj}, t) = p^j_{mes}(t), & s \in V_j^-, \text{ if } V_j^+ = \varnothing, \end{cases} \quad i \in V_{qp}^f \subset V_q^f, \qquad (7)$$

$$\begin{cases} q^i(t) = q^{is}(0, t) = q^i_{mes}(t), & s \in V_i^+, \text{ if } V_i^- = \varnothing, \\ q^j(t) = q^{sj}(l^{sj}, t) = q^j_{mes}(t), & s \in V_j^-, \text{ if } V_j^+ = \varnothing, \end{cases} \quad j \in V_{pq}^f \subset V_p^f. \qquad (8)$$

The inverse problem under consideration involves finding the leak locations $\xi = (\xi^{(k_{i_1}, k_{j_1})}, ..., \xi^{(k_{i_z}, k_{j_z})})$, $\xi^{(k_{i_s}, k_{j_s})} \in (0; l^{k_{i_s}, k_{j_s}})$ and the corresponding values of water leaks $v(t) = (v^{k_{i_1}, k_{j_1}}(t), ..., v^{k_{i_z}, k_{j_z}}(t))$, $(k_{i_s}, k_{j_s}) \subset E^{loss}$, for given $s = 1, ..., L$ $t \in [t_0, T]$, using the provided mathematical model and additional observed information (7), (8).

We will consider these conditions as additional information used to construct a residual functional dependent on initial conditions from the set Γ:

$$\Phi(\xi, v) = \int_{\Gamma} \sum_{i \in \tilde{V}_q^f} \int_{\tau}^{T} \left([q^i(t; \xi, v(t), \gamma) - q_{mes}^i(t)]^2 + \Re(\xi, v) \right) dt \rho_{\Gamma}(\gamma) d\gamma \to \min, \tag{9}$$

where $q^i(t; \xi, v(t), \gamma)$, $i \in \tilde{V}_q^f$ are the computed flow values at the observed points resulting from solving the direct problem under any possible initial conditions $\varphi^{ij}(x, \gamma)$, $\gamma \in \Gamma$ and specified allowable locations and volumes of leaks $(\xi, v(t))$; $[\tau, T]$ is the time interval for monitoring the process, the regimes of which no longer depend on the initial conditions.

In the case where the set of possible initial conditions is finite, given by (4) with a uniform distribution, then the functional (9) will take the following form:

$$\Phi(\xi, v) = \frac{1}{N_{U^0}} \sum_{j=1}^{N_{U^0}} \left(\sum_{i \in \tilde{V}_q^f} \int_{\tau}^{T} [q^i(t; \xi, v(t), \gamma^j) - q_{mes}^i(t)]^2 dt + \Re(\xi, v) \right) \to \min, \tag{10}$$

3 Formulas to Solve the Problem

To solve the formulated optimization problem numerically we'll use both effective numerical methods of first-order optimization, so pay special attention to solving the boundary value problem with unseparated conditions. So we'll derive formulas for the components of the gradient of the minimized objective functional with respect to the identifiable parameters.

For the components of the gradient of the functional (9) we obtain the following formulas:

$$grad_{v^{ij}} \Phi(\xi, v) = \int_{\Gamma} \left\{ c^2 \frac{\rho}{S^{ij}} \psi^{ij}(\xi^{ij}, t; \gamma) + 2\varepsilon(v^{ij}(t) - \tilde{v}^{ij}(t)) \right\} \rho_{\Gamma}(\gamma) d\gamma, \, t \in [t_0, T],$$

$$grad_{\xi^{ij}} \Phi(\xi, v(t)) = \int_{\Gamma} \left\{ c^2 \frac{\rho}{S^{ij}} \int_{t_0}^{T} v^{ij}(t) (\psi^{ij}(x, t; \gamma))'_x |_{x=\xi^{ij}} dt + 2\varepsilon_2 (\xi^{ij} - \hat{\xi}^{ij}) \right\} \rho_{\Gamma}(\gamma) d\gamma$$

Here, the functions $\varphi^{ij}(x, t) = \varphi^{ij}(x, t; \hat{q}, \hat{p})$ and $\psi^{ij}(x, t) = \psi^{ij}(x, t; \hat{q}, \hat{p})$, $(i, j) \in E$ are the solutions to the following adjoint problem:

$$\begin{cases} -\frac{\partial \varphi^{ij}(x,t)}{\partial x} = \frac{\partial \psi^{ij}(x,t)}{\partial t}, \\ -\frac{\partial \varphi^{ij}(x,t)}{\partial t} = c^2 \frac{\partial \psi^{ij}(x,t)}{\partial x} - 2a^{ij} \varphi^{ij}(x, t), \end{cases} x \in (0, l^{ij}), \, j \in V_i^+, \, i \in V, \, t \in [t_0, T],$$

$$\varphi^{ij}(x, T) = 0, \quad \psi^{ij}(x, T) = 0, \quad x \in [0, l^{ij}], j \in V_i^+, i \in V,$$

$$\psi(l^i, t) = \begin{cases} -2S^{ij}[q^i(t; v, \hat{q}, \hat{p}) - q^i_{mes}(t)]/(\rho c^2), & j \in V_i^-, \ i \in \tilde{V}_q^f, \ t \in [\tau, T], \\ 0, & i \in V^f/\tilde{V}_q^f \ \text{for} \ t \in [t_0, T] \quad \text{and} \ i \in V^f \ \text{for} \ t \in [t_0, \tau], \end{cases}$$

$$\psi(0, t) = \begin{cases} -2S^{ij}[q^i(t; u, \hat{q}, \hat{p}) - q^i_{mes}(t)]/(\rho c^2), & j \in V_i^+, \ i \in \tilde{V}_q^f, \ t \in [\tau, T], \\ 0, & i \in V^f/\tilde{V}_q^f \ \text{for} \ t \in [t_0, T] \quad \text{and} \ i \in V^f \ \text{for} \ t \in [t_0, \tau], \end{cases}$$

$$\sum_{j \in V_i^+} \varphi^{ij}(l^{ij}, t) - \sum_{j \in V_i^-} \varphi^{ji}(0, t) = 0, \quad i \in V^{\text{int}}, t \in [t_0, T],$$

$$\frac{1}{S^{k_ik}} \psi^{k_ik}(l^{k_ik}, t) = \frac{1}{S^{kk_j}} \psi^{kk_j}(0, t), k_i \in V_k^+, k_j \in V_k^-, k \in V^{\text{int}}, t \in [t_0, T],$$

$$\psi(l^i, t) = \psi(0, t) = 0, i \in V^f, t \in [t_0, \tau].$$

To identify the unknown locations of leaks, it is necessary to address the aforementioned problem for various sets of areas suspected of having leaks. The solution to the problem corresponds to the problem in which the functional (9) attains its smallest value. Enumerating through such possibilities in networks of intricate structures becomes a labor-intensive task, especially in cases involving a substantial number of leaks or a large number of segments. In practical applications, segments suspected of potential leaks are typically assigned with reasonable plausibility by experts (specialists) relying on their experience, information regarding segment operation duration and modes, and other relevant considerations.

It is evident that if, upon solving problem (1)–(9), a sufficiently small value of the functional is obtained such that $|v^{ks}(t)| \leq \varepsilon, t \in [\tau, T]$, ε is a sufficiently small number, it indicates the absence of raw material leakage in that specific segment, as well as in the entire pipeline. Typically, the consequence of the absence of a leak in the (k, s)-th segment is a notably large minimum value of the functional or, in cases where the obtained value ξ^{ks} satisfies $0 \leq \xi^{ks} \leq \varepsilon$ or $l^{ks} - \varepsilon \leq \xi^{ks} \leq l^{ks}$.

4 Findings from Computational Experiments

We examine a specifically designed test problem for a liquid distribution network comprising 5 nodes, as illustrated in Fig. 1. Here $N = 6, M = 5, V^f = \{1, 4, 3, 6\}, N^f = 4, N^{\text{int}} = 2$. In this configuration, there are no external inflows or outflows within the network.

We assume the observation of the liquid transportation process (pumping plant operation mode at the ends of the sections) over a duration of 30 min, considering a kinematic viscosity of $v = 1.5 \cdot 10^{-4}$ (m^2/s) and density $\rho = 920$ (kg/m^3) ($2a = 0.017$ specific to the case under consideration, with the sound velocity in the liquid being 1200 (m/s)). The pipeline sections have a diameter of 530 mm and vary in lengths:

$l^{(1,2)} = 100$ (km), $l^{(5,2)} = 30$ (km), $l^{(3,2)} = 70$ (km), $l^{(5,4)} = 100$ (km), $l^{(5,6)} = 60$ (km).

Assume there was a state in the pipes at the initial time instance $t = 0$ with the following values of pressure and flow rate in the pipes:

$\hat{p}^{1,2}(x) = 2300000 - 5.8955x$ (Pa), $\hat{p}^{5,2}(x) = 1745669 - 1.17393x$ (Pa), $\hat{p}^{3,2}(x) = 1827844 - 1.677043x$ (Pa), $\hat{p}^{5,4}(x) = 1827844 - 2.35786x$ (Pa), $\hat{p}^{5,6}(x) = 1827844 - 0.94415x$ (Pa).

$\hat{q}^{1,2}(x) = 300 \ (m^3/hour), \ \hat{q}^{5,2}(x) = 200 \ (m^3/hour), \ \hat{q}^{3,2}(x) = 100 \ (m^3/hour),$
$\hat{q}^{5,4}(x) = 120 \ (m^3/hour), \ \hat{q}^{5,6}(x) = 80 \ (m^3/hour).$

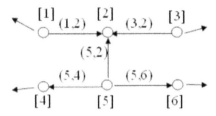

Fig. 1. The scheme of liquid distribution network with 5 nodes

Suppose the flow rates at the ends of this pipeline section are defined by the functions:

$\tilde{p}_0^1(t) = 2000000 + 300000 \ e^{-0.0003t}$ (Pa), $\tilde{p}_0^3(t) = 1900000 - 72156 \ e^{-0.0004t} =$ (Pa),
$\tilde{p}_l^4(t) = 1800000 - 66571 \ e^{-0.0007t}$ (Pa), $\tilde{p}_0^6(t) = 1600000 + 86372 \ e^{-0.0002t}$ (Pa).

Assuming that the point of leakage is located at point $\xi = 30$ (km) of the first section of the pipeline network, and the rate of leakage is determined by the function $v^{loss}(t) = 50 - 10e^{-0.0003t} \ (m^3/h)$, we numerically solved the boundary-value problem (1)–(4). This allowed us to determine the numerical values of pressure at the ends of section $p^i(t)$, $i \in V_p^f$. Subsequently, by utilizing a set of uniformly distributed random numbers, these values were altered within a range of 2% (simulating measurement errors) and employed as the observed states of the process. Notably, the point and rate of leakage $\xi, v^{loss}(t)$ were not considered ("forgotten") in this scenario.

To ascertain $\xi, v^{loss}(t)$ we employed the method of the projection of conjugate gradients. The numerical solution of the boundary-value problem (1)–(4) was achieved utilizing the sweep method scheme introduced in [18], implemented on grids with steps $h_x = 10$ m and $h_t = 100$ (sec).

Table 1 displays the results obtained from minimizing functional (9) with varying initial values of the identified parameters $(\xi, v^{loss}(t))^0$. Additionally, it provides the necessary number of iterations (one-dimensional minimizations) using the method of conjugate gradients projection. The presented results stem from solving the problem under the circumstances where observed flow rate values at the network ends incorporate measurement errors.

In this experiment, to generate observations for the inverse problem, we introduce noise terms $\eta \chi_i q^m(t_i)$ to the values $q^m(t_i) = q^m(t_i; \xi, v)$, $t_i = ih_t$, $m \in \tilde{V}_q^f$, $i = 1, \ldots, N_t$ obtained by solving the direct problem. Here, χ_i is a random variable uniformly distributed on a segment $[-1, 1]$. The variable η assumes values equal to 0.0, 0.005, and 0.01, corresponding to noise levels of 0% (without noise), 0.5%, and 1% from the measured value when assessing flow rates in vertices $\tilde{V}_q^f$, respectively.

In Table 1, the notations are as follows: $\tilde{\xi}^{(1,2)}$ – the obtained leak location value, $\tilde{\Im}$– the resulting optimal value of the functional, $\Im_0$– the initial value of the functional, N_{iter}– the number of iterations (one-dimensional minimizations) required by the conjugate

Table 1. The findings from computational experiments

η	$\xi_0^{(1,2)}$ / $v_0^{(1,2)}(t)$	60 / $90-10e^{-0.0003t}$	20 / $20-10e^{-0.0003t}$	90 / $30-10e^{-0.0003t}$	10 / $66+20e^{-0.0003t}$	45,685 / $66+20e^{-0.0003t}$
0%	$\mathfrak{I}_0$	76.104	16.704	11.664	42.48	57.636
	$\tilde{\mathfrak{I}}$	$5.73 \cdot 10^{-7}$	$1.26 \cdot 10^{-7}$	$3.19 \cdot 10^{-6}$	$1.85 \cdot 10^{-6}$	$7.43 \cdot 10^{-7}$
	$\tilde{\xi}^{(1,2)}$	30.003	29.998	30.008	29.994	29.998
	N_{iter}	6	5	16	14	8
	$\delta\xi^{(1,2)}$	0.00009	0.00006	0.0003	0.0002	0.00006
	$\delta v^{(1,2)}$	0.0003	0.0006	0.0003	0.0002	0.0003
0.5%	$\mathfrak{I}_0$	76.352	16.837	11.683	42.148	57.732
	$\tilde{\mathfrak{I}}$	0.023	0.014	0.024	0.017	0.020
	$\tilde{\xi}^{(1,2)}$	29.841	30.332	30.068	29.654	29.796
	N_{iter}	6	5	14	12	7
	$\delta\xi^{(1,2)}$	−0.005	0.011	0.002	−0.011	−0.006
	$\delta v^{(1,2)}$	0.043	0.030	0.049	0.041	0.042
1%	$\mathfrak{I}_0$	77.119	16.924	11.832	43.744	57.413
	$\tilde{\mathfrak{I}}$	0.067	0.071	0.073	0.062	0.065
	$\tilde{\xi}^{(1,2)}$	28.527	29.392	29.923	30.597	29.839
	N_{iter}	6	5	14	12	7
	$\delta\xi^{(1,2)}$	−0.049	−0.020	−0.002	0.020	−0.005
	$\delta v^{(1,2)}$	0.109	0.092	0.107	0.094	0.093

gradient projection method, $\delta\xi^{(1,2)} = |\xi^{(1,2)*} - \tilde{\xi}^{(1,2)}|/\xi^{(1,2)}$, – relative error at the location of leakage and $\delta v^{(1,2)} = \max_{t\in[t_0,T]} |v^{(1,2)*}(t) - \tilde{v}^{(1,2)}(t)|/|v^{(1,2)}(t)|$ – relative error in its volume.

As indicated in Table 1, an improvement in measurement accuracy has the most pronounced impact on the precision of determining the volume of leaks. Generally, the error order of the obtained values for the identified parameters aligns with the order of the measurement error.

Figure 2 depicts the graphs of the actual leak function and the resultant loss functions obtained when solving the problem under the assumption of leakage presence.

Suppose that at $t = 0$, the states of the regimes for all segments are not accurately specified when the monitoring of boundary conditions begins. However, based on technological considerations, it is known that the flow amount and pressure in the segments

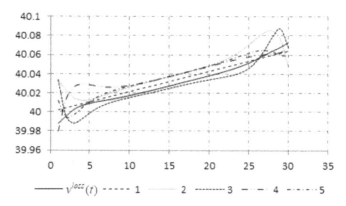

Fig. 2. The exact and experimental time dependences of liquid leakage.

can assume values within the following ranges:

$$300 \leq q_0^{(1,2)n}(x) \leq 320;\ 1.7 \cdot 10^6 \leq p_0^{(1,2)n}(x) \leq 2.7 \cdot 10^6,$$
$$200 \leq q_0^{(5,2)n}(x) \leq 220;\ 1.6 \cdot 10^6 \leq p_0^{(5,2)n}(x) \leq 2.1 \cdot 10^6,$$
$$100 \leq q_0^{(3,2)n}(x) \leq 120;\ 1.6 \cdot 10^6 \leq p_0^{(3,2)n}(x) \leq 2.2 \cdot 10^6, \quad (11)$$
$$120 \leq q_0^{(5,4)n}(x) \leq 140;\ 1.6 \cdot 10^6 \leq p_0^{(5,4)n}(x) \leq 2.2 \cdot 10^6,$$
$$80 \leq q_0^{(5,6)n}(x) \leq 100;\ 1.7 \cdot 10^6 \leq p_0^{(5,6)n}(x) \leq 2.2 \cdot 10^6.$$

We utilize a set of initial conditions that satisfy the constraints (11) rather than exact initial conditions.

Specifically, let admissible initial conditions for the flow amount and pressure in the segments be established on a finite set, derived by uniformly dividing the value range into ν points with a step size of $h_\nu = 20/\nu$, as determined by (11). We define the set of admissible initial conditions presuming a steady regime in all segments of the network at $t_0 = 0$. The flow amount and pressure assumed the following values:

$$\{q_0^{(1,2)n}(x) = 300 + nh_\nu;\ q_0^{(5,2)n}(x) = 200 + nh_\nu;$$
$$q_0^{(3,2)n}(x) = 100 + nh_\nu;\ q_0^{(5,4)n}(x) : 120 + nh_\nu;$$
$$q_0^{(5,6)n}(x) = 80 + nh_\nu\};$$
$$\{p_0^{(1,2)n}(x) = 2.3 \cdot 10^6 - 5.8955x + 2 \cdot 10^4 nh_\nu;$$
$$p_0^{(5,2)n}(x) = 1.7 \cdot 10^6 - 1.174x + 2 \cdot 10^4 nh_\nu;$$
$$p_0^{(3,2)n}(x) = 1.8 \cdot 10^6 - 1.67x + 2 \cdot 10^4 nh_\nu;$$
$$p_0^{(5,4)n}(x) = 1.8 \cdot 10^6 - 2.358x + 2 \cdot 10^4 nh_\nu;$$
$$p_0^{(5,6)n}(x) = 1.8 \cdot 10^6 - 0.944x + 2 \cdot 10^4 nh_\nu\},$$
$$x \in [0, l],\ n = 1, 2, \ldots, \nu$$

Table 2 presents the results achieved through the minimization of the functional for different initial values of the identified parameters $(\xi, v(t))$ and varying numbers ν of initial conditions.

Table 3 presents the results from numerical experiments attained by solving the problem without noise in the measurements. However, it includes a leak at the point

Table 2. The results of solving problem 1, concerning leakage in section (1,2), for various values of v and different initial states $(\xi^{(1,2)}, v^{(1,2)}(t))_0^i, i = 1, ..., 5$, where $\alpha = e^{-0.0003t}$.

v		60	20	90	10	45.685
	$\xi_0^{(1,2)}$					
	$v_0^{(1,2)}(t)$	$90-10\alpha$	$20-10\alpha$	$30-10\alpha$	$66+20\alpha$	$66+20\alpha$
3	$\mathfrak{I}_0$	155.248	22.323	22.323	84.303	118.809
	$\widetilde{\mathfrak{I}}$	$8.2 \cdot 10^{-4}$	$2.7 \cdot 10^{-4}$	$1.0 \cdot 10^{-3}$	$2.2 \cdot 10^{-4}$	$7.4 \cdot 10^{-5}$
	$\widetilde{\xi}^{(1,2)}$	30.145	29.996	30.190	30.022	30.029
	$\delta \xi^{(1,2)}$	0.004	0.0001	0.006	0.0007	0.0005
	$\delta v^{(1,2)}$	0.0005	0.0004	0.0001	0.0009	0.0005
5	$\mathfrak{I}_0$	155.390	33.608	22.443	84.086	118.854
	$\widetilde{\mathfrak{I}}$	$1.6 \cdot 10^{-4}$	$3.9 \cdot 10^{-4}$	$4.3 \cdot 10^{-4}$	$2.8 \cdot 10^{-4}$	$6.4 \cdot 10^{-5}$
	$\widetilde{\xi}^{(1,2)}$	29.963	30.093	30.100	29.985	29.995
	$\delta \xi^{(1,2)}$	0.001	0.003	0.003	0.0005	0.0001
	$\delta v^{(1,2)}$	0.0006	0.0003	0.0005	0.0004	0.0002

$\xi^{(5,6)*} = 15$ km in segment (5,6), with the leak amount specified by the function $v^{(5,6)*}(t) = 10 + 10e^{3\alpha t}$ $m^3/hour$.

Table 3. The outcomes of addressing problem 1, related to leakage in section (5,6), are illustrated for various initial states.

$\xi_0^{(5,6)}$	$v_0^{(5,6)}(t)$	$\widetilde{\xi}^{(5,6)}$	$\mathfrak{I}_0$	$\widetilde{\mathfrak{I}}$	N_{iter}
25	$5+10\alpha$	15.219	2.556	$3.55 \cdot 10^{-5}$	165
10	$30+10\alpha$	14.781	44.615	$3.55 \cdot 10^{-5}$	204
22.56	$15+10\alpha$	15.222	3.467	$3.59 \cdot 10^{-5}$	197

5 Conclusion

In this paper, we present a numerical solution to an inverse problem in the water distribution network with a complex loopback structure. The objective is to identify the locations and volumes of leaks based on additional observations of non-stationary fluid flow regimes at specific points in the pipeline. We transform the problem into a parametric optimal control problem involving unknown initial conditions and non-separated boundary conditions. Subsequently, we employ numerical methods of first-order optimization to address the problem. The paper concludes with the presentation of results from numerical experiments.

References

1. Oyedeko, K.F.K., Balogun, H.A.: Modeling and simulation of a leak detection for oil and gas pipelines via transient model: a case study of the Niger Delta. J. Energy Technol. Policy **5**, 1 (2015)
2. Datta, S., Sarkar, S.: A review on different pipeline fault detection methods. J. Loss Prev. Process Ind. **41**, 97–106 (2016)
3. Colombo, A.F., Lee, P., Karney, B.W.: A selective literature review of transient-based leak detection methods. J. Hydro-Environ. Res. **2**, 212–227 (2009)
4. Al-Khomairi, A.M.: Leak detection in long pipelines using the least squares method. J. Hydraul. Res. **46**(3), 392–401 (2008)
5. Shamloo, H., Haghighi, A.: Optimum leak detection and calibration of pipe networks by inverse transient analysis. J. Hydraul. Res. **48**(3), 371–376 (2010)
6. Kapelan, Z.S., Savic, D.A., Walters, G.A.: A hybrid inverse transient model for leakage detection and roughness calibration in pipe networks. J. Hydraul. Res. **41**(5), 481–492 (2003)
7. Lee, P.J., Vitkovsky, J.P., Lambert, M.F., Simpson, A.R., Liggett, J.A.: Leak location in pipelines using the impulse response function. J. Hydraul. Res. **45**(5), 643–652 (2007)
8. Ferrante, M., Brunone, B., Meniconi, S., Karney, B.W., Massari, C.: Leak size, detectability and test conditions in pressurized pipe systems. Water Resour. Manag. **28**(13), 4583–4598 (2014)
9. Vitkovsky, J., Lambert, M., Simpson, A., Liggett, J.A.: Experimental observation and analysis of inverse transients for pipeline leak detection. J. Water Resour. Plan. Manag. **133**(6), 519–530 (2007)
10. Wichowski, R.: Hydraulic transients analysis in pipe networks by the method of characteristics (MOC). Arch. Hydro Eng. Environ. Mech. **53**(3), 267–291 (2006)
11. Adamkowski, A.: Analysis of transient flow in pipes with expanding or contracting sections. J. Fluids Eng. **125**(4), 716–722 (2003)
12. Samarskii, A.A., Vabishchevich, P.N.: Numerical Methods for Solving Inverse Problems in Mathematical Physics. LKI, Moscow (2009). (in Russian)
13. Aida-zade, K.R., Ashrafova, E.R.: Numerical leak detection in a pipeline network of complex structure with unsteady flow. Comput. Math. Math. Phys. **57**, 1919–1934 (2017)
14. Aida-zade, K.R., Ashrafova, E.R.: Localization of the points of leakage in an oil main pipeline under non-stationary conditions. J. Eng. Phys. Therm. **85**(5), 1148–1156 (2012)
15. Charnyi, I.A.: Unsteady Flows of Real Fluids in Pipelines. Nedra, Moscow (1975). (in Russian)
16. Chaudhry, H.M.: Applied Hydraulic Transients. Van Nostrand Reinhold, New York (1988)
17. Ashrafova, E.R.: Numerical investigation of the duration of the effect exerted by initial regimes on the process of liquid motion in a pipeline. J. Eng. Phys. Therm. **88**(5), 1–9 (2015)
18. Aida-zade, K.R., Ashrafova, Y.R.: Solving systems of differential equations of block structure with nonseparated boundary conditions. J. Appl. Ind. Math. **9**(1), 1–10 (2015)

Numerical Solution to a Problem of Optimizing Placement and Flow Rates of Wells

Arzu Bagirov[1], Tatiana Gunkina[2], and Alexander Handzel[2(✉)]

[1] Institute of Control Systems, Baku, Azerbaijan
[2] North-Caucasus Federal University, Stavropol, Russia
akhandzel@ncfu.ru

Abstract. The paper proposes an approach to the numerical solution of the problem of optimizing wells' placement and the values of their flow rates. The selection of rational placement of wells is the first and extremely important task that oil and gas companies face when designing the development of hydrocarbon fields. Typically, standard patterns are used for well placement, in which wells are located at the nodal points of various regular grids. The method proposed in this paper for determining the optimal placement of wells and their flow rates is based on considering an oil reservoir as a dynamic system with distributed parameters and on application of the methods for optimal control of such systems. The optimization goal is to achieve a uniform distribution of reservoir pressure over the area of the oil deposit after withdrawing the planned volume of oil from it. Fulfillment of this criterion ensures, in particular, a uniform rise of bottom water and prevents early water flooding of production wells. The state of the reservoir is described by the equation of non-stationary two-dimensional filtration of weakly compressible oil. Thus, the problem under consideration is a problem of optimal control of distributed systems, the solution of which is carried out by reducing the original problem to a discrete problem for which the gradient of the objective functional is determined, which allows to use standard first-order optimization methods. The paper presents the results of numerical experiments and their analysis.

Keywords: Oil Field · Filtration · Placement of Wells · Flow Rate · Distributed System · Optimal Control

1 Introduction

As is known, determining the optimal (rational) placement of wells is one of the most important tasks that arise when designing the development of oil and gas fields. The following approaches are most often used to solve these problems: heuristic rules developed by the practice of developing hydrocarbon deposits [1, 2], multivariate computer calculations of well placement schemes proposed by experts, et al. [3, 4]. In this work, the determination of the optimal placement of wells and their flow rates is performed by solving a problem of optimal control of a distributed system with lumped sources, in which the control parameters are the coordinates of the location of the sources and their power [5, 6].

2 Statement of the Problem

Let us formulate the problem of optimal placement of wells and optimization of their operating modes. To describe the process of two-dimensional filtration of weakly compressible oil in a porous medium, we use the following boundary value problem [7, 8]:

$$c(x)\frac{\partial p}{\partial t} - div(a(x)\nabla p) + \sum_{i=1}^{L} q^i(t)\delta\left(x - x^i\right) = 0,$$

$$x \in \Omega^0 \subset E^2, t \in (0, T]; \tag{1}$$

$$p(x, 0) = p_0(x), x \in \Omega; \tag{2}$$

$$p(x, t)|_{x \in \Gamma_1} = p_1(x, t), \left.\frac{dp(x, t)}{dn}\right|_{x \in \Gamma_2} = 0, t \in (0, T]; \tag{3}$$

$$a(x) = \frac{k(x)h(x)}{\mu}, c(x) = h(m\beta_l + \beta_r),$$

here $p = p(x, t)$ is the pressure at point $x \in \Omega$ at time instance t; Ω is the filtration domain with boundary Γ, which consists of two nonintersecting parts Γ_1 and Γ_2; $\Omega^0 = \Omega \backslash \Gamma$; $k(x)$ is the permeability, $h(x)$ is the reservoir thickness; μ is the liquid viscosity; m is the porosity; β_l, β_r are liquid and rock compressibility factors; $x^i = (x_1^i, x_2^i)$ are the coordinates of i th well placement, $q^i(t)$ is the flow rate of i th well, $i = 1, 2, \ldots L$; L is the number of wells, $\delta(\cdot)$ is a generalized two-dimensional Dirac function, T is the duration of the planning period.

In the problem considered below, it is assumed that all functions, reservoir and oil parameters involved in the initial boundary value problem (1)–(3), including well coordinates x^i, $i = 1, \ldots, L_1$ are given, whereas L_2 wells with unknown coordinates $x^i, i = L_1 + 1, \ldots, L_1 + L_2 = L$ must be put into operation, observing the following operating and technological, geological, and planned constraints:

$$\left(x_1^i, x_2^i\right), t \in \Omega, i = L_1 + 1, \ldots, L; \tag{4}$$

$$\|x^i - x^j\| \geq D, i, j = 1, \ldots, L, i \neq j; \tag{5}$$

$$0 \leq \underline{q^i} \leq q^i(t) \leq \overline{q^i}, i = 1, \ldots, L; \tag{6}$$

$$\sum_{l=1}^{L} \int_0^T q^l(t)dt \geq q^{target}, \tag{7}$$

here $\|\cdot\|$ is the Euclidean norm on the plane, D is the minimal admissible distance between wells; q^{target} is the target production.

Depending on the production situation in the field, various indicators such as minimization of reservoir energy losses, maximization of hydrocarbon production and others, including their combinations and multi-criteria cases, can act as an optimality criterion [9–11].

In this work, we will consider the problem of minimizing the deviation of reservoir pressure from its weighted average value. Such a statement of the problem can arise if the oil deposit has an aquifer. Let it be required to ensure a uniform rise of bottom water and a uniform distribution of pressure throughout the entire field of the deposit. In this case, the expression of the corresponding objective functional for the problem of well placement and control of their flow rates can be written in the form:

$$J(X, Q; p) = \iint_\Omega \left[p(x, T) - \bar{p}\right]^2 dx + \varepsilon_1 \sum_{i=1}^{L} \int_0^T \left[q^i(t)\right]^2 dt + \varepsilon_2 \sum_{i=L_1+1}^{L} \|x^i\|^2, \tag{8}$$

here the second and third terms in (8) are introduced to regularize the functional; ε_1, $\varepsilon_2 > 0$ are regularization parameters, $\bar{p}$ is the average pressure in the reservoir.

In addition to the above conditions of the problem, we assume that the optimized well flow rates are piecewise constant functions in time:

$$q^i(t) = Q_k^i = const, t \in [\tau_{k-1}, \tau_k), k = 1, \ldots, n, \tau_0 = 0, \tau_n = T, l = 1, \ldots, L. \tag{9}$$

The stated problem (1)–(9) is a parametric problem of optimal control of systems with distributed parameters [12–14], in which the following vectors are optimized:

$$Q = (Q_1^1, \ldots, Q_n^1, \ldots, Q_1^L, \ldots, Q_n^L) \in E^{Ln},$$

$$X = (x_1^{L_1+1}, x_2^{L_1+1}, \ldots, x_1^L, x_2^L) \in E^{2L_2}$$

The overall dimension of the problem is $Ln + 2L_2$. To solve it numerically, the following scheme is used. Applying finite-difference approximation, the original problem is reduced to a finite-dimensional mathematical programming problem with constraints of the type of equalities and inequalities of a special form. The obtained problem belongs to the class of optimization problems of a network structure [15, 16].

3 Numerical Solution of the Problem

Using the approach described in [5, 6, 13, 14], we obtain calculation formulas for the analytical determination of the gradient of the target functional of a finite-dimensional mathematical programming problem, with the help of which first-order methods and their combinations can be used to solve this problem (the projection method of conjugate gradient, the penalty functions method, etc.).

Let us introduce a grid in the domain $\Omega \times [0, T]$. For the sake of simplicity, we assume that domain Ω is a rectangle: $[0, a] \times [0, b]$, and $\Gamma_1 = \emptyset$:

$x_{1i} = ih_1, x_{2j} = jh_2, t_s = sh_t, i = 0, \ldots, N_1, j = 0, \ldots, N_2, s = 0, \ldots, N_t, h_t = T/N_t, h_1 = a/N_1, h_2 = b/N_2$; and N_t, N_1, and N_2 are preset positive numbers defined by the selected pattern of the approximation of the given problem and the accuracy of its solution.

$$\omega = \{(i, j, s) : i = 0, ..., N_1, j = 0, ..., N_2, s = 0, ..., N_t\}.$$

Let us denote

$$p(x_{1i}, x_{2j}, t_s) = p_{ijs}, \quad q^l(t_s) = Q^l_\chi \equiv const, \quad s = s_{\chi-1}, ..., s_\chi - 1, \quad \chi = 1, ..., n. \tag{10}$$

Using any pattern of grid method [17], we write the approximated boundary-value problem (1)–(3) in a general form

$$p_{ijs} = F_{ijs}(\Re_{ijs}, q_{ijs}, h_{ts}(\tau)), \quad (i, j, s) \in \omega, \tag{11}$$

$$q_{ijs} = \sum_{l=1}^{L} q^l_{ijs}. \tag{12}$$

Relation (11) is determined by the function F_{ijs} and the set of its arguments

$$\Re_{ijs} = \{p_{\xi\eta\chi} : (\xi, \eta, \chi) \in \omega_{ijs} \subset \omega\}. \tag{13}$$

$\Re_{ijs}$ depends on a specific applied pattern of grid method, and index set ω_{ijs} is determined by the selected approximation model.

To approximate two-dimensional Dirac function when calculating the influence of point sources (14), it is possible to use linear function, Gauss function, and other functions [7, 10]. Two functions: linear function and Gauss function were analyzed. For linear function, the influence of a given point source affects only the values of the pressure in the nodes of the grid surrounding the point source. As for Gauss function, the approximation of sources by means of this function can be expressed by the following formula:

$$q^l_{ijs} = \frac{1}{2\pi\sigma^2} Q^l_\chi \exp\left\{ -\frac{\left[(x_{i1} - x^l_1)^2 + (x_{2j} - x^l_2)^2\right]}{2\sigma^2} \right\}, \tag{14}$$

for $i = 0, ..., N_1, j = 0, ..., N_2, s = s_{\chi-1}, ..., s_\chi - 1, \chi = 1, ..., n$.

When Gauss function is applied the action of each point source is distributed across all the nodal points of the grid and it turns out to be an essential advantage of this function.

Constraints (6), (7) and objective function (8) are approximated in the following way:

$$\underline{q}^l \leq Q^l_\chi \leq \overline{q}^l, \chi = 1, \ldots, n, l = 1, \ldots, L, \tag{15}$$

$$\sum_{l=1}^{L} \sum_{\chi=1}^{n} Q^l_\chi (\tau_\chi - \tau_{\chi-1}) \geq q^{plan}, \tag{16}$$

$$I(X, Q, p) = h_1 h_2 \sum_{i=0}^{N_1} \sum_{j=0}^{N_2} \beta_{ij} [p_{ijN_t} - \overline{p}]^2 + \varepsilon_1 \sum_{l=1}^{L} \sum_{s=1}^{n} Q^l_s (\tau_s - \tau_{s-1}) + \varepsilon_2 \sum_{l=L_1+1}^{L} \|X^l\|, \tag{17}$$

$$\bar{p} = \sum_{i=0}^{N_1} \sum_{j=0}^{N_2} p_{ijN_t}/(N_1 N_2).$$

Here β_{ij} are the factors of the quadrature formula applied to calculate integral in (8).

Thus, we obtained the problem of finite-dimensional mathematical programming (4), (5), (11), (15)–(17), in which the equality-type constraints (11) are the main ones. Using the structure of these constraints, we will derive the formulas for the gradient of function (17) with respect to the optimized parameters:

$$\nabla I(X, Q; p) = (\nabla_Q I(X, Q; p), \nabla_x I(X, Q; p)) \tag{18}$$

Let us introduce an auxiliary impulse-vector:

$$\psi_{ijs} = \frac{dI(X, Q; p)}{dp_{ijs}}, \ (i, j, s) \in \omega, \tag{19}$$

Let us introduce an index set ω_{ijs}^- adjoint to the set ω_{ijs}:

$$\omega_{ijs}^- = \{(\xi, \eta, \chi): (i, j, s) \in \omega_{\xi\eta\chi}\}, (i, j, s) \in \omega. \tag{20}$$

Using index set ω_{ijs}^- formula (19) may be rewritten as:

$$\begin{aligned}\psi_{ijs} &= \frac{\partial I(X,Q;p)}{\partial p_{ijs}} + \sum_{(\xi,\eta,\chi)\in\omega_{ijs}^-} \frac{dI(X,Q;p)}{dp_{\xi\eta\chi}} \frac{\partial p_{\xi\eta\chi}}{\partial p_{ijs}} \\ &= \frac{\partial I(X,Q;p)}{\partial p_{ijs}} + \sum_{(\xi,\eta,\chi)\in\omega_{ijs}^-} \frac{\partial p_{\xi\eta\chi}}{\partial p_{ijs}} \psi_{\xi\eta\chi}, (i, j, s) \in \omega.\end{aligned} \tag{21}$$

System of Eqs. (21) makes it possible to determine vectors $\psi_{ijs}, (i, j, s) \in \omega$. The components of the gradient of the objective function (17) with respect to the coordinates and the flow rates of the wells can be calculated by means of [15, 16]:

$$\frac{dI}{dQ_\chi^l} = \frac{\partial I}{\partial Q_\chi^l} + \sum_{s=s_{\chi-1}}^{s_\chi-1} \sum_{(i,j)\in\omega} \frac{\partial p_{ijs}}{\partial Q_\chi^l} \psi_{ijs}, \chi = 1, \ldots, n, l = 1, \ldots, L; \tag{22}$$

$$\frac{dI}{dx_k^l} = \sum_{s=0}^{N_t} \sum_{(i,j)\in\omega} \frac{\partial p_{ijs}}{\partial x_k^l} \psi_{ijs}, l = l_1, \ldots, L, k = 1, 2. \tag{23}$$

Thus, to apply the proposed approach, we have to obtain the solutions of two systems of Eqs. (11) and (21) and afterwards we can calculate the gradient of (17) using (22), (23) that provide us with a tool to use efficient first order optimization methods [18].

4 Results of Numerical Experiments

Let us consider the results of numerical experiments to find the optimal placement of wells and optimal values of their flow rates. The considered problem was numerically solved for the following input data:

$$\Omega = [0; 3000 \text{ (m)}] \times [0; 6000 \text{ (m)}], dp(x,t)/dn = 0, x \in \Gamma,$$

$$h(x) = const = 5\,(m),\ p(x, 0) = p_0(x) = const = 150\,(atm)$$

$$\beta_l = 0.000225(\text{atm}^{-1}),\ \beta_r = 0.000010\left(\text{atm}^{-1}\right),\ m_0 = 0.2,$$

$$\mu = 2.5(\text{centipoise}),\ k = 0.4(\text{darcy}).$$

The number of intervals of flow rates constancy is 5. The range of variation of each well flow rate (values $\underline{q}^l$ and $\overline{q}^l$) is set to 25% of its average value:

$$\underline{q}^l = 0.75 q_{ave}^l,\ \overline{q}^l = 1.25 q_{ave}^l,\ l = 1, \ldots, L.$$

Let us mention that when calculating the influence of point sources (oil wells), Gauss function was used to model Dirac function since this option resulted in obtaining the solutions presented below while an attempt to use linear approximation of Dirac function lead to the failure in solving the considered problem.

During the research, two series of numerical calculations were carried out.

In the first series of experiments, the coordinates of the initial location of wells were very different from the pattern which is usually applied when planning oil field development. The results of these experiments are shown in Figs. 1, 2.

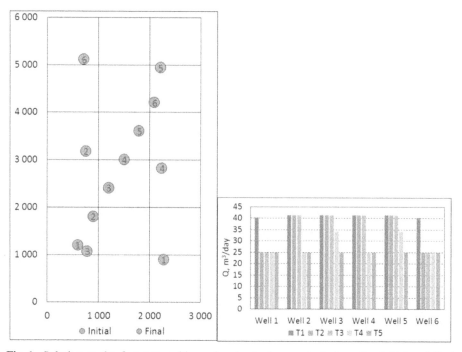

Fig. 1. Solution to the first test problem of optimal placement of six wells: initial and final placement of wells (left), variation of wells' flow rates for intervals of constancy T_1–T_5 (right).

At the first test problem, the wells were initially placed approximately diagonally across the field; at the second test problem, the wells were placed in two groups. After calculation, the wells are placed approximately uniformly across the field, but slightly shifted from an absolutely uniform placement. After solving the first problem three left wells are located slightly higher than the right ones, for the solution of the second test problem the opposite situation is observed: the left points are located lower than the right ones.

As for the flow rates, it is clear that the algorithm sets the flow rates for all wells to the maximal value at the initial interval of constant flow rates T_1 and to the minimum value at the final interval of constant flow rates T_5.

This distribution of flow rates over intervals of their constancy can be associated with a reduction to the minimum impact of wells on the formation at the final moment of time, at which the pressure distribution in the formation, according to the criterion, should be the most uniform.

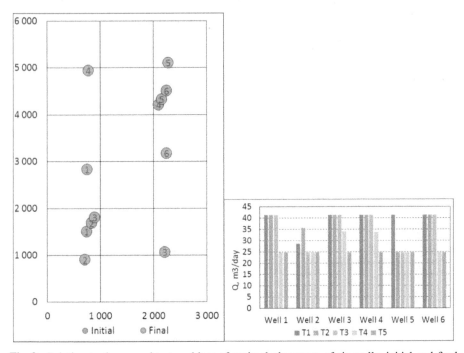

Fig. 2. Solution to the second test problem of optimal placement of six wells: initial and final placement of wells (left), variation of wells' flow rates for intervals of constancy T_1–T_5 (right).

Based on the results of the experiments described above, it seems that the optimal pattern for the problem under consideration is a uniform placement of wells, but the algorithm used cannot obtain this optimal pattern for some reason. In this regard, additional numerical experiments were performed in which the initial pattern was set to the points uniformly distributed over the area of the field.

Numerical Solution to a Problem of Optimizing Placement and Flow Rates 355

At first, the case of uniform well placement was considered. The flow rates of all the wells were equal and constant in time. The results of calculations are given in Figs. 3, 4 and 5: initial well positions, and in Fig. 6 (left): map of isobars.

Then three test problems were considered and solved.

The third test problem. The wells are initially arranged by uniform pattern. The range of flow rate deviation is 25%

$$\underline{q}^l = 0.75 q_{ave}^l, \overline{q}^l = 1.25 q_{ave}^l, l = 1, \ldots, L.$$

The calculation results are shown in the Figs. 3 and 6 (right). It can be easily seen that there is a small shift of wells' coordinates from initial uniform pattern. As for flow rates, the algorithm sets the values of the flow rates of all the wells to the minimal value at the last time interval of flow rate constancy T_5.

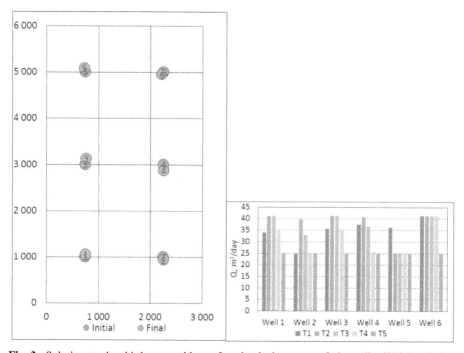

Fig. 3. Solution to the third test problem of optimal placement of six wells: initial and final placement of wells (left), variation of wells' flow rates for intervals of constancy T_1–T_5 (right).

The fourth test problem. Again, the wells are initially arranged by uniform pattern. The range of flow rate is deviation 50%

$$\underline{q}^l = 0.50 q_{ave}^l, \overline{q}^l = 1.50 q_{ave}^l, l = 1, \ldots, L.$$

The calculation results are shown in the Figs. 4 and 7 (left). Obviously, there is a greater shift of wells' coordinates from initial uniform pattern. As for flow rates, the

356 A. Bagirov et al.

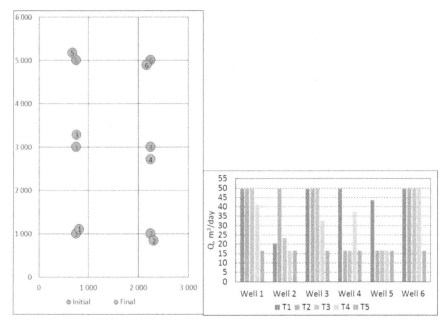

Fig. 4. Solution to the fourth test problem of optimal placement of six wells: initial and final placement of wells (left), variation of wells' flow rates for intervals of constancy T_1–T_5 (right).

algorithm sets the values of the flow rates of all the wells to the minimal value at the last time interval of flow rate constancy T_5, as it was already observed in test problem 3.

The fifth test problem. Once again, the wells are initially arranged by uniform pattern, but the range of flow rate deviation is 75%:

$$\underline{q}^l = 0.25\, q^l_{ave}, \quad \overline{q}^l = 1.75\, q^l_{ave}, l = 1, \ldots, L.$$

The calculation results are shown in the Figs. 5 and 7 (right). For this case, there is an essential shift of wells' coordinates from the initial uniform pattern. Analyzing Fig. 5, one can draw a conclusion that there is a symmetry in wells' final placement. As for flow rates, the algorithm tends to set the initial values of the flow rates of all the wells to the maximal value at the first time intervals of flow rate constancy T_1, T_2, T_3, and the algorithm minimizes the value of all wells flow rate at the last time interval of flow rate constancy T_5.

Furthermore, the values of total production of identically located wells are almost equal. The values of total production of 1st and 6th wells are 102.47 m^3 and 102.00 m^3, of 3rd and 4th are 190.15 m^3 both, and of 2nd and 5th are 203.51 m^3 and 203.80 m^3.

Numerical Solution to a Problem of Optimizing Placement and Flow Rates 357

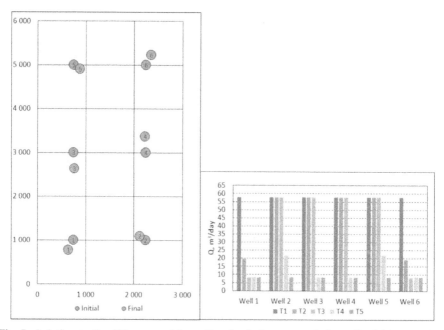

Fig. 5. Solution to the fifth test problem of optimal placement of six wells: initial and final placement of wells (left), variation of wells' flow rates for intervals of constancy T_1–T_5 (right).

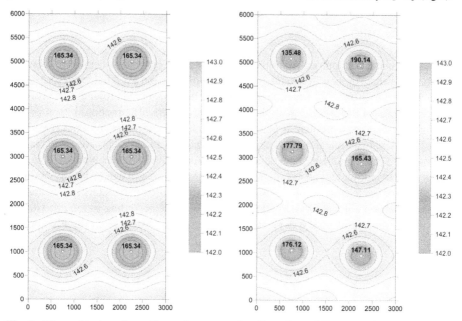

Fig. 6. Maps of isobars for uniform placement of wells and equal flow rates for each well and each time interval (left) and for the third test problem (right). Regular figures are isobars, bold figures are the values of a given well total production.

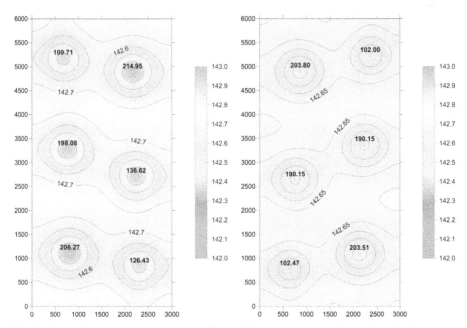

Fig. 7. Maps of isobars for the fourth (left) and fifth test problems (right). Regular figures are isobars, bold figures are the values of a given well total production.

5 Conclusion

The paper presents an approach to numerical determining optimal placement of oil wells and optimal values of their flow rates. The process of filtration in oil field is described by a parabolic type differential equation thus turning the considered problem to a problem of optimal control of a system with distributed parameters. The proposed approach of solving the considered problem is based on the reduction of initial continuous optimal control problem to a finite dimensional mathematical programming problem and obtaining formulas for the gradient of the objective function of that problem. Numerical experiments are carried out and analyzed. The results of the research can be briefly outlined as follows: for some cases of oil fields with uniform filtration parameters (permeability, porosity, reservoir thickness) optimal placement of wells according to the criterion of uniform distribution of reservoir pressure differs to some extent from usually applied regular well patterns.

References

1. Zheltov, Y.: Oil Fields Development. Nedra, Moscow (1986)
2. Bershchansky, Ya.V., Kulibanov, V.N., Meerov, M.V., Permin, O.Yu.: Control of Oil Fields Development. Nedra, Moscow (1983)
3. Ermolaev, A.I., Abdikadirov, B.A.: Optimizing placement of wells at oil fields by means of integer programming. Control Sci. (6), 45–49 (2007)

4. Liu, D., Sun, J.: The Control Theory and Application for Well Pattern Optimization of Heterogeneous Sandstone Reservoirs. Petroleum Industry Press, Beijing (2017)
5. Aida-zade, K.R., Handzel, A.V.: Optimization of oil field wells' placement. In: III International Conference "Mathematical and Computational Applications" (ICMCA-2002), Konya, Turkey, pp. 108–110 (2002)
6. Aida-zade, K.R., Bagirov, A.G.: On the problem of spacing of oil wells and control of their production rates. Autom. Remote Control **67**(1), 44–53 (2006)
7. Charny, I.A.: Underground Hydro-Gas-Dynamics. GNTI, Moscow (1963)
8. Aziz, K., Settari, A.: Petroleum Reservoir Simulation. Nedra, Moscow (1982)
9. Akhmetzyanov, A.V., Kulibanov, V.N.: On the problem of optimal control of oil fields. Autom. Remote Control (6), 5–13 (1999)
10. Akhmetzyanov, A.V., Kulibanov, V.N.: Optimal placement of sources for steady scalar fields. Autom. Remote Control (6), 50–58 (1999)
11. Abdullaev, F.M., Aida-zade, K.R., Kalaushin, M.A.: Control system of Yamburg field. Instrum. Control Syst. (2), 32–37 (1990)
12. Butkovsky, A.G.: Methods of Control of Systems with Distributed Parameters. Nauka, Moscow (1975)
13. Aida-zade, K.R., Handzel, A.V.: An approach to lumped control synthesis in distributed systems. Appl. Comput. Math. **6**(1), 69–79 (2007)
14. Aida-zade, K.R., Abdullaev, V.M.: Optimizing placement of the control points at synthesis of the heating process control. Autom. Remote Control **78**(9), 1585–1599 (2017)
15. Aida-zade, K.R.: Investigation and the numerical solution of finite-dimensional approximation of control problems of distributed systems. Comput. Math. Math. Phys. (3), 122–130 (1989)
16. Aida-zade, K.R.: Investigation of non-linear optimization problems of net structure. Autom. Remote Control (4), 63–71 (1989)
17. Samarsky, A.A.: The Theory of Difference Schemes. Nauka, Moscow (1977)
18. Polyak, B.T.: Introduction to Optimization. Nauka, Moscow (1983)

Time Dilation Principle to Solve Game Problems of Control

Arkadii Chikrii, Greta Chikrii, and Viktor Kuzmenko(✉)

V.M. Glushkov Institute of Cybernetics NASU, 40, pr. Akademika Glushkova, Kyiv 03187, Ukraine
kvn_ukr@yahoo.com

Abstract. This paper outlines a method for solving the game problem of convergence of the generalized quasi-linear non-stationary system trajectory with the cylindrical terminal set whose body part depends on time. The method of resolving functions stands for the theoretical basis of investigation. We examine the situation when Pontryagin's condition, reflecting an advantage in control resources of one of the opposing sides, does not hold. Because of this, with the help of a certain scalar function, we construct the modified condition. This function is called the function of time dilation. Fulfillment of this condition provides an opportunity for the pursuer, based on the Filippov-Castaing theorems on measurable choice, to build control which guarantees successive solution of the problem at hand. In so doing, the technique of set-valued mapping and their selections is applied. The process of convergence consists of two parts – active and passive. Control of the first player is constructed in view of his opponent's control in the past, namely with a certain time delay depending on the function of time dilation. We obtain sufficient conditions for convergence in a finite time in the class of quasi-strategies, and, under additional conditions, in the class of stroboscopic strategies with delay.

Keywords: conflict-controlled process · set-valued mapping · Pontryagin's condition · the function of time dilation · measurable choice · Minkowski' geometric difference · Aumann integral

1 Introduction

The problems of taking solutions and control choice in conditions of conflict and uncertainty are among the most difficult in the theory of extremum problems. Attempts to build optimal strategies for counter-acting sides lead to great mathematical difficulties.

Really, in the time-continuous model, in the course of the process of conflict counteracting at each instant of time one of the sides maximizes and another one minimizes the quality criterion. Therefore, the problem of optimization of the conflict-controlled process comes down to finding the continuum min-max or max-min depending on the character of current information availability. The mentioned peculiarities of formal realization relate to the ideology of dynamic programming and therefore to the main equation of the theory of differential games – the Hamilton-Jacobi-Bellman-Isaacs equation (HJBI) [1]. These ideas are expressed in the form of sets in the Pontryagin-Pshenichnyi procedures [2, 3], the method of alternated integral, and the method of semi-group T_ε-operators.

In view of the above-mentioned, a number of efficient mathematical methods for taking solutions in dynamic games have been proposed, which provide guaranteed results without account of optimality. These are, first of all, Pontryagin's First Direct Method and the Method of Resolving Functions. It is on these methods we will focus our attention in this paper.

It should be emphasized that information on the current state or control of the adversary, as a rule, is available to the side, taking solution with delay in time. The method of reducing the problem of approach with delayed information to the equivalent game with complete information has been developed [4].

To count on success in counteracting conflict, a certain advantage in control resources and dynamic capabilities is a requisite. Classic Pontryagin's condition stands for it in the above-mentioned methods. There are various modifications to this condition.

One way is to multiply the body part of the terminal set by a certain scalar function, whose integral is equal to the unit, and include it into Pontryagin's condition.

Another one deals with annexing some used part of the evader's control resource. The latter is multiplied by a certain matrix function and subtracted from the body part of the terminal set. In this case, Pontryagin's condition breaks down into two conditions providing the non-empty of corresponding sets. This technique is effective in the case of objects with different inertia.

However, for some special problems, in particular, the problem of "soft landing" ("soft meeting") and problems of meeting for oscillatory systems, the above-mentioned conditions are not satisfied. To overcome this difficulty, a special function, later named the function of time dilation, and corresponding modified Pontryagin's condition were introduced in the papers [5, 6]. This approach is called the principle of time dilation. Its application made it possible, on the basis of the First Direct Method, to obtain the solution of the problem of "soft meeting" for a wide range second-order systems [7].

In this paper, to realize the principle of time dilation in the game approach problem we use the Method of Resolving Functions. An important feature of this method is that, in the case of strictly convex control domains and relatively simple dynamics of the conflict-controlled processes, the resolving function appears as the greatest root of the quadratic equation and, consequently, can be found in analytic form. This feature is of decisive importance when solving specific game problems.

Also, an advantage of the Method of Resolving Functions is that for objects with simple motions and spherical control domains, this method provides full justification of the methods of the Euler chase, parallel pursuit, proportional navigation, and pursuit along ray. These methods are used by the designers of rocket and airspace techniques.

The wide possibilities of the method have been demonstrated on the problems of group and alternate pursuit [8], the problems with state constraints. A number of examples from the monograph [1] were solved with the use of resolving functions.

The cumulative principle inherent in the method looks quite natural. Its universality allows encompassing in the unique scheme the conflict-controlled processes, described by the difference-differential, partial derivative, and fractional derivative equations, impulse systems, as well as by non-stationary processes and stochastic systems.

The peculiarity of non-stationary conflict-controlled processes lies in the fact that their parameters (dynamics, control domains, terminal set) depend on time and it is necessary, because of the character of these dependences, to establish the relationships that ensure opportunity for the pursuer to terminate the game in a finite time, under certain information availability to the players.

2 Statement of the Game Problem

To encompass as wide as possible range of dynamic processes functioning in conditions of conflict and uncertainty and build design schemes of approach, we consider the generalized quasi-linear non-stationary conflict-controlled processes.

Let the motion of the object in the finite-dimensional Euclidean space be given by the relation

$$z(t) = \omega(t, t_0) + \int_{t_0}^{t} \Omega(t, \tau)\phi(\tau, u(\tau), v(\tau))d\tau, \ t \geq t_0 \geq 0. \tag{1}$$

Here function $\omega(t, t_0)$, acts from R_{t_0} to R^n, $R_{t_0} = \{t : t \geq t_0\}$, and represents itself a block of initial data: $\omega(t_0, t_0) = z(t_0) = z_0$. It is Lebesgue measurable in t and bounded for $t > t_0$.

The matrix function $\Omega(t, \tau)$, $t \geq \tau \geq t_0$, is supposed to be measurable in t and summable in τ for each $t \in R_{t_0}$.

The control block is given by the function $\phi(t, u, v)$, which meets the Caratheodori conditions, it is measurable in t and jointly continuous in (u, v) for each $t \in R_{t_0}$.

The control parameters of the players are chosen at each instant of time from the control domains $U(t)$ and $V(t)$, respectively. Herewith $U(t)$ and $V(t)$ are measurable compact-valued mappings for $t \in [t_0, +\infty)$. By admissible controls of the players are meant measurable in t selections of set-valued mappings $U(t)$ and $V(t)$, respectively. They exist by the measurable choice theorems [9].

In addition, we suppose that there exists a locally summable function $c(t)$, which satisfies the following inequality

$$\|\phi(t, u, v)\| \leq c(t), u \in U(t), v \in V(t), t \in [t_0, +\infty).$$

The pair $(t, z(t))$, $t \geq t_0$ designates the state of system (1) at each instant of time t.

Except for the system (1) that describes the dynamics of conflict-controlled process, the terminal set $M^*(t)$ is given. We suppose that it has a cylindrical form:

$$M^*(t) = M_0 + M(t), t \in [t_0, +\infty). \tag{2}$$

Here M_0 is a linear subspace from R^n, and $M(t)$ – a measurable compact-valued mapping whose direct images belong to the orthogonal supplement of L to M_0 in R^n.

Goal of the first player (the pursuer) is, by means of the choice of admissible control parameter $u(t)$, $u(t) \in U(t)$, to bring the trajectory of the process (1) to the terminal set (2) in the shortest time, for arbitrary admissible counter-actions of the adversary.

Goal of the second player (the evader) – by means of the choice of admissible control parameter $v(t)$, $v(t) \in V(t)$, to avoid meeting of the trajectory of process (1) with the terminal set $M^*(t)$ at a finite moment of time or if it is impossible to maximally postpone the moment of such meeting.

To fully formulate the approach problem, it is necessary to determine the level of the players' awareness in the process of conflict counteracting. Let us take the side of the pursuer and find out what result he can guarantee for himself. We assume that the evader chooses as controls the measurable selections of the set-valued mapping $V(t)$. Denote Ω_E the set of such selections.

Suppose that the pursuer, at the moment of taking solution t, $t \geq t_0$, uses information about the initial state of the process – (t_0, z_0) together with the evader's control pre-history

$$v_t(\cdot) = \{v(s) : v(s) \in V(s), s \in [t_0, t]\},$$

i.e. $u(t) = u(t_0, z_0, v_t(\cdot)) \in U(t)$. Then we say that he employs a quasi-strategy that designates the control $u(t)$. If the pursuer uses the delayed control pre-history we say about the delayed quasi-strategy.

In the case at the moment t, $t \geq t_0$, the pursuer uses information on the initial process state and instantaneous evader control $v(t)$, $v(t) \in V(t)$, i.e. $u(t) = u(t_0, z_0, v(t)) \in U(t)$, we say about counter-control by Krasovskii [10], which is prescribed by the stroboscopic strategy by Hajek [11]. If in the course of the game, the pursuer employs delayed information about the evader state then we say about the stroboscopic strategy with delay.

In this paper, the pursuer uses delayed information about the evader control. This is conditioned by the fact that we consider the case when Pontryagin's condition does not hold. The introduction of the function of time dilation makes it feasible to formulate the modified Pontryagin's condition and on its basis to construct the desired control using the technique of resolving functions. This provides a solution of the game approach problem in hand.

3 Pontryagin's Condition. Function of Time Dilation

Denote π the orthogonal projector that acts from R^n onto L. We put

$$\phi(\tau, U(\tau), v) = \{\phi(\tau, u, v) : u \in U(\tau)\}, v \in V(\tau), \tau \geq t_0.$$

Consider the following set-valued mappings.

$$W(t, \tau, v) = \pi \Omega(t, \tau) \phi(\tau, U(\tau), v), \quad W(t, \tau) = \bigcap_{v \in V(\tau)} W(t, \tau, v).$$

By virtue of the assumptions about parameters of the conflict-controlled process (1) and in view of the theorems on the direct image [9], the mapping $\phi(\tau, U(\tau), v)$ is measurable in τ and continuous in v by the Hausdorff metrics [12].

In its turn, the set-valued mapping $W(t, \tau, v)$ is also measurable in τ. In what follows, we will consider it compact-valued. Then the mapping $W(t, \tau)$ is measurable in τ and closed-valued [9]. Denote

$$\Delta = \{(t, \tau) : t_0 \leq \tau \leq t < +\infty\}.$$

Condition 1 (classic Pontryagin's condition). Set-valued mapping $W(t, \tau), (t, \tau) \in \Delta$, has non-empty direct images.

Let us study the mapping

$$P(\omega(\cdot, t_0)) = \{t \geq t_0 : \pi\omega(t, t_0) \in M(t) - \int_{t_0}^{t} W(t, \tau)d\tau\}. \tag{3}$$

The integral of set-valued mapping in formula (3) represents itself as the Aumann integral [9]. If the inclusion in parenthesis does not hold for all $t \geq t_0$, then we set $P(\omega(\cdot, t_0)) = \emptyset$.

Theorem 1. Let for the conflict-controlled process (1), (2) Condition 1 be fulfilled, $P \in P(\omega(\cdot, t_0)) \neq \emptyset$, and $W(P, \tau) = coW(P, \tau)$. Then at the moment P the trajectory of process (1) can be brought to the set $M^*(t)$ by a certain counter-control.

Proof. From the inclusion (3) and assumptions of the theorem follows the relationship

$$\pi\omega(P, t_0) \in M(P) - \int_{t_0}^{P} W(P, \tau)d\tau.$$

It testifies to the existence of point m, $m \in M(P)$ and measurable Pontryagin's selection $\gamma(P, \tau)$, $\gamma(P, \tau) \in W(P, \tau)$, such that

$$\pi\omega(P, t_0) = m - \int_{t_0}^{P} \gamma(P, \tau)d\tau. \tag{4}$$

Let us examine the selection.

$$U(\tau, v) = \{u \in U(\tau) : \pi\Omega(P, \tau)\phi(\tau, u, v) - \gamma(P, \tau) = 0\}, v \in V(\tau), \tau \in [t_0, P]. \tag{5}$$

The relationship (4) demonstrates that the set-valued mapping $U_*(\tau) = U(\tau, v(\tau))$ is measurable and compact-valued for arbitrary measurable selection $v(\tau)$ of the set-valued mapping $V(\tau)$. Therefore, by the theorem on measurable choice, it has a measurable selection $u_*(\tau)$. We assign it as the pursuer's control. Then from formulas (3)–(5) there follows that $\pi z(P) = m \in M(P)$.

In the case of fulfillment of Pontryagin's condition the game, as was shown above, can be terminated in a finite time according to the scheme of the First Direct Method. The same is valid for various classes of strategies when the scheme of the Method of Resolving Functions [8] is applied. However, Condition 1 does not hold for many problems of complicated nature.

The goal of this paper is to propose and describe the procedure of control in conflict situation in the case when Condition 1 fails. Based on the principle of time dilation [5–7], it makes it feasible to weaken the classic Pontryagin's condition and to widen class the class of conflict-controlled processes for which the problem of approach in a finite time can be solved on the basis of the method of resolving functions.

Let the control block in system (1) be structured as follows

$$\phi(t, u, v) = f(t, u) + g(t, v),$$

i.e. the controls of counter-acting sides are separated. By the way, this automatically ensures fulfillment of the condition for the existence of the saddle point in the "small game" [10].

Denote by

$$X \underset{*}{*} Y = \{z : z + Y \subset X\} = \bigcap_{y \in Y} (X - y)$$

Minkowski's geometric difference of sets X and Y [2].

Condition2 (modification of Pontryagin's condition)
There exists a differentiable monotonically increasing function $I(t)$, $I(t) \geq t$, $t \geq t_0$, $I(t_0) = t_0$, such that the set-valued mapping

$$W_I(t, \tau) = \pi \Omega(I(t), q(t, \tau)) f(q(t, \tau), U(q(t, \tau)))$$
$$\underset{*}{*} \pi \Omega(I(t), \psi(t, \tau)) g(\psi(t, \tau), V(\psi(t, \tau))) \dot{I}(t - \tau + t_0)$$

has non-empty images for $(t, \tau) \in \Delta$. Here

$$q(t, \tau) = I(t) - t + \tau, \psi(t, \tau) = I(t) - I(t - \tau + t_0) + t_0,$$

and the mappings $U(q(t, \tau))$ and $V(\psi(t, \tau))$ are compact-valued and measurable in τ.

For better understanding, ahead of events, let us explain the meaning of designations, appearing in Condition 2. The time t and the function of time dilation $I(t)$ concern the time duration, and τ is a current time. By function $q(t, \tau)$ we denote the moment of the pursuer control choice based on the evader control at the time $\psi(t, \tau)$. It is easy to see that $q(t, \tau) \geq \psi(t, \tau)$, $(t, \tau) \in \Delta$, in view of the inequalities $I(t) \geq t \geq t_0$. We see that there is a certain delay of information in the pursuer control choice concerning the evader control [4].

In Condition 2 the mapping $W_I(t, \tau)$ is measurable in τ, $\tau \in [t_0, t]$. Also, since the set-valued mapping $\pi \Omega(I(t), q(t, \tau)) f(q(t, \tau), U(q(t, \tau)))$ is closed-valued, it is closed-valued as well. Therefore, by theorem on measurable choice [9], it contains a measurable in τ selection $\gamma_I(t, \tau)$. Let us fix it for later.

Have a look at the function

$$N(t, u_0(\cdot)) = \pi \omega(I(t), t_0) + \int_{t_0}^{q(t,t_0)} \pi \Omega(I(t), \tau) f(\tau, u_0(\tau)) d\tau + \int_{t_0}^{t} \gamma_I(t, \tau) d\tau.$$

It connects the initial data, the pursuer starting control $u_0(\tau)$, $u_0(\tau) \in U(\tau)$, $\tau \in [t_0, q(t, t_0)]$, and the chosen selection $\gamma_I(t, \tau)$.

Let us apply the method of resolving functions to solving the problem of approaching a cylindrical terminal set (2) by the trajectory of (1) in condition of conflict [8]. To this end, we analyze the following two set-valued mappings

$$W_I(t, \tau, v) = \pi \Omega(I(t), q(t, \tau)) f(q(t, \tau), U(q(t, \tau)))$$
$$- \pi \Omega(I(t), \psi(t, \tau)) g(\psi(t, \tau), v) \dot{I}(t - \tau + t_0)$$

and

$$R_I(t, \tau, v, u_0(\cdot)) = \{\rho \geq 0 : [W_I(t, \tau, v) - \gamma_I(t, \tau)] \cap \rho[M(I(t)) - N(t, u_0(\cdot))] \neq \emptyset\}$$
$$= \{\rho \geq 0 : 0 \in W_I(t, \tau, v) - \gamma_I(t, \tau) - \rho[M(I(t)) - N(t, u_0(\cdot))]\} \quad (6)$$

We see that the second mapping is presented in two equivalent forms. Here $(t, \tau) \in \Delta$, $v \in V(\psi(t, \tau))$.

By Condition 2, the set-valued mapping $R(t, \tau, v, u_0(\cdot))$ has non-empty images on the semi-axis R_+. We name its support function in the direction $+1$ by the resolving function [8]:

$$\rho_I(t, \tau, v, u_0(\cdot)) = \sup\{\rho : \rho \in R_I(t, \tau, v, u_0(\cdot))\}, (t, \tau) \in \Delta, v \in V(\psi(t, \tau)).$$

From formula (6) there follows that if for some $u_0(\cdot), t \geq t_0, N(t, u_0(\cdot)) \in M(I(t))$, then

$$R_I(t, \tau, v, u_0(\cdot)) = [0, +\infty), \rho_I(t, \tau, v, u_0(\cdot)) = +\infty \text{ for } v \in V(\psi(t, \tau)), \tau \in [t_0, t].$$

Consider the set

$$T_I(\omega(\cdot, t_0), \gamma_I(\cdot, \cdot)) = \{t \geq t_0 : \sup_{u_0(\cdot) \in U_0^t} \inf_{v(\cdot) \in V^t} \int_{t_0}^{t} \rho_I(t, \tau, v, u_0(\cdot)) d\tau \geq 1\}, \quad (7)$$

where

$$U_0^t = \{u_0(\tau) : u_0(\tau) \in U(\tau), \tau \in [t_0, q(t, t_0)]\},$$
$$V^t = \{v(\tau) : v(\tau) \in V(\psi(t, \tau)), \tau \in [t_0, t]\}.$$

If inequality in (7) does not hold for all t, $t > t_0$, then we set $T_I(\omega(\cdot, t_0), \gamma_I(\cdot, \cdot)) = \emptyset$.

To simplify further treatment, we assume that the exact upper bound in $u_0(\cdot)$ in the relationship (7) is achieved.

Before moving on to the formulation of the main result, let us make some transformations over the solution representation – system (1) – with the account of the players' information availability.

Consider the projection of the process (1):

$$\pi z(t) = \pi \omega(t, t_0) + \int_{t_0}^{t} \pi \Omega(t, \tau) f(\tau, u(\tau)) d\tau + \int_{t_0}^{t} \pi \Omega(t, \tau) g(\tau, v(\tau)) d\tau$$

and present $\pi z(I(t))$ in the form

$$\pi z(I(t)) = \pi \omega(I(t), t_0) + \int_{t_0}^{q(t,t_0)} \pi \Omega(I(t), \tau) f(\tau, u_0(\tau)) d\tau$$

$$+ \int_{q(t,t_0)}^{I(t)} \pi \Omega(I(t), \tau) f(\tau, u(\tau)) d\tau + \int_{t_0}^{I(t)} \pi \Omega(I(t), \tau) g(\tau, v(\tau)) d\tau. \qquad (8)$$

To level the integration limits in the last two terms of the above expression we change variable τ for $q(t, \tau)$ and $\psi(t, \tau)$, respectively. As a result, we obtain the following representation

$$\pi z(I(t)) = \pi \omega(I(t), t_0) + \int_{t_0}^{q(t,t_0)} \pi \Omega(I(t), \tau) f(\tau, u_0(\tau)) d\tau$$

$$+ \int_{t_0}^{t} \pi \Omega(I(t), q(t, \tau)) f(q(t, \tau), u(q(t, \tau))) d\tau$$

$$+ \int_{t_0}^{t} \pi \Omega(I(t), \psi(t, \tau)) g(\psi(t, \tau), v(\psi(t, \tau))) \dot{I}(t - \tau + t_0) d\tau. \qquad (9)$$

Here is taken into account that for $q(t, t_0) \le \tau \le I(t)$ the inequality $t_0 \le q(t, \tau) \le t$ is true and for $t_0 \le \tau \le I(t)$ the inequality $t_0 \le \psi(t, \tau) \le t$ is also true. Thus, as a result of the change of variables in representation (8), the limits of integration in the last two terms of formula (9) became the same, i.e. $[t_0, t]$. Thus the choice of control $u(\tau)$, $\tau \in [q(t, t_0), I(t)]$, is reduced to the choice of control $u(q(t, \tau))$, $\tau \in [t_0, t]$.

The following result is valid.

Theorem 2 Let the control block in the game problem (1), (2) have the form $\phi(t, u, v) = f(t, u) + g(t, v)$ and let the mapping $M(t)$, $t \ge t_0$, be convex-valued. Also, suppose that Condition2 is fulfilled for some function of time dilation $I(t)$, $t \ge t_0$. Then, if for the given block of initial data $\omega(t, t_0)$, $t \ge t_0$, there exist a measurable selection $\gamma_I(t, \tau)$, of set-valued mapping $W_I(t, \tau)$, $(t, \tau) \in \Delta$, and time T, such that

$$T \in T_I(\omega(\cdot, t_0), \gamma_I(\cdot, \cdot) \ne \emptyset, \qquad (10)$$

then the trajectory of the process (1) can be brought to set $M^*(t)$ at the moment $I(T)$ using the pursuer quasi-strategy with delay.

Proof. We will study the game problem on the interval $[t_0, I(T)]$. Let us choose control of the pursuer (1) $u_0(\tau)$, $u_0(\tau) \in U_0^T$, from the condition for maximum of the following expression

$$\inf_{v(\cdot) \in V^T} \int_{t_0}^{T} \rho_I(T, \tau, v(\tau), u_0(\cdot)) d\tau \qquad (11)$$

Let $N(T, u_0(\cdot)) \notin M(I(T))$. We introduce the so-called test function.

$$k_I(t) = 1 - \int_{t_0}^{t} \rho_I(T, \tau, v(\tau), u_0(\cdot))d\tau, \quad v(\tau) \in V(\psi(T, \tau)).$$

Function $k_I(t)$, as the function of the integral upper bound, is continuous in t. Therefore, by the definition of time T (10), from formula (7), in view of the known theorems of the mathematical analysis, there follows the existence of a time instant t_*, $t_* \in (t_0, T]$, such that $k_I(t_*) = 0$. We will call the sections of time $[q(T, t_0), q(T, t_*)]$ and $(q(T, t_*), I(T)]$ by active and passive, respectively. Representation of solution without separating active and passive sections is given by formulas (8), (9) at $t = T$.

Let us consider the set-valued mapping

$$U_I(\tau, v) = \{u \in U(q(T, \tau)) : \pi\Omega(I(T), q(T, \tau))f(q(T, \tau), u)$$
$$- \pi\Omega(I(T), \psi(T, \tau))g(\psi(T, \tau), v)\dot{I}(T - \tau + t_0) - \gamma_I(T, \tau) \in$$

$$\in \bar{\rho}_I(T, \tau, v, u_0(\cdot))[M(I(T)) - N(T, u_0(\cdot))]\}, \quad \tau \in [t_0, T], \quad v(\tau) \in V(\psi(T, \tau)),$$
where

$$\bar{\rho}_I(T, \tau, v, u_0(\cdot)) = \begin{cases} \rho_I(T, \tau, v, u_0(\cdot)), & \tau \in [t_0, t_*] \\ 0, & \tau \in (t_*, T]. \end{cases}$$

The mapping $U_I(\tau, v)$ is $L \times B$ – measurable, by the theorem on inverse image [9], and also is closed-valued. Consequently, by the theorem on measurable choice [9], it has $L \times B$ – measurable selection.

$$u_I(\tau, v) = u(q(T, \tau), v(\psi(T, \tau)), \quad \tau \in [t_0, T]. \tag{12}$$

Let us choose control of the pursuer in the form (12) both on the active and passive sections. The control (12) presents itself a superposition measurable function. In other words, function $u(\tau) = u_I(\tau, v(\tau))$ is measurable for arbitrary selection $v(\tau)$, $v(\tau) \in V(\psi(T, \tau))$.

In the case $N(T, u_0(\cdot)) \in M(I(T))$ we choose control of the pursuer analogously, but with the zero resolving function. It is generally accepted that this case corresponds to Pontryagin's First Direct Method.

Upon separation of the active and passive sections of time, from formula (9) for $t = T$ we obtain the relationship

$$\pi z(I(T)) = \pi\omega(I(T), t_0) + \int_{t_0}^{q(T,t_0)} \pi\Omega(I(T), \tau)f(\tau, u_0(\tau))d\tau + \int_{t_0}^{T} \gamma_I(T, \tau)d\tau$$

$$+ \int_{t_0}^{t_*} \pi\Omega(I(T), q(T, \tau))f(q(T, \tau), u(q(T, \tau)))d\tau$$

$$+ \int_{t_*}^{T} \pi\Omega(I(T), q(T, \tau))f(q(T, \tau), u(q(T, \tau)))d\tau$$

$$-\int_{t_0}^{T} \pi\Omega(I(T), \psi(T,\tau))g(\psi(T,\tau), v(\psi(T,\tau)))\dot{I}(T-\tau+t_0)d\tau \int_{t_0}^{T} \gamma_I(T,\tau)d\tau.$$
(13)

Formula (13) for $N(T, u_0(\cdot)) \notin M(I(T))$, with the account of the laws of control choice, yields the relationship

$$\pi z(I(t)) \in N(T, u_0(\cdot)) + \int_{t_0}^{T} \overline{\rho}_I(T, \tau, v(\tau), u_0(\cdot))[M(I(T)) - N(T, u_0(\cdot))]d\tau$$

$$= N(T, u_0(\cdot))[1 - \int_{t_0}^{T} \overline{\rho}_I(T, \tau, v(\tau), u_0(\cdot))]d\tau$$

$$+ \int_{t_0}^{T} \overline{\rho}_I(T, \tau, v(\tau), u_0(\cdot))M(I(T))d\tau = M(I(T))$$

The above formula takes into account that the test function turns into zero at the moment t_* and the resolving function equals zero on the passive section.

If $N(T, u_0(\cdot)) \in M(I(T))$ then, as a result of control choice by the law (12) with the zero resolving function, from formula (13) it directly follows the inclusion $\pi z(I(T)) \in M(I(T))$, which is equivalent to the inclusion $z(I(T)) \in M^*(I(T))$.

Theorem 3. Let in the dynamics of the conflict-controlled process (1), controls of counter-acting sides are separated ($\phi(t, u, v) = f(t, u) + g(t, v)$), Condition 2 hold with some function of time dilation $I(t)$. Also, suppose that the set-valued mappings $M(t)$, $t \geq t_0$, and $R_I(t, \tau, v, u_0(\cdot))$, $(t, \tau) \in \Delta$, $v \in V(\psi(t, \tau))$, $u_0(\tau) \in U(\tau)$, are convex-valued. Then, if for given function $\omega(t, t_0)$, $t \geq t_0$, there exist a measurable selection $\gamma_I(t, \tau)$, $\gamma_I(t, \tau) \in W_I(t, \tau)$, $(t, \tau) \in \Delta$, and the moment T, such that

$$T \in T_I(\omega(\cdot, t_0), \gamma_I(\cdot, \cdot) \neq \emptyset,$$

then the trajectory of process (1) can be brought to the terminal set (2) at the time T using some stroboscopic strategy with delay.

Proof. On the section $[t_0, q(T, t_0)]$, we choose control $u_0(\tau)$, $u_0(\tau) \in U_0^T$, from the condition for the maximum of expression (11). Denote

$$\rho(T) = \int_{t_0}^{T} \inf_{v(\cdot) \in V^T} \rho(T, \tau, v(\tau), u_0(\cdot))d\tau.$$

We set

$$\rho^*(T, \tau) = [^1/_{\rho(T)}] \inf_{v(\cdot) \in V^T} \rho(T, \tau, v(\tau), u_0(\cdot)). \quad (14)$$

Since $\rho(T) \geq 1$, then, in view of the control choice $u_0(\cdot)$ and formula (14), we infer that $\rho^*(T, \tau) \leq \rho^*(T, \tau, v(\tau), u_0(\cdot))$, $\tau \in [t_0, T]$, $v(\tau) \in V(\psi(T, \tau))$, and function $\rho^*(T, \tau)$ is a measurable selection for each of the mappings $R_I(T, \tau, v, u_0(\cdot))$, $v(\tau) \in V(\psi(T, \tau))$, $\tau \in [t_0, T]$.

Consider the set-valued mapping

$$U^*(\tau, v) = \{u \in U(q(T, \tau)) : \pi\Omega(I(T), q(T, \tau))f(q(T, \tau), u)$$
$$- \pi\Omega(I(T), \psi(T, \tau))g(\psi(T, \tau), v)\dot{I}(T - \tau + t_0) - \gamma_I(T, \tau) \in$$
$$\in \rho^*(T, \tau)[M(I(T)) - N(T, u_0(\cdot))]\}, \tau \in [t_0, T], v \in V(\psi(T, \tau)). \quad (15)$$

This mapping is compact-valued and $L \times B$ – measurable. Then, by the theorem on measurable choice, it has $L \times B$ – measurable selection $u^*(\tau, v)$, which is a superposition measurable function. Let us set control of the pursuer equal $u^*(\tau) = u^*(\tau, v(\tau))$, $v(\tau) \in V(\psi(T, \tau))$, $\tau \in [t_0, T]$.

In the case $N(T, u_0(\cdot)) \in M(I(T))$ we choose control in the analogous way and set $\rho^*(\tau, v) = 0$ in (15).

Now we show that the trajectory of process (1) hits the set $M^*(t)$ at the moment $I(T)$.

If $N(T, u_0(\cdot)) \notin M(I(T))$ then from the relationship (9), with account of the law of control choice, we deduce:

$$\pi z(I(T)) \in N(T, u_0(\cdot))[1 - \int_{t_0}^{T} \rho^*(T, \tau)d\tau] + \int_{t_0}^{T} \rho^*(T, \tau)M(I(T))d\tau. \quad (16)$$

Since $M(I(T))$ is a convex compact, function $\rho^*(T, \tau)$ is non-negative and $\int_{t_0}^{T} \rho^*(T, \tau)d\tau = 1$, then $\int_{t_0}^{t} \rho^*(T, \tau)M(I(T))d\tau = M(I(T))$. Then formula (16) yields the inclusion $\pi z(I(T)) \in M(I(T))$.

Analogous inclusion can be deduced in the case $N(T, u_0(\cdot)) \in M(I(T))$.

Remark. The principle of time dilation has been applied to solving stationary approach problems on the basis of the First Direct Method (see, for example, [7, 13]). Its idea can be also applied in the case of the matrix resolving functions [14].

4 Conclusion

We develop a method of convergence of the trajectory of the conflict-controlled process with the terminal set when the condition for the advantage of one of the counter-acting sides fails. The gist of the method consists in the introduction of the so-called function of time dilation, which ensures fulfillment of the analog of Pontryagin's condition. This makes it possible to build control of the pursuer with account of the evader control in the past. Using this control the pursuer achieves its goal in a finite time.

The study is carried out in the frames of the method of resolving functions. We deduce sufficient conditions for the approach in a finite time for various cases of information availability to the players.

References

1. Isaacs, R.F.: Differential Games. Wiley Inter-science, New York (1965), 479 p.
2. Pontryagin, L.S.: Selected Scientific Works. V.2. Nauka, Moscow (1988), 576 p. (in Russian)
3. Pshenichnyi, B.N., Ostapenko, V.V.: Differential Games. Naukova Dumka, Kyiv (1992), 260 p. (in Russian)
4. Chikrii, G.T.: Using the effect of information delay in differential pursuit games. Cybern. Syst. Anal. **43**(2), 233–245 (2007). https://doi.org/10.1007/s10559-007-0042-x
5. Nikolskii, M.S.: Application of the first direct method in the linear differential games. Izvestia Acad. Nauk SSSR, vol. 10. pp. 51–56 (1972). (in Russian)
6. Zonnevend, D.: On one method of pursuit. DAN SSSR, vol. 24, no. 6, pp. 1296–1299 (1972). (in Russian)
7. Chikrii, G.Ts.: Principle of time stretching for motion control in condition of conflict controlled systems: theory and applications. In: Series in Automation, Control and Robotics, pp. 53–82. River Publishers (2021). https://doi.org/10.1201/9781003337010-4
8. Chikrii, A.A.: Conflict Controlled Processes. Springer, Dordrecht (2013), 424 p.
9. Aubin, J.-P., Frankowska, H.: Set-Valued Analysis. Birkhauser, Boston (1990), 461 p.
10. Krasovskii, N.N., Subbotin, A.I.: Positional Differential Games. Nauka, Moscow (1974), 455 p. (in Russian)
11. Hajek, O.: Pursuit Games, p. 266. Academic Press, New York (1975)
12. Kuratovskij, K.: Topology, vol. 1. Mir, Moscow (1966), 596 p. (in Russian)
13. Chikrii, G.Ts., Kuzmenko, V.M.: Solving the soft convergence problem for controlled oscillatory systems based on the time dilation principle. Cybern. Syst. Anal. **59**(3), 428–438 (2023). https://doi.org/10.1007/s10559-023-00577-z
14. Chikrii, A.A., Chikrii, G.Ts.: Matrix resolving functions in game dynamic problems. In: Kondratenko, Y.P., Kreinovich, V., Pedrycz, W., Chikrii, A., Gil-Lafuente, A.M. (eds.) Artificial Intelligence in Control and Decision-making Systems. SCI, vol.1087. Springer, Cham (2023). https://doi.org/10.1007/978-3-031-25759-9_5

Synthesis of Power Control of Moving Sources with Optimization of Measurement Points Location on Heating of the Rod

Vugar Hashimov(✉)

Institute of Control Systems of the Ministry of Science and Education of Republic of Azerbaijan, Baku, Azerbaijan
vugarhashimov@gmail.com

Abstract. The article discusses controlling the power of moving heat sources to heat a rod as per given rules and trajectories. The heat source control values are determined based on temperature measurements taken at nearby measuring points. An optimization approach is suggested for feedback coefficients of heat sources and measurement points in a rod's state. Current values of the powers of moving lumped sources are distributed along the rod in a certain neighbourhood of their current location. The problem of control synthesis is reduced into the parametric optimal control problem for a system with distributed parameters. To solve the problem using numerical methods, first-order optimization methods are proposed. Formulas were derived for the gradient of the objective functional, considering feedback parameters and the coordinates of measurement points. The test problem can be solved numerically using first-order optimization methods with the help of these formulas. The report will present the results of numerical experiments.

Keywords: Rod Heating · Feedback Control · Moving Sources · Points of Temperature Measurement · Parameters of Feedback

1 Introduction

The problem of optimizing power synthesis for controlling a heat source when heating a rod that moves along given trajectories and follows given rules is considered. The current source powers are determined based on temperature measurements at the installation points of the measuring devices [1–9].

The work presents original results in synthesizing optimal control of systems with distributed parameters, particularly in the synthesis of control power for moving sources; to optimize the locations of measurement points, a proposed formula for linear dependence on the measured temperature values is used to set the current values of source powers; the current control synthesis problem has been reduced to an optimization problem with a finite-dimensional.

A specific feature of the problem under consideration is that the current values of the powers of moving lumped sources are distributed along the rod in a certain neighbourhood of their current location.

In this work, formulas have been derived to determine the components of the objective functional's gradient. These formulas consider both feedback parameters and the location of measurement points. The resulting formulas can be used in numerical gradient-type optimization methods to solve the control synthesis problem.

The proposed method can be applied to regulation and control systems for objects described by various types of initial-boundary value problems.

2 Statement of the Problem

Let's consider the process of heating the rod, which is described by the following initial-boundary value problem for the parabolic partial differential equation [10]:

$$u_t(x,t) = a^2 u_{xx}(x,t) - \lambda_0[u(x,t) - \theta] + \sum_{i=1}^{N_c} q_i(t)\delta_\sigma(x; z_i(t)), \quad (x,t) \in (0,l) \times (0,T], \tag{1}$$

$$u(x,0) = b(x) = b = \text{const}, \quad x \in [0,l], \tag{2}$$

$$u_x(0,t) = \lambda[u(0,t) - \theta], \quad u_x(l,t) = \lambda[u(l,t) - \theta], \quad t \in (0,T], \tag{3}$$

Here: $u(x,t)$ is the temperature of the rod at a point $x \in [0,l]$ at the moment t; a, λ, λ_0 are given parameters of the heating process; θ is ambient temperature.

The function $\delta_\sigma(x; \hat{z}), \hat{z} \in [\sigma, l - \sigma]$ is continuously differentiable with respect to x for a given positive parameter σ, determine the distribution of the instantaneous value of the $q(t)$ power of the source at time t, lumped at the point $\hat{z}$, in a domain $(\hat{z} - \sigma, \hat{z} + \sigma)$:

$$\delta_\sigma(x; \hat{z}) \begin{cases} \geq 0, & \text{if } x \in (\hat{z} - \sigma, \hat{z} + \sigma), \\ = 0, & \text{if } x \notin (\hat{z} - \sigma, \hat{z} + \sigma). \end{cases}$$

In addition, it is required that these functions meet the following conditions:

$$\int_{\hat{z}-\sigma}^{\hat{z}+\sigma} \delta_\sigma(x; \hat{z}) dx = 1.$$

It is clear that for σ approaches to 0 the function $\delta_\sigma(x; 0)$ tends to the δ-function of Dirac [11]. However, the problem considered in this work is closer to real applications.

The process of heating is being carried out by N_c sources moving along given trajectories, powers $q = q(t) = (q_1(t), q_2(t), \ldots, q_{N_c}(t))$, which the controls that are piece-wise continuous should meet the given constraints:

$$q_i(t) \in Q_i = \left[\underline{q_i}, \overline{q_i}\right], \quad i = 1, 2, \ldots, N_c, \quad t \in [0,T], \tag{4}$$

where values $\underline{q_i}, \overline{q_i}, i = 1, \ldots, N_s$ are given. Continuous functions $z_i(t) \in [\sigma, l - \sigma]$ are given, they determine the position of ith power source at the moment t on the rod, $i = 1, 2, \ldots, N_c$.

We assume that the initial temperature in (2) is the same throughout the rod, but not precisely defined, and belongs to a given set $B \subset R$ with a known density function $\rho_B(b)$:

$$\rho_B(b) \geq 0,\ b \in B,\ \int_B \rho_B(b)db = 1.$$

Ambient temperature θ remains constant, and its values belong to the set $\Theta \subset R$ is described by a density function $\rho_\Theta(\theta)$ with specified properties

$$\rho_\Theta(\theta) \geq 0,\ \theta \in \Theta,\ \int_\Theta \rho_\Theta(\theta)d\theta = 1.$$

The formulated problem is to determine the values of the sources power $q = q(t) = (q_1(t), q_2(t), \ldots, q_{N_c}(t)) \in Q$ which minimize the following objective functional:

$$J(q) = \int_B \int_\Theta I(q; b, \theta) \rho_B(b) \rho_\Theta(\theta) d\theta db, \tag{5}$$

$$I(q; b, \theta) = \int_0^l \mu(x)[u(x, T) - U(x)]^2 dx + \varepsilon \|q(t) - \hat{q}(t)\|^2_{L_2^{N_c}[0,T]}. \tag{6}$$

Here: function $u(x, t) = u(x, t; q, b, \theta)$ is the solution to the initial-boundary value problem (1)–(3) at $u(x, 0) = b$, ambient temperature θ, values of sources power $q(t)$; function $U(x)$, $x \in [0, l]$ is the desired state of the rod at $t = T$; $\mu(x) \geq 0$, $x \in [0, l]$ is the weight function; ε, $\hat{q}(t)$ are regularization parameters of the objective functional of the problem [12].

Let at the given N_o points of the rod $\xi_j \in [\sigma, l - \sigma]$, $j = 1, 2, \ldots, N_c$ during the process of heating temperature measurements continuously in time are carried out:

$$\hat{u}_j(t) = \int_{\xi_j - \sigma}^{\xi_j + \sigma} u(x, t) v_\sigma(x; \xi_j) dx,\ j = 1, 2, \ldots, N_o,\ t \in [0, T]. \tag{7}$$

The relations (7) show that measuring the current state at any given point ξ_j affects the states of its neighboring points: $(\xi_j - \sigma, \xi_j + \sigma)$ with weight function $v_\sigma(x; \xi_j)$, $j = 1 \ldots, N_c$.

The weight function $v_\sigma(x; \hat{\xi})$ has certain properties which are as follows:

$$v_\sigma(x; \hat{\xi}) \begin{cases} \geq 0, \text{ if } x \in \left(\hat{\xi} - \sigma, \hat{\xi} + \sigma\right), \\ = 0, \text{ if } x \notin \left(\hat{\xi} - \sigma, \hat{\xi} + \sigma\right), \end{cases}$$

$$\int_{\hat{\xi} - \sigma}^{\hat{\xi} + \sigma} v_\sigma(x; \hat{\xi}) dx = 1.$$

It is clear that, if σ tends to 0 the value of $\int_{\hat{\xi}-\sigma}^{\hat{\xi}+\sigma} f(x) v_\sigma(x; \hat{\xi}) dx$ tends to the value of function $f(x)$ at point $\hat{\xi}$.

Using the following relationship, we determine the power values $q_i(t)$, $i = 1, 2, \ldots, N_c$ based on the obtained measurement:

$$q_i(t) = \sum_{j=1}^{N_o} \alpha_i^j \left[\int_{\xi_j-\sigma}^{\xi_j+\sigma} u(x,t) v_\sigma(x;\xi_j) dx - \beta_i^j \right], \quad i = 1, 2, \ldots, N_c, \ t \in [0, T], \quad (8)$$

where $\alpha_i^j, \beta_i^j, \xi_j$ are coefficients of feedback, $i = 1, 2, \ldots, N_c, j = 1, 2, \ldots, N_o$ [13, 14].

The value in brackets is equal to the deviation of the temperature at the jth measuring point from the value β_i^j that is nominal relative to the ith source for the jth measuring point, α_i^j are the coefficients of the amplification.

The number of feedback coefficients or parameters in formulas (8) is $\mathcal{N} = 2N_c N_o + N_o$ needs to be determined.

Since the function $u(x, t)$ is the solution of problem (1)–(3) is continuous by t at $t \in [0, T]$, then the power $q(t)$, obtained from formula (8), in measurements (7) are continuous functions.

When we apply formula (8) in Eq. (7) with continuous feedback (1), the result is:

$$u_t(x,t) = a^2 u_{xx}(x,t) - \lambda_0[u(x,t) - \theta]$$
$$+ \sum_{i=1}^{N_c} \delta_\sigma(x; z_i(t)) \left[\sum_{j=1}^{N_o} \alpha_i^j \int_{\xi_j-\sigma}^{\xi_j+\sigma} u(\gamma, t_k) v_\sigma(\gamma; \xi_j) d\gamma - \beta_i^j \right], \quad (9)$$
$$x \in (0, l), \ t \in (0, T].$$

Equations (9) is an integro-differential equation or an equation with integral loading. The initial-boundary value problems that correspond to many works have been thoroughly studied (the bibliography can be found in [15]). By approximating the integral, the equation is reduced to a point-loaded equation. The study of these equations was conducted [15–19].

The objective is to find feedback coefficients $\alpha = \left((\alpha_i^j)\right), \beta = \left((\beta_i^j)\right), \xi = (\xi_j)$, $i = 1, 2, \ldots, N_c, j = 1, 2, \ldots, N_o$ that minimize the objective function under given constraints (4). For the parameters optimized in the problem, we introduce the notation $y = (\alpha, \beta, \xi) \in \mathbb{R}^\mathcal{N}$, and functional (5) and (6) of the considered problem can be written as follows:

$$J(y) = \int_B \int_\Theta I(y; b, \theta) \rho_B(b) \rho_\Theta(\theta) d\theta db, \quad (10)$$

$$I(y; b, \theta) = \int_0^l \mu(x)[u(x, T) - U(x)]^2 dx + \varepsilon \|y - \hat{y}\|_{\mathbb{R}^\mathcal{N}}^2. \quad (11)$$

Here: $u(x, t) = u(x, t; y, b, \theta)$ solution of the initial-boundary value problem concerning Eqs. (9), (2), (3) for the given parameters $y = (\alpha, \beta, \xi)$, with the initial condition $u(x, 0) = b$ and ambient temperature θ.

Constraints (4) on the power of sources with continuous feedback (8) will become joint constraints on the parameters y and the phase state temperature at the measuring points ξ_j in integral form, $j = 1, 2, \ldots, N_o$.

$$\underline{q_i} \leq \sum_{j=1}^{N_o} \alpha_i^j \int_{\xi_j-\sigma}^{\xi_j+\sigma} u(x,t) v_\sigma(x;\xi_j) dx - \beta_i^j \leq \overline{q_i}, \quad (12)$$
$$t \in [0, T], \ i = 1, 2, \ldots, N_c.$$

We can express these limitations in an equivalent manner

$$g_i(t; y) = \left|g_i^0(t; y)\right| - \frac{\overline{q_i} - \underline{q_i}}{2} \leq 0, t \in [0, T], i = 1, 2, \ldots, N_c, \tag{13}$$
$$g_i^0(t; y) = \frac{\overline{q_i} + \underline{q_i}}{2} - q_i(t; y).$$

For constraints of the location of measurement points, we will use the following natural constraints:

$$\sigma \leq \xi_j \leq l - \sigma, j = 1, 2, \ldots, N_o. \tag{14}$$

The problem resulting from (9), (2)–(4), (10), and (1) belongs to the class of control synthesis problems for systems with distributed parameters. The objective is to find a finite-dimensional vector $y \in \mathbb{R}^{\mathcal{N}}$.

The problem at hand involves integrodifferential equations that describe the process being studied. The objective functional values depend on the set of solutions of the initial-boundary value problem. This is because the initial condition of the rod and the ambient temperature can take values from certain sets B and Θ. Note that the dimension of the optimization problem resulting from reducing the original problem to a finite-dimensional one is determined by multiplying the number of sources and sampling points. This dimension is acceptable given current computing technology and optimization methods.

3 Approach and Formulas for Solving the Problem

First of all, it is evident that the optimization problem (10), (11) with continuous feedback (8) lacks convexity in terms of optimized feedback parameters y, given the convexity of the $q(t)$ source optimal control problem (1)–(6). Note that the permissible parameter range y, which is defined by formulas (12), is non-convex. This follows the non-linearity of the dependence of the solution to the initial-boundary value problem $u(x, t)$ from parameters $y = (\alpha, \beta, \xi)$. The formulas below for functional gradients can be useful for solving the problem of determining locally optimal feedback parameters or for locally determining their results specified by an expert. In general, global optimization methods can be used in combination with local conditional gradient optimization methods to optimize feedback parameters.

To solve the resulting finite-dimensional parametric optimal control problem (1)–(6), we apply the penalty function method to take into account constraints (13). Taking into account, as indicated above, the multi-extremal nature of the problem in terms of feedback parameters for both continuous and discrete feedback, the solution to the problem can be solved by the method of penalty functions with different initial search points relative to the parameter vector y.

The external penalty functional with respect to functional (10), (11) is written as:

$$J_{\mathcal{R}}(y) = \int_B \int_\Theta I_{\mathcal{R}}(y; b, \theta) \rho_B(b) \rho_\Theta(\theta) d\theta db, \tag{15}$$

$$I_{\mathcal{R}}(y; b, \theta) = \int_0^l \mu(x)[u(x, T) - U(x)]^2 dx + \varepsilon \|y - \hat{y}\|_{\mathbb{R}^{\mathcal{N}}}^2 + \mathcal{R}G(y), \tag{16}$$
$$G(y) = \sum_{i=1}^{N_c} \int_0^T \left[g_i^+(t; y)\right]^2 dt.$$

Here the coefficient $\mathcal{R}$ tends to infinity, and the function $g_i^+(t; y) = 0$, if $g_i(t; y) \leq 0$, and $g_i^+(t; y) = g_i(t; y)$, if $g_i(t; y) > 0$.

To minimize the external penalty functional (15), (16), and the linearity constraint (14), we also use the gradient projection method [12]:

$$y^{k+1} = \mathcal{P}_{(14)}[y^k - \omega_k \text{grad}_y J_{\mathcal{R}}(y^k)], k = 0, 1, 2, \ldots.$$
$$\omega_k = \underset{\omega \geq 0}{\text{argmin}} J_{\mathcal{R}}\left(\mathcal{P}_{(14)}[y^k - \omega_k \text{grad}_y J_{\mathcal{R}}(y^k)]\right). \quad (17)$$

Here ω_k is the minimization step of one-dimensional function, y^0 is an arbitrary starting point of the search from R^N; $\mathcal{P}_{(14)}[\cdot]$ is the projecting operator of point ξ_j into the $[\sigma, l - \sigma]$, $j = 1, 2, \ldots, N_o$ defined by constraints (14). Considering the linearity of constraints (14), the operator $\mathcal{P}_{(14)}[\cdot]$ is easy to construct constructively [12].

Theorem 1. With continuous feedback (8), the functional $J_{\mathcal{R}}(y)$ of problem (9), (2), (3), (4), (15), (16) for each value of the coefficient $\mathcal{R}$ differentiable by $y = (\alpha, \beta, \xi)$, and the components of its gradient have the formulas

$$\frac{\partial J_{\mathcal{R}}(y)}{\partial \alpha_i^j} = \int_B \int_\Theta \left\{ -\int_0^T \left[\int_{z_i(t)-\sigma}^{z_i(t)+\sigma} \psi(x, t) \delta_\sigma(x; z_i(t)) dx + 2\mathcal{R}g_i^+(t; y) \text{sgn}(g_i^0(t; y)) \right] \right.$$
$$\left. \times \left[\int_{\xi_j-\sigma}^{\xi_j+\sigma} u(\gamma, t) v_\sigma(\gamma; \xi_j) d\gamma - \beta_i^j \right] dt + 2\varepsilon \left(\alpha_i^j - \hat{\alpha}_i^j \right) \right\} \rho_B(b) \rho_\Theta(\theta) d\theta db, \quad (18)$$

$$\frac{\partial J_{\mathcal{R}}(y)}{\partial \beta_i^j} = \int_B \int_\Theta \left\{ \alpha_i^j \int_0^T \left[\int_{z_i(t)-\sigma}^{z_i(t)+\sigma} \psi(x, t) \delta_\sigma(x; z_i(t)) dx \right. \right.$$
$$\left. \left. + 2\mathcal{R}g_i^+(t; y) \text{sgn}(g_i^0(t; y)) \right] dt + 2\varepsilon \left(\beta_i^j - \hat{\beta}_i^j \right) \right\} \rho_B(b) \rho_\Theta(\theta) d\theta db, \quad (19)$$

$$\frac{\partial J_{\mathcal{R}}(y)}{\partial \xi_j} = \int_B \int_\Theta \sum_{i=1}^{N_o} \left\{ -\int_0^T \left[\int_{z_i(t)-\sigma}^{z_i(t)+\sigma} \psi(x, t) \delta_\sigma(x; z_i(t)) dx \right. \right.$$
$$\left. + 2\mathcal{R}g_i^+(t; y) \text{sgn}(g_i^0(t; y)) \right] \left[\int_{\xi_j-\sigma}^{\xi_j+\sigma} u_\gamma(\gamma, t) v_\sigma(\gamma; \xi_j) d\gamma \right] dt \quad (20)$$
$$\left. + 2\varepsilon (\xi_j - \hat{\xi}_j) \right\} \rho_B(b) \rho_\Theta(\theta) d\theta db,$$

$i = 1, 2 \ldots, N_c, j = 1, 2 \ldots, N_o$. Here the function $\psi(x, t) = \psi(x, t; y, b, \theta, \mathcal{R})$ is the solution for the following conjugate initial-boundary value problem:

$$\psi_t(x, t) = -a^2 \psi_{xx}(x, t) + \lambda_0 \psi(x, t) - \sum_{j=1}^{N_o} v_\sigma(x; \xi_j)$$
$$\times \sum_{i=1}^{N_c} \left\{ \alpha_i^j \int_{z_i(t)-\sigma}^{z_i(t)+\sigma} \psi(\gamma, t) \delta_\sigma(\gamma; z_i(t)) d\gamma + 2\mathcal{R}g_i^+(t; y) \text{sgn}(g_i^0(t; y)) \right\} dt, \quad (21)$$

$$x \in (0, l), \ t \in [0, T),$$

$$\psi(x, T) = -2\mu(x)[u(x, T) - U(x)], \ x \in [0, l], \quad (22)$$

$$\psi_x(0, t) = \lambda \psi(0, t), \ \psi_x(l, t) = -\lambda \psi(l, t), \ t \in [0, T). \quad (23)$$

Proof. From the independence of the initial temperature of the rod and the temperature of the external environment on the vector of its parameters it follows:

$$\text{grad}_y J_{\mathcal{R}}(y) = \text{grad} \int_B \int_\Theta I_{\mathcal{R}}(y; b, \theta) \rho_\Theta(\theta) \rho_B(b) d\theta db$$
$$= \int_B \int_\Theta \text{grad}_y I_{\mathcal{R}}(y; b, \theta) \rho_\Theta(\theta) \rho_B(b) d\theta db.$$

Taking this into account, we will further study the differentiability of functional (11) at given temperatures b and θ.

In Eq. (9) we use the notation:

$$F(x, t; u, y) = \sum_{i=1}^{N_c} \left[\sum_{j=1}^{N_o} \alpha_i^j \int_{\xi_j - \sigma}^{\xi_j + \sigma} u(\gamma, t) v_\sigma(\gamma; \xi_j) d\gamma - \omega^i \right] \delta_\sigma(x; z_i(t)), \quad (24)$$

and Eq. (9) is written as follows:

$$u_t(x, t) = a^2 u_{xx}(x, t) - \lambda_0 [u(x, t) - \theta] + F(x, t; u, y). \quad (25)$$

Next, we will apply the incremental method to the independent variables y. Let us assume that the parameters $y = (\alpha, \beta, \xi)$ have been incremented Δy, let denote $\tilde{y} = y + \Delta y = (\alpha + \Delta\alpha, \beta + \Delta\beta, \xi + \Delta\xi)$. It is clear that the $F(x, t; u, y)$ will be increased:

$$\Delta F(x, t; u, y) = F(x, t; u, \tilde{y}) - F(x, t; u, y) \quad (26)$$

and also phase function

$$\Delta u(x, t; y) = u(x, t; \tilde{y}) - u(x, t; y). \quad (27)$$

Then the function $\Delta u(x, t; y)$ is the solution to the problem:

$$\Delta u_t(x, t) = a^2 \Delta u_{xx}(x, t) - \lambda_0 \Delta u(x, t) + \Delta F(x, t; u, y), \quad (28)$$

$$x \in (0, l), \ t \in (0, T],$$

$$\Delta u(x, 0) = 0, \ x \in [0, l], \quad (29)$$

$$\Delta u_x(0, t) = \lambda \Delta u(0, t), \ \Delta u_x(l, t) = -\lambda \Delta u(l, t), \ t \in (0, T], \quad (30)$$

For the first term of the functional increment (16):

$$\Delta I_\mathcal{R}(y; b, \theta) = \Delta I_T(y; b, \theta) + \mathcal{R}\Delta G(y), \quad (31)$$

considering (28)–(30), we have:

$$\begin{aligned} \Delta I_\mathcal{R}(y; b, \theta) &= I_\mathcal{R}(\tilde{y}; b, \theta) - I_\mathcal{R}(y; b, \theta) \\ &= \int_0^l 2\mu(x)[u(x, T) - U(x)]\Delta u(x, T)dx + 2\varepsilon\langle y - \hat{y}, \Delta y\rangle. \end{aligned} \quad (32)$$

Synthesis of Power Control of Moving Sources 379

Let's multiply both sides of Eq. (28) by a still arbitrary function $\psi(x,t)$, integrate over x, $x \in [0,l]$ and over t, $t \in [0,T]$, and move all terms to the left. Adding the resulting relation to (32), we obtain

$$\Delta I_R(y; b, \theta) = \int_0^l 2\mu(x)[u(x,T) - U(x)]\Delta u(x,T)dx + 2\varepsilon\langle y - \hat{y}, \Delta y\rangle \\ + \int_0^T \int_0^l \psi(x,t)\big(\Delta u_t(x,t) - a^2 \Delta u_{xx}(x,t) + \lambda_0 \Delta u(x,t) - \Delta F(x,t;u,y)\big)dxdt. \quad (33)$$

Integrating by parts we get:

$$\Delta I_R(y; b, \theta) = \int_0^l 2\mu(x)[u(x,T) - U(x)]\Delta u(x,T)dx + 2\varepsilon\langle y - \hat{y}, \Delta y\rangle \\ + \int_0^l \psi(x,T)\Delta u(x,T)dx - \int_0^l \psi(x,0)\Delta u(x,0)dx - a^2 \int_0^T \psi(l,t)\Delta u_x(l,t)dt \\ + a^2 \int_0^T \psi(0,t)\Delta u_x(0,t)dt + a^2 \int_0^T \psi_x(l,t)\Delta u(l,t)dt - a^2 \int_0^T \psi_x(0,t)\Delta u(0,t)dt \\ - \int_0^T \int_0^l \big(\psi_t(x,t) + a^2 \psi_{xx}(x,t) - \lambda_0 \psi(x,t)\big)\Delta u(x,t)dxdt \\ - \int_0^T \int_0^l \psi(x,t)\Delta F(x,t;u,y)dxdt. \quad (34)$$

Bracket $\langle\cdot,\cdot\rangle$ denotes scalar product operation.
For the last term (34), considering (24), (26), after transformations we obtain:

$$\int_0^T \int_0^l \psi(x,t)\Delta F(x,t;u,y)dxdt = \int_0^T \int_0^l \psi(x,t)\big[F(x,t;u,\tilde{y}) - F(x,t;u,y)\big]dxdt \\ = \sum_{i=1}^{N_c}\sum_{j=1}^{N_o} \Delta\alpha_i^j \int_0^T \int_{z_i(t)-\sigma}^{z_i(t)+\sigma} \psi(x,t)\delta_\sigma(x;z_i(t))dx \int_{\xi_j-\sigma}^{\xi_j+\sigma} u(\gamma,t)v_\sigma(\gamma;\xi_j)d\gamma dt \\ - \sum_{i=1}^{N_c}\sum_{j=1}^{N_o} \Delta\beta_i^j \int_0^T \int_{z_i(t)-\sigma}^{z_i(t)+\sigma} \psi(x,t)\delta_\sigma(x;z_i(t))dxdt \\ + \sum_{j=1}^{N_o} \Delta\xi^j \sum_{i=1}^{N_c} \alpha_i^j \int_0^T \int_{z_i(t)-\sigma}^{z_i(t)+\sigma} \psi(x,t)\delta_\sigma(x;z_i(t))dx \int_{\xi_j-\sigma}^{\xi_j+\sigma} u_\gamma(\gamma,t)v_\sigma(\gamma;\xi_j)d\gamma dt \\ + \sum_{j=1}^{N_o}\sum_{i=1}^{N_c} \alpha_i^j \int_0^T \int_{z_i(t)-\sigma}^{z_i(t)+\sigma} \psi(x,t)\delta_\sigma(x;z_i(t))dx \int_{\xi_j-\sigma}^{\xi_j+\sigma} \Delta u(\gamma,t)v_\sigma(\gamma;\xi_j)d\gamma dt.$$

For the second term in formula (31), after simple transformations we have:

$$\Delta G(y) = G(\tilde{y}) - G(y) = \sum_{i=1}^{N_c} \int_0^T \Big\{\big[g_i^+(t;\tilde{y})\big]^2 - \big[g_i^+(t;y)\big]^2\Big\}dt \\ = \sum_{i=1}^{N_c} \int_0^T \left\{\left[|g_i^0(t;\tilde{y})| - \frac{\overline{q}-\underline{q}}{2}\right]^2 - \left[|g_i^0(t;y)| - \frac{\overline{q}-\underline{q}}{2}\right]^2\right\}dt \\ = -2\sum_{i=1}^{N_c}\sum_{j=1}^{N_o} \Delta\alpha_i^j \int_0^T g_i^+(t;y)\mathrm{sgn}(g_i^0(t;y))(\int_{\xi_j-\sigma}^{\xi_j+\sigma} u(\gamma,t)v_\sigma(\gamma;\xi_j)d\gamma - \beta_i^j)dt \\ + 2\sum_{i=1}^{N_c}\sum_{j=1}^{N_o} \Delta\beta_i^j \int_0^T g_i^+(t;y)\mathrm{sgn}(g_i^0(t;y))dt \\ - 2\sum_{j=1}^{N_o} \Delta\xi^j \sum_{i=1}^{N_c} \alpha_i^j \int_0^T g_i^+(t;y)\mathrm{sgn}(g_i^0(t;y)) \int_{\xi_j-\sigma}^{\xi_j+\sigma} u_\gamma(\gamma,t)v_\sigma(\gamma;\xi_j)d\gamma dt \\ - 2\sum_{j=1}^{N_o}\sum_{i=1}^{N_c} \alpha_i^j \int_0^T g_i^+(t;y)\mathrm{sgn}(g_i^0(t;y)) \int_{\xi_j-\sigma}^{\xi_j+\sigma} \Delta u(\gamma,t)v_\sigma(\gamma;\xi_j)d\gamma dt.$$

It's clear that:

$$\Delta I_{\mathcal{R}}(y; b, \theta) = \int_0^l 2\mu(x)[u(x, T) - U(x)]\Delta u(x, T)dx$$
$$+ \int_0^l \psi(x, T)\Delta u(x, T)dx - \int_0^l \psi(x, 0)\Delta u(x, 0)dx - a^2 \int_0^T \psi(l, t)\Delta u_x(l, t)dt$$
$$+ a^2 \int_0^T \psi(0, t)\Delta u_x(0, t)dt + a^2 \int_0^T \psi_x(l, t)\Delta u(l, t)dt - a^2 \int_0^T \psi_x(0, t)\Delta u(0, t)dt$$
$$- \int_0^T \int_0^l (\psi_t(x, t) + a^2\psi_{xx}(x, t) - \lambda_0\psi(x, t))\Delta u(x, t)dxdt$$
$$- \sum_{i=1}^{N_c} \sum_{j=1}^{N_o} \Delta \alpha_i^j \{\int_0^T \left[\int_{z_i(t)-\sigma}^{z_i(t)+\sigma} \psi(x, t)\delta_\sigma(x; z_i(t))dx + 2\mathcal{R}g_+^i(t; y)\text{sgn}(g_i^0(t; y))\right]$$
$$\times \left[\int_{\xi_j-\sigma}^{\xi_j+\sigma} u(\gamma, t)\hat{\omega}_\sigma(\gamma; \xi_j)d\gamma\right]dt + 2\varepsilon\left(\alpha_i^j - \hat{\alpha}_i^j\right)\}$$
$$+ \sum_{i=1}^{N_c} \sum_{j=1}^{N_o} \Delta \beta_i^j \{\int_0^T \left[\int_{z_i(t)-\sigma}^{z_i(t)+\sigma} \psi(x, t)\delta_\sigma(x; z_i(t))dx + 2\mathcal{R}g_+^i(t; y)\text{sgn}(g_i^0(t; y))\right]dt$$
$$+ 2\varepsilon\left(\beta_i^j - \hat{\beta}_i^j\right)\}$$
$$- \sum_{j=1}^{N_o} \Delta \xi^j \{\sum_{i=1}^{N_c} \int_0^T \left[\int_{z_i(t)-\sigma}^{z_i(t)+\sigma} \psi(x, t)\delta_\sigma(x; z_i(t))dx + 2\mathcal{R}g_+^i(t; y)\text{sgn}(g_i^0(t; y))\right]$$
$$\times \left[\int_{\xi_j-\sigma}^{\xi_j+\sigma} u_\gamma(\gamma, t)v_\sigma(\gamma; \xi_j)d\gamma\right]dt + 2\varepsilon(\xi^j - \hat{\xi}^j)\}$$
$$- \sum_{i=1}^{N_c} \sum_{j=1}^{N_o} \{\int_0^T \left[\int_{z_i(t)-\sigma}^{z_i(t)+\sigma} \psi(x, t)\delta_\sigma(x; z_i(t))dx + 2\mathcal{R}g_+^i(t; y)\text{sgn}(g_i^0(t; y))\right]$$
$$\times \int_{\xi_j-\sigma}^{\xi_j+\sigma} \Delta u(\gamma, t)v_\sigma(\gamma; \xi_j)d\gamma dt\}. \tag{35}$$

It is known [12] that the components of the functional gradient are determined from (35) by the linear parts of its increment with respect to each of the components. Consequently, we have formulas (18)–(20) and the conjugate problem (21)–(23). The theorem has been proven.

4 The Result of Numerical Experiments

Let us present the results of solving a test problem of rod heating process control with continuous feedback. The following parameter values were used in the problem:

$$a^2 = 1, \lambda_0 = 0.01, \lambda = 0.001, T = 1, l = 1, N_c = 2, N_o = 3;$$
$$\mu(x) = 1, U(x) = 50; \mathcal{R} = 1, \varepsilon = 1;$$
$$\sigma = 0.03; \xi_j \in [0.03; 0.997], j = 1, 2, 3;$$
$$B = [12.5; 13; 13.5], \rho_B(b) = \tfrac{1}{3}; \Theta = [11; 12; 13; 14], \rho_\Theta(\theta) = \tfrac{1}{4};$$
$$z_1(t) = 0.4 - 0.35\sin(2\pi t), z_2(t) = 0.6 + 0.35\sin(2\pi t), t \in [0; 1];$$
$$-10 \leq q_1(t) \leq 90, -10 \leq q_2(t) \leq 80, t \in [0, 1].$$

One common approach for solving the synthesis problem for vector parameters y involves iterating the optimization process with the selected penalty coefficient $\mathcal{R}$ and regularization parameters $\varepsilon, \hat{y}$. At each iteration (17), the algorithm calculates the optimized parameters y^k by exploring all possible values of $b \in B$ and $\theta \in \Theta$, ultimately refining the solution through a series of calculated steps within the given constraints:

1. solve the direct initial-boundary value problem (9), (2), (3);
2. solve the conjugate boundary-value problem (21)–(23);
3. calculate components (18)–(20) of the gradient of the penalty function (15), (16);

4. carry out one-dimensional minimization along to $\omega \geq 0$ in the direction of the resulting anti-gradient of the functional for positional constraints (14).

These steps are repeated until that stopping criterion is met.

The implicit scheme of Crank-Nicolson method [20] is a suitable approach for solving the direct (9), (2), (3) and conjugate (21)–(23) loaded initial-boundary value problems given the specified spatial and time variable steps. With $h_x = 0.01$ and $h_t = 0.01$ this method can efficiently handle the numerical computation of the system based on the provided parameters.

For the functions $\delta_\sigma(x; z)$ or $\nu_\sigma(x; \xi)$ had used a Gaussian type function like [21]:

$$\delta_\sigma(x; z) = \begin{cases} 0, & |x - z| > \sigma, \\ \frac{1}{\sigma\sqrt{2\pi}} \exp\left(-\frac{(x-z)^2}{2\sigma^2}\right), & |x - z| \leq \sigma, \end{cases}$$

where

$$\sigma = \frac{1}{\sqrt{2\pi}} \int_{z-\sigma}^{z+\sigma} \exp\left(-\frac{(x-z)^2}{2\sigma^2}\right) dx,$$

It is can be easily checked

$$\int_{z-\sigma}^{z+\sigma} \delta_\sigma(x; z) dx = 1.$$

For numerical calculations the value σ was used equal to h_x.

Figure 1 shows the graphs of lumped heat sources power for the synthesized optimal vector of feedback parameters y^*.

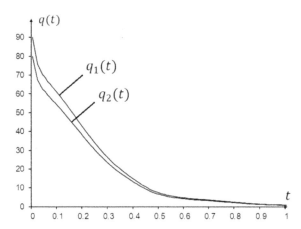

Fig. 1. Graphs of heat sources power for the vector of feedback parameters y_1^*.

Table 1 and 2 shows the results of solving the problem at different iterations from initial points y_1^0 and y_2^0 with the value of the penalty equal to $\mathcal{R} = 10$ and the values of the regularization parameters $\varepsilon = 0.001$ obtained.

Table 1. Results of solving the problem obtained at iterations from initial vector y_1^0 and optimal vector y_1^*.

k	α			β			ξ	$J_\mathcal{R}(y)$
0	−0.2000	−0.2000	−0.2000	50.000	50.000	50.000	0.2534	131.78
	−0.2000	−0.2000	−0.2000	50.000	50.000	50.000	0.5045	
							0.7589	
1	−1.0139	−1.0237	−1.0152	49.835	49.914	49.823	0.3500	11.374
	−0.8535	−1.0511	−0.9530	49.650	49.650	49.650	0.5017	
							0.6500	
2	−0.6856	−0.9184	−0.7342	50.456	50.605	50.443	0.1939	1.7846
	−0.5092	−0.8908	−0.6759	50.227	50.521	50.366	0.5034	
							0.8344	
3	−0.6480	−0.8889	−0.6918	50.5267	50.702	50.518	0.2875	0.0846
	−0.4997	−0.8878	−0.6741	50.2550	50.570	50.404	0.4752	
4	−0.8198	−0.9674	−0.7606	51.577	51.773	52.572	0.2258	0.0026
	−0.5508	−0.9451	−0.7822	51.275	51.606	52.432	0.5726	
							0.7925	
5	−0.8593	−0.8476	−0.9372	50.757	50.937	51.002	0.2075	0.0001
	−0.7818	−0.8951	−0.8024	50.726	50.879	50.932	0.5487	
							0.8599	

Figure 2 shows the graphs of function

$$J_\mathcal{R}(t; y_1^*) = \int_B \int_\Theta \int_0^l \mu(x)[u(x, t; y_1^*, b, \theta) - U(x)]^2 dx \rho_B(b) \rho_\Theta(\theta) d\theta db,$$

where $u(x, t; y_1^*, b, \theta)$ is the solution of the initial–boundary value problem (9), (2), (3) for optimal feedback parameters y_1^* obtained from the numerical problem solution.

Synthesis of Power Control of Moving Sources

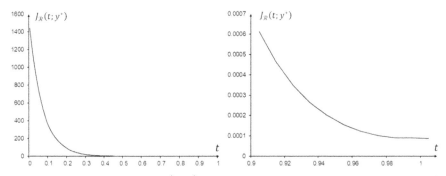

Fig. 2. Graphs of the function $J_{\mathcal{R}}(t; y_1^*)$, $t \in [0; 1]$ for the vector of feedback parameters y_1^*.

Table 2. Results of solving the problem obtained at iterations from initial vector y_2^0 and optimal vector y_2^*.

k	α			β			ξ	$J_{\mathcal{R}}(y)$
0	−0.3500	−0.3500	−0.3500	49.500	49.500	49.500	0.1887	26.204
	−0.3500	−0.3500	−0.3500	49.500	49.500	49.500	0.4539	
							0.6586	
1	−0.9250	−0.9224	−0.9205	49.519	49.516	49.519	0.1671	0.3768
	−0.9540	−0.9551	−0.9528	49.519	49.519	49.5190	0.4604	
							0.5500	
2	−0.7867	−0.8372	−0.8183	49.566	49.569	49.568	0.3500	0.0278
	−0.6274	−0.7285	−0.7533	49.554	49.561	49.563	0.6452	
							0.6500	
3	−0.7288	−0.8213	−0.7994	49.946	49.976	49.965	0.3573	0.0145
	−0.6074	−0.7396	−0.7655	49.901	49.965	49.981	0.5321	
							0.6599	
4	−0.6627	−0.8728	−0.8065	50.397	50.459	50.443	0.4329	0.0035
	−0.5214	−0.7816	−0.7774	50.341	50.485	50.520	0.5861	
							0.7285	
5	−0.6618−	−0.8737	−0.8065	50.396	50.459	50.442	0.2537	0.0001
	0.52052	−0.7824	−0.7774	50.342	50.486	50.521	0.5254	
							0.8975	

5 Conclusion

The article examines the problem of synthesizing power control of moving sources when the rod is heated. The current power values of the sources are distributed in the vicinity of their location points. To synthesize the source power values, their dependence on the rod temperature values measured at the measurement points was constructed. At the same time, the problem also optimizes the location of measurement points. Formulas are proposed for the gradient components through feedback parameters and coordinates of measurement points. The resulting formulas will allow you to apply effective first-order optimization methods for the numerical solution of the problem.

The proposed approach can be conveniently adapted to handle optimal control problems associated with lumped sources described by diverse initial-boundary value problems.

References

1. Butkovskii, A.G., Pustyl'nikov, L.M.: Teoriya podvizhnogo upravleniya sistemami s raspredelennymi parametrami. Nauka, Moscow (1980). (in Russian)
2. Butkovskii, A.G.: Metody upravleniya sistemami s raspredelennymi parametrami. Nauka, Moscow (1984). (in Russian)
3. Deineka, V.S., Sergienko, I.V.: Optimal'noye upravleniye neodnorodnimi raspredelennimi sistemami. Naukova Dumka, Kiev (2003). (in Russian)
4. Polyak, B.T., Khlebnikov, M.V., Rapoport, L.B.: Mathematical Theory of Automatic Control. LENAND, Moscow (2019). (in Russian)
5. Ray, W.H.: Advanced Process Control. McGraw-Hill Book Company, New York (1980)
6. Sergienko, I.V.: Deineka, V.S.: Optimal Control of Distributed Systems with Conjugation Conditions. Kluwer Acad. Publ., New York (2005)
7. Sirazetdinov, T.K.: Optimizatsiya sistem s raspredelennymi parametrami. Nauka, Moscow (1977). (in Russian)
8. Utkin, V.I.: Skol'zjashhie rezhimy v zadachah optimizacii i upravlenija. Nauka, Moscow (1981). (in Russian)
9. Egorov, A.I.: Osnovy teorii upravleniya. Fizmatlit, Moscow (2004). (in Russian)
10. Aida-zade, K.R., Hashimov, V.A.: Feedback control of locally lumped stabilizers for damping membrane oscillations with optimization of points of stabilizers placement and points of measuring. J. Mod. Technol. Eng. **5**(2), 111–128 (2020)
11. Mardanov, M.J., Sharifov, Y.A., Zeynalli, M.F.: Existence and uniqueness of the solutions to impulsive nonlinear integro-differential equations with nonlocal boundary conditions. Proc. Inst. Math. Mech. Natl. Acad. Sci. Azerb. **45**(2), 222–233 (2019)
12. Vasil'ev, F.P.: Metody optimizatsii. Faktorial Press, Moscow (2002). (in Russian)
13. Aida-zade, K.R., Abdullaev, V.M.: On an approach to designing control of the distributed parameter processes. Autom. Remote Control **73**(9), 1443–1455 (2012)
14. Guliyev, S.Z.: Synthesis of zonal controls for a problem of heating with delay under nonseparated boundary conditions. Cybern. Syst. Anal. **54**(1), 110–121 (2018)
15. Nakhushev, A.M.: Nagruzhennye uravneniya i ikh prilozheniya. Nauka, Moscow (2012). (in Russian)
16. Abdullaev, V.M., Aida-zade, K.R.: Numerical method of solution to loaded nonlocal boundary value problems for ordinary differential equations. Comput. Math. Math. Phys. **54**, 1096–1109 (2014)

17. Abdullaev, V.M., Aida-zade, K.R.: On the numerical solution of loaded systems of ordinary differential equations. Comput. Math. Math. Phys. **44**(9), 1505–1515 (2004)
18. Abdullayev, V.M.: Numerical solution to optimal control problems with multipoint and integral conditions. Proc. Inst. Math. Mech. Natl. Acad. Sci. Azerb. **44**(2), 171–186 (2018)
19. Alikhanov, A.A., Berezgov, A.M., Shkhanukov-Lafishev, M.X.: Boundary value problems for certain classes of loaded differential equations and solving them by finite difference methods. Comput. Math. Math. Phys. **48**, 1581–1590 (2008)
20. Samarskii, A.A.: The Theory of Difference Schemes. Marcel Dekker, New York (2001)
21. Aida-zade, K.R., Bagirov, A.G.: On the problem of spacing of oil wells and control of their production rates. Autom. Remote Control **67**(1), 44–53 (2006)

"Schedule" System for Universities Under the Bologna Education Process

Gulchohra Mammadova[1], Reshad Ismibayli[1], and Sona Rzayeva[2](✉)

[1] University of Architecture and Construction, Baku, Azerbaijan
rector@azmiu.edu.az
[2] Institute of Control Systems, Baku, Azerbaijan
sonarza@yahoo.com

Abstract. The accession of universities to the Bologna education system requires new approaches to the organization of the educational process, including the need to develop a timetable of classes with new criteria.

The article provides a comparative analysis of the traditional education system and the education system based on the credit-module system. The Bologna model takes into account an important feature of the credit system, namely, the individual organization of the educational process, which provides students with the opportunity to draw up individual curricula, freely determine the sequence of studying disciplines, independently draw up a personal semester schedule of classes, and select teachers in academic subjects.

An approach to scheduling is presented, consisting of two levels: teacher schedules and individual schedules for each student are formed. Each type of schedule has its own "hard" restrictions and "soft" requirements.

The developed software is based on dialogue between the user and the system, combining automation of the functions of assigning classes, monitoring restrictions and requirements, and monitoring the process of assigning classrooms, with the ability of the program user to influence the progress of scheduling, including to improve its efficiency.

The result of the program is two schedules: individual schedules of teachers and all subgroups, including students enrolled in them, meeting all the "hard" restrictions and, to the maximum extent possible, "soft" requirements for the class schedule.

Keywords: Bologna process · class schedule · credit system · asynchronous learning system · restrictions and requirements for the schedule · interactive mode

1 Introduction

The Bologna system of higher education is a system of unified educational standards that unites the educational processes of education in universities of different countries. The name is associated with the University of Bologna, where in 1999 29 European countries signed an agreement to create common educational standards. The outcome of the meeting was the Bologna Declaration on the creation of a "European Higher Education Area". Currently, 48 countries are parties to this agreement [1–4].

An important element of the Bologna process is the European Credit Transfer and Accumulation System (ECTS). This is a pan-European system for recording a student's level of knowledge in the process of mastering an educational program. A credit is a conventional unit in quantitative terms that measures the volume of academic workload at the first two levels of higher education - bachelor's and master's degrees. ECTS credits are based on the workload required by students to achieve the expected learning outcomes. The European Credit Transfer and Accumulation System is designed to increase transparency and facilitate the transfer of credits across different European institutions. In other words, it allows students to more easily move between countries to study without losing their credits. Thus, universities automatically accept credits previously obtained at another university. All this increases student mobility and makes national education systems more compatible at the international level [5–9].

Both under the traditional education system and under the Bologna system, the academic semester at any university begins with drawing up a class schedule. Scheduling is the procedure of distributing activities into fixed time slots and spaces subject to various restrictions. The complexity of this task is associated with many factors, the main ones being the following:

- a large amount of information coming from many departments of the educational institution: dean's offices, departments, educational department, personnel department, administrative department;
- lack of classroom resources, especially specialized ones;
- organizational, methodological, pedagogical, and medical requirements;
- the wishes of teachers and students, which often may not coincide.

In addition, many universities have their own requirements for the class schedule related to the use of their existing classroom fund and the organization of the educational process in general.

Classes should be scheduled at predetermined time slots to, firstly, give students some flexibility in choosing courses; secondly, to allow lecturers and teachers to carry out other administrative and research activities and, finally, to guarantee a good level of use of the infrastructure [10–13].

Unlike the traditional education system, class scheduling under the Bologna education system is an iterative process. Each iteration consists of three stages. In the first stage, based on information from departments and dean's offices, a schedule of teachers is drawn up. The second stage is student registration for courses. The third stage is the analysis of both schedules. If a number of teachers have not formed groups - not enough students have registered with them, or many students have not been able to register due to the lack of free places in the subject they are interested in, then the work on creating a schedule returns to the first stage and the process is repeated. To achieve this, the initial schedule of teachers is adjusted. This may be a change in the timing of some classes and an increase in the number of places for classes that have caused increased interest among students [14–18].

Analyzing which courses and which teachers are filling up too quickly and which are under–filling can help determine how to schedule these classes in the future. However, very few institutions comprehensively track the performance indicators of the previous year's schedule, which often results in a gap between the needs of students and the actual proposals of faculty leadership.

It should be noted that even for a traditional system, the created algorithms do not guarantee that all criteria and requirements are met in full and the solution has the best possible optimization cost.

The task of scheduling training sessions is a multi-criteria problem of a combinatorial type, the characteristic feature of which is its large dimension and large time costs, i.e. refers to NP-hard. Currently, there are no universal methods for solving such problems. Many attempts have been made using various computational techniques to obtain optimal solutions to the scheduling problem.

2 Problem Statement

Creating a schedule "manually" is a complex, painstaking, and lengthy process, which does not exclude overlaps in the assignment of classrooms and the distribution of teachers. In addition, it is necessary to have qualified specialists who are well aware of both the scheduling process itself and the specifics of the educational institution. "Manual" scheduling does not always guarantee its admissibility, not to mention its optimality. Often after a schedule has been created, it becomes necessary to change some details. For example, change the time of some classes, replace the teacher, or change the location of the class. However, this may violate the restrictions, firstly, the user may simply not notice, and secondly, re-checking the entire schedule, which takes time and may not result in an error-free lesson schedule. Moreover, manual review does not guarantee a conflict-free schedule. In most cases, a number of problems arise due to planning errors. Conflicts in the schedule affect the normal course of the educational process. Classes are postponed until conflicts are resolved. The most serious mistake is when a student who has successfully completed registration is forced to refuse enrollment in the chosen discipline due to a change in the schedule.

All these problems can be avoided by automating scheduling.

There are many commercial college scheduling programs on the market with their own advantages and disadvantages, but they all operate within a traditional education system.

The credit education system, which is part of the Bologna process, differs from the traditional one. The scheduling process, as well as the requirements and restrictions placed on it, are changing [19–21].

The main difference is that two schedules must be created: the teacher's schedule and the individual schedule of each student, which he draws up independently, taking into account his capabilities, preferences, and needs. Students make their schedule after the teachers' schedule is ready, i.e. it is input information for them.

The subjects studied are divided into three categories:

- mandatory ones that cannot be missed;
- compulsory with a choice, these subjects are usually grouped into several thematic blocks, and the student must select several subjects from each block;
- free choice items.

With the traditional education system, the established curriculum is quite stable and mandatory for everyone, so there are no sudden changes when drawing up the schedule, so you can use last year's schedule and make a small "fine-tuning" if necessary.

Under the conditions of a credit education system, if in the first years of study, students try to take compulsory subjects, then in subsequent years of study, the spread of the subjects studied can be much wider. In addition, disciplines and students from other faculties or specialties can be added. The composition of students changes every academic semester, and their preferences in choosing the subjects they study also change. Therefore, there is no possibility of transferring last year's schedule even with minor changes and it is necessary to draw up a new schedule every year.

Taking all this into account, it is clear that previously developed software applications for scheduling classes for the traditional education system will not work in universities that have joined the Bologna process.

This research is aimed at developing a software application that will take into account all the features of the Bologna education system and cover the entire process from registering students for courses to creating two schedules: a teacher's schedule and an individual schedule for each student.

2.1 Synchronous and Asynchronous Learning Systems

The traditional, sometimes also called classical, education system is a synchronous education system. Synchronicity means coordination of actions. The classical education system is designed for group learning.

In synchronous learning, a group of students studies at the same time. All students who entered the university at the same time for the same specialty study all subjects at the same time: according to the same curriculum, divided into semesters, according to the same class schedule for all. Groups are formed on an ongoing basis and, with rare exceptions, students in a group who started their studies at the university together finish it at the same time.

In contrast to rigid synchronous learning, asynchronous learning is learning that occurs at a pace set by the student [22, 23].

Asynchronous learning mode is a form of learning in which students learn educational material at a comfortable pace and at a time convenient for them. Asynchronous learning allows students to develop their own study schedules within the overall program of their chosen major. Each student chooses subjects and time based on his capabilities, preferences, and needs. Students who entered a university at the same time for the same specialty can study at the university for a different number of years, rarely intersecting in classes, and sometimes not even intersecting. The composition of the groups is temporary; they are formed only to study a specific subject with a specific teacher.

The Bologna system implies student independence in matters of study, choice of discipline for study, schedule of classes and exams, and a supervisor for a diploma.

Both synchronous and asynchronous forms of learning can take place online and offline.

3 Stages of Scheduling Classes with a Credit Education System

As noted, with the credit education system, two schedules are created: the teacher's schedule and the individual schedule of each student. This happens in the first two stages. At the third stage, the generated schedule is evaluated and, if necessary, the process begins from the first stage, i.e. this process is iterative.

After students are enrolled at the university, the dean's office presents the entire curriculum for the chosen specialty, indicating the timing and cost (in credit units) of each subject.

First stage. The dean's office analyzes applications submitted by students with their chosen subjects for the current semester. Whenever possible, students' wishes are taken into account. Taking into account these requests, the departments formulate the teaching load of each teacher for the current semester. Then the information agreed with the management of the dean's office goes to the person responsible for drawing up the schedule, who begins to compile it. A student wishing to follow his own pattern of study may register for any new subject only after completing the prerequisites. When drawing up a schedule, in addition to the wishes of students, the presence of a sufficient number of teaching staff, the number of applications received for a given discipline, the permissible size of each formed temporary group, and the schedule of the educational institution are taken into account. For a discipline taught offline, an audience is assigned that meets the requirements of this lesson: capacity, availability of necessary technical means, and other equipment.

If there are more applicants for a particular subject than previously planned, the number of places can be increased either by adding an additional teacher to teach this subject or by increasing the number of students in the group of teachers already included in the schedule.

Second stage. The generated schedule of teachers appears on the university website, after which students are allocated a "window", i.e. time to register for their chosen subjects. If all places in a temporary group formed to study a subject have already been filled, then registration is no longer possible. An additional list is formed from students who do not fall into the limit allocated for the chosen subject. If someone from the main list refuses from the additional list, in order of priority, the student takes a place in the group formed to study this subject. Changes may occur in subsequent iterations.

In addition, at the registration stage, control is carried out regarding the possibility of studying the discipline in the current semester. If the control is not passed, the student will not be enrolled in this subject even if there are free places.

Third stage. Evaluation of the created schedule. If there are teachers with unformed groups and a large number of students who are unable to enroll in their chosen disciplines, then there are several ways to get out of this situation:

- change the schedule of a teacher who is left without a lesson due to an unformed group;
- include an additional teacher in the schedule or increase, if possible, the size of an already formed group.

If this is done, then the next iteration begins from the first stage.

4 Restrictions and Requirements for the Class Schedule

Hard restrictions are restrictions on time and space that must be complied with. If any strict requirement is not met, then such a schedule is not accepted for execution. A feasible schedule is a schedule that satisfies all stringent requirements. Soft requirements are not mandatory. A schedule is still considered valid even if it does not satisfy the soft requirements, provided that all hard restrictions are satisfied. But it is desirable that soft requirements be met to the greatest extent possible. The quality of the schedule depends on the fulfillment of soft requirements criteria, i.e. its optimality. The schedule must be drawn up within the work schedule of the educational institution [24–26].

The composition of soft requirements and their priorities in different educational institutions may be different.

Hard restrictions include events that should not overlap in time:

- no teacher can teach more than one lesson at the same time interval;
- no more than one lesson may be held in any one classroom at the same time interval;
- at the same time interval, the total number of assigned classrooms cannot exceed the university's classroom fund.

Another group of hard restrictions are spatial restrictions:

- the lesson cannot be scheduled in a classroom whose capacity is less than the number of students;
- a lesson requiring special equipment should be scheduled in specialized classrooms;
- a lesson cannot be scheduled in classrooms assigned to departments and faculties without their consent.

Another type of hard restrictions is related to the implementation of the teaching load - all classes planned for each teacher must be included in the schedule.

Soft requirements are preferences that do not involve time or space conflicts. When drawing up a schedule for a credit education system, wishes and preferences can be on the part of both teachers and students.

Soft requirements from teachers:

- minimizing transitions between academic buildings;
- taking into account the wishes of teachers who want to have time for scientific work and preparation for classes.

Soft requirements from students:

- balanced (even) assignment of hours for each subject;
- non-overlapping of compulsory subjects in time to increase the possibility of their choice.

After the teachers' schedules appear on the university website and the "window" for registration opens, students begin to draw up their schedules. Successfully registering for a subject of interest means including that subject in their individual schedule.

A student may successfully register for any subject if the following requirements are met:

- in the same time interval, each student can register no more than once;
- compliance of the selected subject with the sequence of study disciplines approved by the curriculum;
- have time to meet the student limit established for studying this subject;
- the number of selected academic disciplines on each school day cannot exceed the established limit and comply with medical recommendations;
- the number of credits gained per semester should not exceed the established standards;
- the chosen discipline must be included in the plan of the specialty in which the student is studying.

Note. With the permission of the dean's office, the last two requirements may be violated for students with good academic performance.

Taking into account the fact that each student draws up an individual schedule personally following his capabilities, needs, and preferences, there are no soft requirements for student schedules.

5 Interactive Schedule Model

The task of creating a class schedule for a university is characterized by a large volume and variety of data, a large number of requirements and restrictions on the schedule, and limited resources. This classifies it as an NP-hard multicriteria optimization problem. Obtaining its exact solution due to the large dimensionality of real scheduling problems is practically impossible in polynomial time [27–32].

To solve planning problems, which include the problem under consideration, several algorithms have been proposed that can be divided into the following groups:

- classical methods (methods of dynamic programming, integer programming, non-linear programming, branch and bound, graph coloring method, exhaustive search method),
- metaheuristic (genetic algorithms, simulated annealing method, greedy algorithms, ant colony method, etc.).

To solve the problem of scheduling at a university, we offer an interactive mode [33]. Interactive scheduling is an idea in which the system shows the user how the schedule is being built, and the user can intervene in the process at certain stages. The algorithm we developed is based on the "traffic light" principle.

The presented algorithm simulates the behavior of the user responsible for scheduling. All labor-intensive work on creating a schedule: distributing classes and monitoring compliance with restrictions and requirements is taken over by the system, and the user has the opportunity to be involved in this process at any given time. At each step, the system presents its solution with the participation of the user. This intermediate solution satisfies all the hard restrictions and should satisfy the soft restrictions as much as possible.

The connection between the user and the developed application is ensured by a convenient interface and a system of hints and help. It seems to us that such a solution is more convenient than a completely closed, fully automated system, which often, after completing the automated part of the program, is followed by manual debugging, which can cause great inconvenience.

Let us describe each of the three stages of creating a schedule, which make up the proposed interactive system for creating a class schedule. All reference information comes from various departments of the educational institution directly into the system database.

In the first stage, the teacher schedule is formed. The information for him is both reference information and operational information, presented by two documents (in this case for a teacher named Isayeva): the planned workload of the teacher (see Fig. 1) and the wishes of the teachers, indicating time intervals undesirable for conducting classes (see Fig. 2).

Fig. 1. Teacher's workload

The system processes each teacher's application one by one. In addition to the application with the teacher's workload, there is a table template on the screen, the number of rows and columns which coincides with the operating mode of the educational institution. Classes already included in the schedule take their place in the table indicating the name of the subject and the details of the teacher teaching this discipline. For each activity selected by the user for inclusion in the activity schedule, the free cells of the table are painted in colors corresponding to the principle of operation of the traffic light. The operation of a traffic light is based on the use of three colors: red, yellow, and green. Red color means prohibited; it is impossible to assign an activity to a red cell due to non-compliance with the most important thing, without which there is no acceptable schedule - hard restrictions. Green color is the ideal choice for including the selected

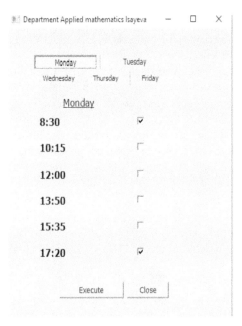

Fig. 2. Teacher's wishes

activity in the schedule. In this case, absolutely all restrictions and requirements, both hard and soft, are met, and the schedule will not only be acceptable but also close to optimal. Yellow cells are associated with soft requirements. In this case, all hard restrictions are met, but one or more soft requirements may be violated.

The selected lesson can be assigned to this cell, and the user receives information about which requirements are being violated and how this affects the optimality of the schedule. If there are alternative options, it is possible to assign a lesson to a yellow cell in which the optimality of the schedule will be higher. The terminology of integer linear programming is used to describe hard restrictions and soft requirements mathematically.

Schedule for a specific teacher (in this case for a teacher named Isayeva) with classes already included (see Fig. 3).

The distribution of classes among table cells is done using the "drag-and-drop" method, with which the user can easily move information between the "Teacher's workload" document and the table with the created schedule.

Classrooms can be assigned during the scheduling process, or the user can return to this issue later.

The system provides the opportunity to obtain information about the availability of free classrooms for the user-selected lesson included in the schedule. For each selected classroom, information is provided about the availability of the necessary equipment for conducting the lesson, who is assigned to this classroom, and its capacity (see Fig. 4).

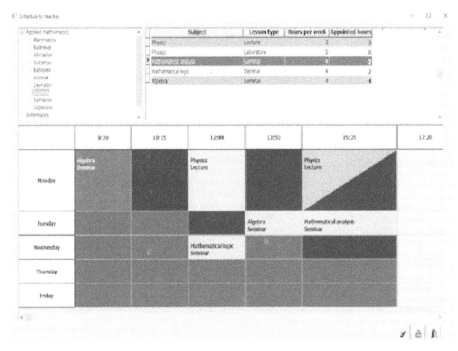

Fig. 3. Teacher's class schedule (Color figure online)

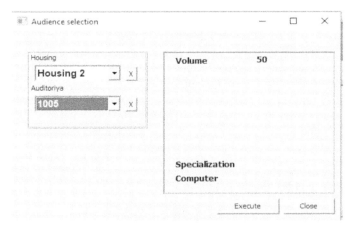

Fig. 4. Audience options

The second stage is the registration of students for selected disciplines and the formation of their individual schedules. Students must register for their chosen subjects within the allotted time slot. If the selected subjects meet the criteria specified in the description of strict restrictions, and the student's application falls within the limit allocated for this lesson, then the student successfully registers, about which he receives the following information (see Fig. 5):

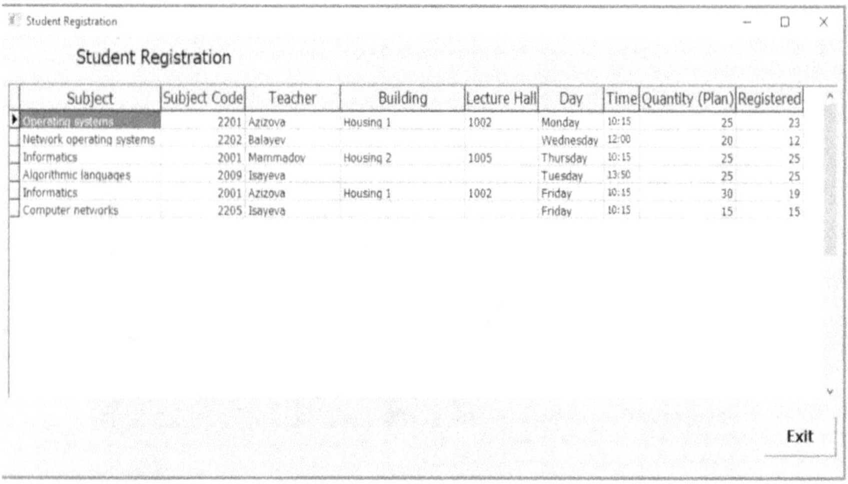

Subject	Subject Code	Teacher	Building	Lecture Hall	Day	Time	Quantity (Plan)	Registered
Operating systems	2201	Azizova	Housing 1	1002	Monday	10:15	25	23
Network operating systems	2202	Balayev			Wednesday	12:00	20	12
Informatics	2001	Mammadov	Housing 2	1005	Thursday	10:15	25	25
Algorithmic languages	2009	Isayeva			Tuesday	13:50	25	25
Informatics	2001	Azizova	Housing 1	1002	Friday	10:15	30	19
Computer networks	2205	Isayeva			Friday	10:15	15	15

Fig. 5. Student's Schedule

If the subjects selected by the student do not pass the test for compliance with strict restrictions, then a message about the reasons is issued immediately during registration. If students were unable to register due to a lack of free places, then they can enroll in an additional list and, if free places become available, have the opportunity to be included in a temporarily formed group.

In the third stage, the generated schedule is assessed. The main indicators for assessing the quality of the schedule are the number of teachers left without formed groups due to the small number of registered students, and the number of students who were unable to enroll in selected subjects. The scheduling process is iterative and consists of three stages. If, after completing the first iteration, a significant number of teachers have not organized groups, then there may be several reasons:

- poorly designed schedule;
- the subject is uninteresting or in little demand.

On the contrary, the large number of students deprived of the opportunity to study the desired subject indicates its popularity.

In this situation, a transition to the first stage of the next iteration is possible. To eliminate the noted problems, the faculty management can take the following actions:

- the time of the lesson was changed if it did not suit the majority of students;
- an unclaimed subject has been replaced with a more in-demand item;
- provision of an additional teacher;
- adding hours for in-demand teachers;
- increasing the composition of groups for in-demand subjects.

If any of the above lists is implemented, then at the first stage the teachers' schedule is adjusted: the time intervals for classes are changed, additional teachers and disciplines are included, and if the number of groups increases, previously assigned classrooms are replaced with classrooms of larger capacity.

Taking into account the results of the first stage, students can change their schedule and join a temporary group to study a subject that was not available to them previously.

Often universities give students a grace period during which they can add and drop classes during the semester. In the first couple of weeks, they check to see if they like the classes and if the timing suits them. Dean's offices, in turn, must monitor the wishes of students in order to give the opportunity to as many students as possible to take all the necessary courses of their choice to the maximum extent possible.

At the end of the last iteration, each teacher and each student receives their own individual schedules.

6 Conclusion

The proposed solution implements a full cycle of creating class schedules, consisting of analyzing the needs and wishes of students in choosing the subjects they study, quickly and conveniently registering them for the selected disciplines, and creating a schedule for both teachers and students. The developed software does not allow conflict situations either in the schedule of teachers or in the schedule of students because compliance with all strict requirements and, to the greatest possible extent, soft requirements is ensured, which completely frees the user from analyzing and monitoring the correctness of the schedule. A user-friendly interface and modern design reduce its compilation to arranging selected activities into valid cells.

The developed system can be included in a single information system university system.

References

1. What is the Bologna Process and Why it is Important in Higher Education? http://www.las.ac/what-is-the-bologna-process-and-why-it-is-important-in-higher-education/. Accessed 13 Apr 2024
2. Kushnir, I.: The role of the Bologna Process in defining Europe. Eur. Educ. Res. J. **15**(6), 664–675 (2016). https://doi.org/10.1177/1474904116657549
3. Zgaga, P.: The Bologna Process in a global setting: twenty years later. Innov.: Eur. J. Soc. Sci. Res. **32**(4), 450-464 (2019). https://doi.org/10.1080/13511610.2019.1674130
4. Mammadova, L., Valiyev, A.: Azerbaijan and European higher education area: students' involvement in Bologna reforms. Res. Educ. Adm. Leadersh. **5**(4), 1083–1121 (2020). https://doi.org/10.30828/real/2020.4.4
5. What is the Academic Credit System in Education? How Does It Benefit International Students? https://www.mastersportal.com/articles/948/what-is-the-academic-credit-system-in-education-how-does-it-benefit-international-students.html. Accessed 13 Apr 2024
6. Saidova, Z.U.: Organization of independent education in the credit-module system. Mod. Sci. Res. Int. Sci. J. **2**(1), 66–70 (2024). https://doi.org/10.5281/zenodo.10498561

7. Gentle-Genitty, C.: Credit is currency: prior learning and conversion to credit (2022). https://doi.org/10.7912/5knf-dk95
8. Wagenaar, R.: International student mobility based on learning outcomes and workload: the European credit transfer and accumulation system. In: Arnold, C., Wilson, M., Bridge, J., Lennon, M.C. (eds.) Learning Outcomes, Academic Credit, and Student Mobility, pp. 141–166. McGill-Queen's University Press (2020). https://doi.org/10.1515/9781553395560-008
9. López-Duarte, C., Maley, J.F., Vidal-Suárez, M.M.: International mobility in higher education: students' attitude to international credit virtual mobility programs. Eur. J. High. Educ. **13**(4), 468–487 (2023). https://doi.org/10.1080/21568235.2022.2068637
10. How To Make A Timetable At The University. https://www.en.scienceforming.com/10862090-how-to-make-a-timetable-at-the-university. Accessed 13 Apr 2024
11. Dvirna, O., Verhal, K., Ivanov, Yi.: The higher educational information system: management of the timetable scheduling and logistics of the educational process. Control Navig. Commun. Syst. **3**, 86–92 (2023). https://doi.org/10.26906/SUNZ.2023.3.086
12. Alghamdi, H., Alsubait, T., Alhakami, H., Baz, A.: A review of optimization algorithms for university timetable scheduling. Eng. Technol. Appl. Sci. Res. **10**(6), 6410–6417 (2020). https://doi.org/10.48084/etasr.3832
13. Kristiansen, S., Stidsen, T. R.: A Comprehensive Study of Educational Timetabling - a Survey. DTU Management Engineering. DTU Management Engineering Report No. 8.2013 (2013)
14. Boltayevich, M.B., Quvondiqovich, I.S.: Mathematical representation of the problem of forming the lesson schedule of higher education institutions in the credit module system. In: 2021 International Conference on Information Science and Communications Technologies (ICISCT), Tashkent, Uzbekistan, pp. 1–3 (2021). https://doi.org/10.1109/ICISCT52966.2021.9670137
15. Ismibayli, R., Rzayeva, S.: University scheduling system on the Bologna form of education. In: 2023 5th International Conference on Problems of Cybernetics and Informatics (PCI), Baku, Azerbaijan, pp. 1–5 (2023). https://doi.org/10.1109/PCI60110.2023.10326003
16. Hahm, S., Kluve, J.: Better with Bologna? Tertiary education reform and student outcomes. Educ. Econ. **27**(4), 425–449 (2019). https://doi.org/10.1080/09645292.2019.1616280
17. Ismibayli, R.E., Rzayeva, S.H.: Creating a schedule of classes in higher education institutions in the conditions of the Bologna system of education (In Russian). Appl. Math. Fundam. Inform. **8**(3), 45–46 (2023)
18. Innovative High School Schedules: Finding the Perfect Timetable for Teachers and Students. https://www.educationadvanced.com/resources/blog/innovative-high-school-schedules-finding-the-perfect-timetable-for-teachers/. Accessed 13 Apr 2024
19. Müller, T., Rudová, H.: Real-life curriculum-based timetabling with elective courses and course sections. Ann. Oper. Res. **239**(1), 1–18 (2014). https://doi.org/10.1007/s10479-014-1643-1
20. Oude Vrielink, R.A., Jansen, E.A., Hans, E.W., van Hillegersberg, J.: Practices in timetabling in higher education institutions: a systematic review. Ann. Oper. Res. **275**(1), 145–160 (2017). https://doi.org/10.1007/s10479-017-2688-8
21. Ajanovski, V.: Integration of a course enrolment and class timetable scheduling in a student information system. Int. J. Database Manag. Syst. (IJDMS) **5**(1), 85–95 (2013). https://doi.org/10.5121/ijdms.2013.5107
22. Synchronous vs. asynchronous learning: what's the difference. https://www.easy-lms.com/knowledge-center/learning-training/synchronous-vs-asynchronous-learning/item10387. Accessed 13 Apr 2024
23. Synchronous vs. Asynchronous Classes: Best Practices and Future Trends in 2024. https://www.research.com/education/synchronous-vs-asynchronous-classes. Accessed 13 Apr 2024
24. Kostadinova, I., Totev, V., Ivanov, I.: A mathematical model for rationality in timetable planning. TEM J. **12**(1), 118–125 (2023). https://doi.org/10.18421/TEM121-16

25. Sigl, B., Golub, M., Mornar, V.: Solving timetable scheduling problem using genetic algorithm. In: Proceedings of the 25th International Conference on Information Technology Interfaces, ITI 2003, Cavtat, Croatia, pp. 519–524. https://doi.org/10.1109/ITI.2003.1225396
26. Rudova, H., Vlk, M.: Multi-criteria soft constraints in timetabling. In: MISTA 2005 - Proceedings of the 2nd Multidisciplinary International Conference on Scheduling, pp. 11–15. Stern School of Business, New York (2005)
27. Algethami, H., Laesanklang, W.: A mathematical model for course timetabling problem with faculty-course assignment constraints. IEEE Access **9**, 11666–111682 (2021). https://doi.org/10.1109/ACCESS.2021.3103495
28. Burke, E., Mccollum, B., Meisels, A., Petrovic, S., Qu, R.: A graph-based hyper-heuristic for educational timetabling problems. Eur. J. Oper. Res. **176**(1), 177–192 (2007). https://doi.org/10.1016/j.ejor.2005.08.012
29. Rappos, E., Thiémard, E., Robert, S., Héche, J.-F.: A mixed-integer programming approach for solving university course timetabling problems. J. Sched. **25**, 391–404 (2022). https://doi.org/10.1007/s10951-021-00715-5
30. Badoni, R.P., et al.: An exploration and exploitation-based metaheuristic approach for university course timetabling problems. Axioms **12**(8) (2023). https://doi.org/10.3390/axioms12080720
31. Abdullah, S., Turabieh, H., McCollum, B., et al.: Meta-heuristic approaches for the university course timetabling problem. J. Heuristics **18**, 1–23 (2012). https://doi.org/10.1007/s10732-010-9154-y
32. Muller, T., Bartak, R.: Interactive timetabling: concepts, techniques and practical results. In: Burke, E., De Causmaecker P. (eds.) The 4th Conference on the Practice and Theory of Automated Timetabling. LNCS, Gent, Belgium, vol. 2740, pp. 58–72 (2002)
33. Ayda-zade, K.R., Rzayeva, S.H., Talibov, S.G.: Automated dialog system for compilation of schedule lessons in high school. Probl. Inf. Technol. **11**, 62–73 (2020, in Russian). https://doi.org/10.25045/jpit.v11.i1.08

Author Index

A

Abasova, Nigar I-125
Abbasov, Majid I-313, II-313
Abdullayev, Vugar Hacimahmud II-87
Abuzarova, Vusala II-87
Ahmedzade, Perviz II-3
Aida-zade, Kamil I-142, II-336
Ajesh, F. II-87
Akbarova, Samira II-18
Akleylek, Sedat I-3, I-91, II-275, II-298
Albayati, Maha Ahmed Abdullah II-32
Aliev, Telman I-17, II-42
Alimhan, Keylan II-184
Aliyev, Almaz I-55
Aliyev, Asif I-210
Aliyeva, Samira II-87, II-133
Alzabidi, Eissa II-52
Andishmand, Saadi I-3
Asadova, Jamila II-325
Asgarov, Taleh II-87
Ashrafova, Yegana II-336
Asil, Ufuk II-67
Azimov, Rustam II-171

B

Bagirov, Arzu II-348
Balametov, Ashraf I-157
Bardadym, Tamara I-363
Bastidas-Arteaga, Emilio II-241
Benghazouani, Salsabila II-76
Brasil, Reyolando II-252

C

Çevik, Nurşah II-275
Chikrii, Arkadii II-360
Chikrii, Greta II-360
Cilden-Guler, Demet II-100

E

Ertürk, Mehmet Fatih I-172
Eua-Anant, Nawapak I-275
Evcil, Mustafa I-28
Fındık, Oğuz II-32, II-52

G

Gunel, Aliyeva I-199
Gunkina, Tatiana II-348
Gurbanov, Majid I-44

H

Hajiyev, Chingiz I-172, II-100, II-144, II-159
Halilov, Elman I-157
Hamani, Ameur II-241
Handzel, Alexander II-348
Hashimov, Vugar II-372
Huseynli, Nicat I-235
Huseynova, Hayat II-111

I

Ibrahimov, Bayram G. I-55
Imanov, Akif II-133
Imanov, Gorkhmaz I-210
Isayev, Mazahir I-44
Isayeva, Tarana I-157
Isgandarova, Sevil I-44
Ismibayli, Reshad I-221, II-386
Ivashenko, Valerian II-198

K

Kasimzade, Azer A. I-235
Katanalp, Burak Yiğit II-3
Katanyukul, Tatpong I-275
Kazim-zada, Aydin II-111
Khanh, Pham Duy II-123
Khankishiyeva Hati, Nargiz I-187
Khasayeva, Natavan I-44

© The Editor(s) (if applicable) and The Author(s), under exclusive license to Springer Nature Switzerland AG 2025
G. Mammadova et al. (Eds.): ITTA 2024, CCIS 2226, pp. 401–403, 2025.
https://doi.org/10.1007/978-3-031-73420-5

Khurshudov, Dursun II-87, II-133
Kinatas, Hasan II-144
Kirci, Orhan II-159
Korablov, Mykola I-350
Kupryianava, Dziana II-213
Kuzmenko, Viktor II-360

L
Luong, Hoang-Chau II-123

M
Mahmudbeyli, Leyla I-44
Mahmudov, Eldaniz I-235
Mammadli, Maryam I-125
Mammadova, Ana II-42
Mammadova, Gulchohra II-18, II-386
Mamyrbayev, Orken II-184
Melikov, Agassi I-78
Mikailzade, Latafat II-87
Mordukhovich, Boris S. II-123
Musaeva, Naila I-17, II-42
Mustafayev, Elshan II-171

N
Nabadova, L. N. I-251
Nabiyev, Vasif II-228
Nagiyeva, Malahat II-133
Nasiboglu, Resmiye I-66
Nasibov, Efendi I-66, II-67
Nematli, Emin I-235
Nouh, Said II-76
Nuraliyev, Jamalladdin II-133
Nuriyeva, Fidan I-260

O
Oralbekova, Dina II-184
Özcan, Caner II-287
Özkan, Merve II-287
Ozkar, Serife I-78

P
Parakarn, Hathaichanok I-275
Pashayev, Adalat I-324
Pertsau, Dmitry II-213
Philip, Felix M. II-87
Poladova, Laman I-78
Polyakova, Lyudmila II-313

Q
Quliyev, Samir I-142, I-287

R
Rafizade, Ulvi R. I-55
Rahimov, Anar I-300
Rana, Huseynova I-199
Rychkov, Andrey I-313
Rzayeva, Narmin II-42
Rzayeva, Sona I-221, II-386

S
Sabir, Mammadov II-87
Sabziev, E. N. I-251
Sabziev, Elkhan I-324
Sahin, Varol II-298
Sandıkkaya, Mehmet Tahir I-28
Satilmis, Hami II-298
Saveliev, Mikhail I-338
Seyhan, Kübra I-91
Shafi, Danyalov II-87
Shunkevich, Daniil II-198
Stetsyuk, Petro I-350
Stovba, Viktor I-350

T
Tang, Yi II-213
Tastan, Murat II-3
Tatur, Mikhail II-213
Tok, Zaliha Yüce I-28
Tran, Dat Ba II-123
Triwiyanto, T. II-87
Turhal, Uğur II-228
Türkyılmaz, Çiğdem Canbay II-3
Türkyılmaz, Emrah II-3

U
Ushakova, Liudmyla I-106
Vasyanin, Volodymyr I-106
Vo, Truc II-123

W
Wangsrimongkol, Buddhathida I-275

Y
Yaqublu, Tofiq I-157
Yazar, Bilge Kagan II-298
Yilmaz, Asuman Günay II-228

Z

Zacchei, Enrico II-241, II-252
Zakrani, Abdelali II-76
Zhao, Di II-213
Zhumazhan, Nurdaulet II-184
Zhurbenko, Mykola I-363
Zverev, Ivan II-262
Zykina, Anna I-338, II-262

Printed in the USA
CPSIA information can be obtained
at www.ICGtesting.com
CBHW051916201024
16144CB00007B/115